Metric Spaces

Satish Shirali and Harkrishan L. Vasudeva

Metric Spaces

With 21 Figures

 Springer

Mathematics Subject Classification (2000): S4E35, 54–02

British Library Cataloguing in Publication Data
Shirali Satish
  Metric spaces
  1. Metric spaces
  I. Title II. Vasudeva, Harkrishan L.
  514.3'2
ISBN 1852339225

Library of Congress Control Number: 2005923525

ISBN 1-85233-922-5
Springer Science+Business Media
springeronline.com

Typeset by SPI Publisher Services, Pondicherry, India
Printed and bound in the United States of America
12/3830-543210 Printed on acid-free paper   SPIN 11334521

Preface

Since the last century, the postulational method and an abstract point of view have played a vital role in the development of modern mathematics. The experience gained from the earlier concrete studies of analysis point to the importance of passage to the limit. The basis of this operation is the notion of distance between any two points of the line or the complex plane. The algebraic properties of underlying sets often play no role in the development of analysis; this situation naturally leads to the study of metric spaces. The abstraction not only simplifies and elucidates mathematical ideas that recur in different guises, but also helps economize the intellectual effort involved in learning them. However, such an abstract approach is likely to overlook the special features of particular mathematical developments, especially those not taken into account while forming the larger picture. Hence, the study of particular mathematical developments is hard to overemphasize.

The language in which a large body of ideas and results of functional analysis are expressed is that of *metric spaces*. The books on functional analysis seem to go over the preliminaries of this topic far too quickly. The present authors attempt to provide a leisurely approach to the theory of metric spaces. In order to ensure that the ideas take root gradually but firmly, a large number of examples and counterexamples follow each definition. Also included are several worked examples and exercises. Applications of the theory are spread out over the entire book.

The book treats material concerning metric spaces that is crucial for any advanced level course in analysis. Chapter 0 is devoted to a review and systematisation of properties which we shall generalize or use later in the book. It includes the Cantor construction of real numbers. In Chapter 1, we introduce the basic ideas of metric spaces and Cauchy sequences and discuss the completion of a metric space. The topology of metric spaces, Baire's category theorem and its applications, including the existence of a continuous, nowhere differentiable function and an explicit example of such a function, are discussed in Chapter 2. Continuous mappings, uniform convergence of sequences and series of functions, the contraction mapping principle and applications are discussed in Chapter 3. The concepts of connected, locally connected and arcwise connected spaces are explained in Chapter 4. The characterizations of connected subsets of the reals and arcwise connected subsets of the plane are also in Chapter 4. The notion of compactness, together with its equivalent characterisations, is included in Chapter 5. Also contained in this chapter are characterisations of compact subsets of special metric spaces. In Chapter 6, we discuss product metric spaces and provide a proof of Tychonoff's theorem.

The authors are grateful to Dr. Savita Bhatnagar for reading the final draft of the manuscript and making useful suggestions. While writing the book we benefited from the works listed in the References. The help rendered by the staff of Springer-Verlag London, in particular, Ms. Karen Borthwick and Ms. Helen Desmond, in transforming the manuscript into the present book is gratefully acknowledged.

Satish Shirali
Harkrishan L. Vasudeva

Contents

0. Preliminaries 1

    0.1. Sets and Functions 1

    0.2. Relations 4

    0.3. The Real Number System 4

    0.4. Sequences of Real Numbers 6

    0.5. Limits of Functions and Continuous Functions 8

    0.6. Sequences of Functions 9

    0.7. Compact Sets 10

    0.8. Derivative and Riemann Integral 11

    0.9. Cantor's Construction 13

    0.10. Addition, Multiplication and Order in R 17

    0.11. Completeness of R 19

1. Basic Concepts 23

    1.1. Inequalities 23

    1.2. Metric Spaces 27

    1.3. Sequences in Metric Spaces 37

    1.4. Cauchy Sequences 44

    1.5. Completion of a Metric Space 54

    1.6. Exercises 58

2. Topology of a Metric Space 64

    2.1. Open and Closed Sets 64

    2.2. Relativisation and Subspaces 78

    2.3. Countability Axioms and Separability 82

    2.4. Baire's Category Theorem 88

    2.5. Exercises 98

3. Continuity ... 103

3.1. Continuous Mappings ... 103
3.2. Extension Theorems ... 109
3.3. Real and Complex-valued Continuous Functions ... 112
3.4. Uniform Continuity ... 114
3.5. Homeomorphism, Equivalent Metrics and Isometry ... 119
3.6. Uniform Convergence of Sequences of Functions ... 123
3.7. Contraction Mappings and Applications ... 132
3.8. Exercises ... 143

4. Connected Spaces ... 156

4.1. Connectedness ... 156
4.2. Local Connectedness ... 163
4.3. Arcwise Connectedness ... 165
4.4. Exercises ... 167

5. Compact Spaces ... 170

5.1. Bounded sets and Compactness ... 171
5.2. Other Characterisations of Compactness ... 178
5.3. Continuous Functions on Compact Spaces ... 182
5.4. Locally Compact Spaces ... 185
5.5. Compact Sets in Special Metric Spaces ... 188
5.6. Exercises ... 194

6. Product Spaces ... 201

6.1. Finite and Infinite Products of Sets ... 201
6.2. Finite Metric Products ... 202
6.3. Infinite Metric Products ... 208
6.4. Cantor Set ... 212
6.5. Exercises ... 215

Index ... 219

0 Preliminaries

We shall find it convenient to use logical symbols such as $\forall, \exists, \ni, \Rightarrow$ and $\Leftrightarrow$. These are listed below with their meanings. A brief summary of set algebra and functions, which will be used throughout this book, is included in this chapter. The words 'set', 'class', 'collection' and 'family' are regarded as synonymous and no attempt has been made to define these terms. We shall assume that the reader is familiar with the set $\mathbf{R}$ of real numbers as a complete ordered field. However, Section 0.3 is devoted to review and systematisation of the properties that will be needed later, The concepts of convergence of real sequences, limits of real-valued functions, continuity, compactness and integration, together with properties that we shall generalise, or that we use later in the book, have been included in Sections 0.4 to 0.8. A sketch of the proof of the Weierstrass approximation theorem for a real-valued continuous function on the closed bounded interval [0,1] constitutes a part of Section 0.8. This has been done for the benefit of readers who may not be familiar with it. The final Sections, Sections 0.9 to 0.11, are devoted to the construction of real numbers from the field $\mathbf{Q}$ of rational numbers (axioms for $\mathbf{Q}$ are assumed). It is a common sense approach to the study of real numbers, apart from the fact that this construction has a close connection with the completion of a metric space (see Section 1.5).

0.1. Sets and Functions

Throughout this book, the following commonly used symbols will be employed:

$\forall$ means "for all" or "for every"
$\exists$ means "there exists"
$\ni$ means "such that"
$\Rightarrow$ means "implies that" or simply "implies"
$\Leftrightarrow$ or "iff" means "if and only if".

The concept of *set* plays an important role in every branch of modern mathematics. Although it is easy and natural to define a set as a collection of objects, it has been shown that this definition leads to a contradiction. The notion of set is, therefore, left undefined, and a set is described by simply listing its elements or by naming its properties. Thus $\{x_1, x_2, \ldots, x_n\}$ is the set whose elements are $x_1, x_2, \ldots, x_n$; and $\{x\}$ is the set whose only element is x. If X is the set of all elements x such that some property $P(x)$ is true, we shall write

$$X = \{x : P(x)\}.$$

The symbol $\varnothing$ denotes the empty set.

We write $x \in X$ if x is a member of the set X; otherwise, $x \notin X$. If Y is a *subset* of X, that is, if $x \in Y$ implies $x \in X$, we write $Y \subseteq X$. If $Y \subseteq X$ and $X \subseteq Y$, then $X = Y$. If $Y \subseteq X$ and $Y \neq X$, then Y is *proper subset* of X. Observe that $\varnothing \subseteq X$ for every set X.

We list below the standard notations for the most important sets of numbers:

N the set of all natural numbers
Z the set of all integers
Q the set of all rational numbers
R the set of all real numbers
C the set of all complex numbers.

Given two sets X and Y, we can form the following new sets from them:

$$X \cup Y = \{x : x \in X \text{ or } x \in Y\},$$
$$X \cap Y = \{x : x \in X \text{ and } x \in Y\}.$$

$X \cup Y$ and $X \cap Y$ are the *union* an *intersection*, respectively, of X and Y. If $\{X_\alpha\}$ is a collection of sets, where α runs through some indexing set Λ, we write

$$\bigcup_{\alpha \in \Lambda} X_\alpha \quad \text{and} \quad \bigcap_{\alpha \in \Lambda} X_\alpha$$

for the union and intersection, respectively, of X_α:

$$\bigcup_{\alpha \in \Lambda} X_\alpha = \{x : x \in X_\alpha \qquad \text{for at least one } \alpha \in \Lambda\},$$
$$\bigcap_{\alpha \in \Lambda} X_\alpha = \{x : x \in X_\alpha \qquad \text{for every } \alpha \in \Lambda\}.$$

If $\Lambda = \mathbf{N}$, the set of all natural numbers, the customary notations are

$$\bigcup_{n=1}^{\infty} X_n \quad \text{and} \quad \bigcap_{n=1}^{\infty} X_n.$$

If no two members of $\{X_\alpha\}$ have any element in common, then $\{X_\alpha\}$ is said to be a *pairwise disjoint collection of sets*.

If $Y \subseteq X$, the *complement* of Y in X is the set of elements that are in X but not in Y, that is,

$$X \backslash Y = \{x : x \in X, \ x \notin Y\}.$$

The complement of Y is denoted by Y^c whenever it is clear from the context with respect to which larger set the complement is taken.

If $\{X_\alpha\}$ is a collection of subsets of X, then *De Morgan's laws* hold:

$$\left(\bigcup_{\alpha\in\Lambda} X_\alpha\right)^c = \bigcap_{\alpha\in\Lambda} (X_\alpha)^c \text{ and } \left(\bigcap_{\alpha\in\Lambda} X_\alpha\right)^c = \bigcup_{\alpha\in\Lambda} (X_\alpha)^c.$$

The *Cartesian product* $X_1 \times X_2 \times \ldots \times X_n$ of the sets $X_1, X_2, \ldots, X_n$ is the set of all ordered n-tuples $(x_1, x_2, \ldots, x_n)$, where $x_i \in X_i$ for $i = 1, 2, \ldots, n$.

The symbol

$$f : X \to Y$$

means that f is a *function* (or *mapping*) from the set X into the set Y; that is, f assigns to each $x \in X$ an element $f(x) \in Y$. The elements assigned to elements of X by f are often called *values* of f. If $A \subseteq X$ and $B \subseteq Y$, the *image* of A and *inverse image* of B are, respectively,

$$f(A) = \{f(x) : x \in A\},$$
$$f^{-1}(B) = \{x : f(x) \in B\}.$$

Note that $f^{-1}(B)$ may be empty even when $B \neq \varnothing$. The *domain* of f is X and the *range* is $f(X)$. If $f(X) = Y$, the function f is said to map X *onto* Y (or the function is said to be *surjective*). We write $f^{-1}(y)$ instead of $f^{-1}(\{y\})$ for every $y \in Y$. If $f^{-1}(y)$ consists of at most one element for each $y \in Y$, f is said to be *one-to-one* (or *injective*). If f is one-to-one, then f^{-1} is a function with domain $f(X)$ and range X. A function that is both injective and surjective is said to be *bijective*.

If $\{X_\alpha : \alpha \in \Lambda\}$ is any family of subsets of X, then

$$f\left(\bigcup_{\alpha\in\Lambda} X_\alpha\right) = \bigcup_{\alpha\in\Lambda} f(X_\alpha)$$

and

$$f\left(\bigcap_{\alpha\in\Lambda} X_\alpha\right) = \bigcap_{\alpha\in\Lambda} f(X_\alpha).$$

Also, if $\{Y_\alpha : \alpha \in \Lambda\}$ is a family of subsets of Y, then

$$f^{-1}\left(\bigcup_{\alpha\in\Lambda} Y_\alpha\right) = \bigcup_{\alpha\in\Lambda} f^{-1}(Y_\alpha)$$

and

$$f^{-1}\left(\bigcap_{\alpha\in\Lambda} Y_\alpha\right) = \bigcap_{\alpha\in\Lambda} f^{-1}(Y_\alpha).$$

If Y_1 and Y_2 are subsets of Y, then

$$f^{-1}(Y_1 \backslash Y_2) = f^{-1}(Y_1) \backslash f^{-1}(Y_2).$$

Finally, if $f: X \to Y$ and $g: Y \to Z$, the *composite* function $g \circ f: X \to Z$ is defined by

$$(g \circ f)(x) = g(f(x)).$$

0.2. Relations

Let X be any set. By a *relation* R on X, we simply mean a subset of $X \times X$. If $(x, y) \in R$, then x is said to be in relation R with y and this is denoted by xRy. Among the most interesting relations are the *equivalence relations*. A relation is said to be an *equivalence relation* if it satisfies the following three properties:

(i) xRx for each $x \in X$ (reflexive);
(ii) if xRy, then yRx (symmetric);
(iii) if xRy and yRz, then xRz (transitive).

Let R be an equivalence relation on a set X. Then the *equivalence class* determined by $x \in X$ is defined by $[x] = \{y \in X : xRy\}$.

It is easy to check that any two equivalence classes are either disjoint or else they coincide. Since $x \in [x]$, it follows that R *partitions* X; that is, there exists a family $\{A_\alpha : \alpha \in \Lambda\}$ of sets such that $X = \bigcup_{\alpha \in \Lambda} A_\alpha$. Conversely, if a pairwise disjoint family $\{A_\alpha : \alpha \in \Lambda\}$ of sets partitions X, that is, $X = \bigcup_{\alpha \in \Lambda} A_\alpha$, then by letting

$$R = \{(x, y) \in X \times X : \exists \alpha \in \Lambda \ni x \in A_\alpha \text{ and } y \in A_\alpha\},$$

an equivalence relation is defined on X whose equivalence classes are precisely A_α.

0.3. The Real Number System

We assume that the reader has familiarity with the set $\mathbf{R}$ of real numbers and those of its basic properties, which are usually treated in an elementary course in analysis, namely, that it satisfies field axioms, the linear ordering axioms and the least upper bound axiom. In the present section, they are listed in detail. Beginning with the set of natural numbers $\mathbf{N}$, it can be shown that there exists a unique set $\mathbf{R}$ that satisfies these properties. The process, though, is lengthy and tedious. Later in the chapter we shall sketch one way of constructing $\mathbf{R}$ from $\mathbf{Q}$.

A. Field Axioms

For all real numbers x, y and z, we have

(i) $x + y = y + x$,
(ii) $(x + y) + z = x + (y + z)$,
(iii) there exists $0 \in \mathbf{R}$ such that $x + 0 = x$,

(iv) there exists a $w \in \mathbf{R}$ such that $x + w = 0$,

(v) $xy = yx$,

(vi) $(xy)z = x(yz)$,

(vii) there exists $1 \in \mathbf{R}$ such that $1 \neq 0$ and $x \cdot 1 = x$,

(viii) if x is different from 0, there exists a $w \in \mathbf{R}$ such that $xw = 1$,

(ix) $x(y + z) = xy + xz$.

The second group of properties possessed by the real numbers has to do with the fact that they are *ordered*. They can be phrased in terms of positivity of real numbers. When we do this, our second group of axioms takes the following form.

B. Order Axioms

The subset P of positive real numbers satisfies the following:

(i) P is closed with respect to addition and multiplication, that is, if x, $y \in P$, then so are $x + y$ and xy,

(ii) $x \in P$ implies $-x \notin P$,

(iii) $x \in \mathbf{R}$ implies $x = 0$ or $x \in P$ or $-x \in P$.

Any system satisfying the axioms of groups A and B is called an *ordered field*, for example, the rational numbers.

In an ordered field we define the notion $x < y$ to mean $y - x \in P$. We write $x \leq y$ to mean $x < y$ or $x = y$.

Absolute value is defined in any ordered field in the familiar manner:

$$|x| = \begin{cases} x & \text{if } x \geq 0, \\ -x & \text{if } x < 0. \end{cases}$$

It can be shown on the basis of this definition that the *triangle inequality*

$$|x + y| \leq |x| + |y|$$

or equivalently,

$$|x - y| \leq |x - z| + |z - x|$$

holds.

The third group of properties of real numbers contains only one axiom, and it is this axiom that sets apart the real numbers from other ordered fields. Before stating this axiom, we need to define some terms. Let X be a nonempty subset of $\mathbf{R}$. If there exists M such that $x \leq M$ for all $x \in X$, then X is said to be *bounded above* and M is said to be an *upper bound* of X. If there exists m such that $x \geq m$ for all $x \in X$, then X is said to be *bounded below* and m is said to be a *lower bound* of X. If X is bounded above as well as below, then it is said to be *bounded*. A number M' is called the *least upper bound* (or *supremum*) of X if it is an upper bound and $M' \leq M$ for each upper bound M of X. The final axiom guarantees the existence of least upper bounds for nonempty subsets of $\mathbf{R}$ that are bounded above.

C. Completeness Axiom

Every nonempty subset of $\mathbf{R}$ that has an upper bound possesses a least upper bound.

We shall denote the least upper bound of X by sup X or by sup $\{x : x \in X\}$ or by $\sup_{x \in X} x$. The *greatest lower bound* or *infimum* can be defined similarly. It follows from C above that every nonempty subset of $\mathbf{R}$ that has a lower bound possesses a greatest lower bound. The greatest lower bound of X is denoted by inf X or by inf $\{x : x \in X\}$ or by $\inf_{x \in X} x$. Note that $\inf_{x \in X} x = -\sup_{x \in X} - (x)$.

The following characterisation of supremum is used frequently.

Proposition 0.3.1. Let X be a nonempty set of real numbers that is bounded above. Then $M = \sup X$ if and only if

(i) $x \leq M$ for all $x \in X$, and
(ii) given any $\varepsilon > 0$, there exists $x \in X$ such that $x > M - \varepsilon$.

There is a similar characterisation of the infimum of a nonempty set of real numbers that is bounded below.

The Cartesian product $\mathbf{R} \times \mathbf{R}$ becomes a field under $+$ and $\cdot$ defined as follows:

$$(x_1, y_1) + (x_2, y_2) = (x_1 + x_2, \ y_1 + y_2),$$
$$(x_1, y_1) \cdot (x_2, y_2) = (x_1 x_2 - y_1 y_2, \ x_1 y_2 + y_1 x_2).$$

It is convenient to denote the ordered pair $(x, 0)$ by x and the ordered pair $(0, 1)$ by i. The reader will check that $(0, \ 1) \cdot (0, 1) = (-1, 0)$, i.e., $i^2 = -1$. The ordered pair (x, y) can now be written as

$$(x, y) = (x, 0) + (0, y) = (x, 0) + (0, 1) \cdot (y, 0) = x + iy.$$

This field is denoted by $\mathbf{C}$ and is called the *field of complex numbers.*

The absolute value of $x + iy$ is defined to be $\sqrt{x^2 + y^2}$. The triangle inequality, as stated above, holds for complex numbers x, y and z.

0.4. Sequences of Real Numbers

Functions that have the set $\mathbf{N}$ of natural numbers as domain play an important role in analysis. Such functions have special terminology and notation, which we describe below.

A *sequence* of real numbers is a map $x : \mathbf{N} \to \mathbf{R}$. Given such a map, we denote $x(n)$ by x_n, and this value is called the nth term of the sequence. The sequence itself is frequently denoted by $\{x_n\}_{n \geq 1}$. It is important to distinguish between the sequence $\{x_n\}_{n \geq 1}$ and its range $\{x_n : n \in \mathbf{N}\}$, which is a subset of $\mathbf{R}$. A real number l is said to be *a limit* of the sequence $\{x_n\}_{n \geq 1}$ if for each $\varepsilon > 0$, there is a positive integer n_0 such that for all $n \geq n_0$, we have $|x_n - l| < \varepsilon$. It is easy to verify that a sequence has at most one limit. When $\{x_n\}_{n \geq 1}$ does have a limit, we denote it by $\lim x_n$. In symbols,

$$l = \lim x_n \text{ if } \forall \varepsilon > 0, \ \exists n_0 \ni n \geq n_0 \Rightarrow |x_n - l| < \varepsilon.$$

A sequence that has a limit is said to *converge* (or to be *convergent*).

A sequence $\{x_n\}_{n \geq 1}$ of real numbers is said to be *increasing* if it satisfies the inequalities $x_n \leq x_{n+1}$, $n = 1, 2, \ldots$, and *decreasing* if it satisfies the inequalities $x_n \geq x_{n+1}$, $n = 1, 2, \ldots$. We say that the sequence is *monotone* if it is either increasing or it is decreasing.

A sequence $\{x_n\}_{n \geq 1}$ of real numbers is said to be bounded if there exists a real number $M > 0$ such that $|x_n| \leq M$ for all $n \in \mathbf{N}$. The following simple criterion for the convergence of a monotone sequence is very useful.

Proposition 0.4.1. A monotone sequence of real numbers is convergent if and only if it is bounded.

Let $\{x_n\}_{n \geq 1}$ be a sequence of real numbers and let $r_1 < r_2 < \ldots < r_n < \ldots$ be a *strictly* increasing sequence of natural numbers. Then $\{x_{r_n}\}_{n \geq 1}$ is called a *subsequence* of $\{x_n\}_{n \geq 1}$.

Proposition 0.4.2. (Bolzano–Weierstrass Theorem) A bounded sequence of real numbers has a convergent subsequence.

The convergence criterion described in Proposition 0.4.1 is restricted to monotone sequences. It is important to have a condition implying the convergence of a sequence of real numbers that is applicable to a larger class and preferably does not require knowledge of the value of the limit. The Cauchy criterion gives such a condition.

A sequence $\{x_n\}_{n \geq 1}$ of real numbers is said to be a *Cauchy sequence* if, for every $\varepsilon > 0$, there exists an integer n_0 such that $|x_n - x_m| < \varepsilon$ whenever $n \geq n_0$ and $m \geq n_0$. In symbols,

$$\forall \varepsilon > 0, \ \exists n_0 \ni (m \geq n_0, \ n \geq n_0) \Rightarrow |x_n - x_m| < \varepsilon.$$

Proposition 0.4.3. (Cauchy Convergence Criterion) A sequence of real numbers converges if and only if it is a Cauchy sequence.

Let $\{x_n\}_{n \geq 1}$ be a bounded sequence in $\mathbf{R}$. Then the limit superior of $\{x_n\}_{n \geq 1}$ is defined by

$$\limsup_{n} x_n = \inf_{n} \{ \sup_{k \geq n} x_k \},$$

and the limit inferior of $\{x_n\}_{n \geq 1}$ is defined by

$$\liminf_{n} x_n = \sup_{n} \{ \inf_{k \geq n} x_k \}.$$

Observe that a sequence $\{x_n\}_{n \geq 1}$ is convergent if and only if

$$\limsup x_n = \liminf x_n.$$

In case $\{x_n\}_{n \geq 1}$ is not bounded from above [respective, below], one defines $\limsup x_n = \infty$ [respective, $\liminf x_n = -\infty$].

0.5. Limits of Functions and Continuous Functions

Mathematical analysis is primarily concerned with limit processes. We have already met one of the basic limit processes, namely, convergence of a sequence of real numbers. In this section, we shall recall the notion of the limit of a function, which is used in the study of continuity, differentiation and integration. The notion is parallel to that of the limit of a sequence. We shall also state the definition of continuity and its relation to limits.

A point $a \in \mathbf{R}$ is said to be a **limit point** of a subset $X \subseteq \mathbf{R}$ if every neighbourhood $(a - \varepsilon, a + \varepsilon)$, $\varepsilon > 0$, of a contains a point $x \neq a$ such that $x \in X$.

Let f be a real-valued function defined on a subset X of $\mathbf{R}$ and a be a limit point of X. We say that $f(x)$ tends to l as x tends to a if, for every $\varepsilon > 0$, there exists some $\delta > 0$ such that

$$|f(x) - l| < \varepsilon \ \forall x \in X \text{ for which } 0 < |x - a| < \delta.$$

The number l is said to be the limit of $f(x)$ as x tends to a, and we write

$$\lim_{x \to a} f(x) = l \qquad \text{or} \qquad f(x) \to l \text{ as } x \to a.$$

Note that $f(a)$ need not be defined for the above definition to make sense. Moreover, the value l of the limit is uniquely determined when it exists.

The following important formulation of limit of a function is in terms of limits of sequences.

Proposition 0.5.1. Let $f: X \to \mathbf{R}$ and let a be a limit point of X. Then $\lim_{x \to a} f(x) = l$ if and only if, for every sequence $\{x_n\}_{n \geq 1}$ in X that converges to a and $x_n \neq a$ for every n, the sequence $\{f(x_n)\}_{n \geq 1}$ converges to l.

Let f be a real-valued function whose domain of definition is a set X of real numbers. we say that f is *continuous at the point* $x \in X$ if, given $\varepsilon > 0$, there exists a $\delta > 0$ such that for all $y \in X$ with $|y - x| < \delta$, we have $|f(y) - f(x)| < \varepsilon$. The function is said to be *continuous on X* if it is continuous at every point of X. If we merely say that a function is 'continuous', we mean that it is continuous on its domain.

It may checked that f is continuous at a limit point $a \in X$ if and only if $f(a)$ is defined and $\lim_{x \to a} f(x) = f(a)$. The following criterion of continuity of f at a point $a \in X$ follows immediately from the preceding criterion and Proposition 0.5.1.

Proposition 0.5.2. Let f be a real-valued function defined on a subset X of $\mathbf{R}$ and $a \in X$ be a limit point of X. Then f is continuous at a if and only if, for every sequence $\{x_n\}_{n \geq 1}$ in X that converges to a and $x_n \neq a$ for every n, $\lim f(x_n) = f(\lim x_n) = f(a)$.

This result shows that continuous functions are precisely those which send convergent sequences into convergent sequences, in other words, they 'preserve' convergence.

The next result, which is known as the *Bolzano intermediate value theorem*, guarantees that a continuous functions on an interval assumes (at least once) every value that lies between any two of its values.

Proposition 0.5.3. Let I be an interval and $f: I \to \mathbf{R}$ be a continuous mapping on I. If $a, b \in I$ and $\alpha \in \mathbf{R}$ satisfies $f(a) < \alpha < f(b)$ or $f(a) > \alpha > f(b)$, then there exists a point $c \in I$ between a and b such that $f(c) = \alpha$.

0.6. Sequences of Functions

Let X be a subset of $\mathbf{R}$. If to every $n = 1, 2, \ldots$ is assigned a real-valued function f_n defined on X, then $\{f_n\}_{n \geq 1}$ is called a sequence of functions on X.

The sequence $\{f_n\}_{n \geq 1}$ is said to *converge pointwise* on X if for each $x \in X$, the sequence $f_1(x)$, $f_2(x)$, $\ldots$ of real numbers is convergent. In this case we define a function f on X by taking $f(x) = \lim f_n(x)$ for every $x \in X$. This function is called the *pointwise limit* of the sequence $\{f_n\}_{n \geq 1}$. Thus a function f which is defined on X is the pointwise limit of the sequence $\{f_n\}_{n \geq 1}$ if, given $x \in X$ and $\varepsilon > 0$, there is an integer n_0 (depending on both x and ε) such that we have $|f(x) - f_n(x)| < \varepsilon$ for all $n \geq n_0$. In symbols,

given $\varepsilon > 0$ and $x \in X$, $\exists$ an integer $n_0 = n_0(x, \varepsilon) \ni n \geq n_0 \Rightarrow |f(x) - f_n(x)| < \varepsilon$.

Recall that a series $\sum_{n=1}^{\infty} x_n$ of real numbers converges to $x \in \mathbf{R}$ if the sequence $\{s_n\}_{n \geq 1}$, where $s_n = \sum_{k=1}^{n} x_k$ (the nth partial sum), converges to x. We write $x = \lim s_n = \sum_{n=1}^{\infty} x_n$ and x is called the *sum of the series*. If $\sum_{n=1}^{\infty} f_n(x)$ converges for every $x \in X$, and if we define $f(x) = \sum_{n=1}^{\infty} f_n(x)$, $x \in X$, the function f is called the sum of the series $\sum_{n=1}^{\infty} f_n$.

A sequence $\{f_n\}_{n \geq 1}$ of functions defined on a set $X \subseteq \mathbf{R}$ is said to *converge uniformly* on X to a function f if, given $\varepsilon > 0$, there is an integer n_0 (depending on ε only) such that for all $x \in X$ and all $n \geq n_0$, we have $|f(x) - f_n(x)| < \varepsilon$. In symbols,

given $\varepsilon > 0$, $\exists$ an integer $n_0 = n_0(\varepsilon) \ni n \geq n_0 \Rightarrow |f(x) - f_n(x)| < \varepsilon \ \forall x \in X$.

The statement '$\{f_n\}_{n \geq 1}$ converges uniformly to f' is written as '$\lim f_n = f$ uniformly' or as '$\lim f_n = f$ (unif)'.

It is clear that every uniformly convergent sequence is pointwise convergent. The converse is, however, not true.

We say that the series $\sum_{n=1}^{\infty} f_n$ converges uniformly on X if the sequence $\{s_n\}_{n \geq 1}$ of functions, where $s_n(x) = \sum_{k=1}^{n} f_k(x)$, $x \in X$, converges uniformly on X. The Cauchy criterion of uniform convergence of sequences of functions is as follows.

Proposition 0.6.1. The sequence of functions $\{f_n\}_{n \geq 1}$ defined on X converges uniformly on X if and only if, given $\varepsilon > 0$, there exists an integer n_0 such that, for all $x \in X$ and all $n \geq n_0$, $m \geq n_0$, we have $|f_n(x) - f_m(x)| < \varepsilon$.

The limit of a uniformly convergent sequence of continuous functions is continuous. More precisely, the following is true:

Proposition 0.6.2. Suppose $\{f_n\}_{n \geq 1}$ is a sequence of continuous functions defined on X that converges uniformly to f. Then f is continuous.

0.7. Compact Sets

The notion of compactness, which is of enormous significance in the study of metric spaces, or more generally in analysis, is an abstraction of an important property possessed by certain subsets of real numbers. The property in question asserts that every open cover of a closed and bounded subset of **R** has a finite subcover. This simple property of closed and bounded subsets has far reaching implications in analysis; for example, a real-valued continuous function defined on [0,1], say, is bounded and uniformly continuous. In what follows, we shall define the notion of compactness in **R** and list some of its characterisations. To begin with, we recall the definition of an open subset of **R**. A subset G of **R** is said to be *open* if for each $x \in G$, there is a neighbourhood $(x - \varepsilon, \ x + \varepsilon)$, $\varepsilon > 0$, of x that is contained in G.

Let X be a subset of **R**. An *open cover* (covering) of X is a collection $C = \{G_\alpha : \alpha \in \Lambda\}$ of open sets in **R** whose union contains X, that is,

$$X \subseteq \bigcup_\alpha G_\alpha.$$

If C' is a subcollection of C such that the union of sets in C' also contains X, then C' is called a *subcover* (or subcovering) from C of X. If C' consists of finitely many sets, then we say that C' is a *finite subcover* (or finite subcovering).

A subset X of **R** is said to be *compact* if every open cover of X contains a finite subcover. The following proposition characterises compact subsets of **R**.

Proposition 0.7.1. (Heine-Borel Theorem) Let X be a set of real numbers. Then the following statements are equivalent:

(i) X is closed and bounded.
(ii) X is compact.
(iii) Every infinite subset of X has a limit point in X.

Proposition 0.7.2. Let f be a real-valued continuous function defined on the closed bounded interval $I = [a, b]$. Then f is bounded on I and assumes its maximum and minimum values on I, that is, there are points x_1 and x_2 in I such that $f(x_1) \leq f(x) \leq f(x_2)$ for all $x \in X$.

For our next proposition we shall need the following definition. Let f be a real-valued continuous function defined on a set X. Then f is said to be *uniformly continuous* on X if, given $\varepsilon > 0$, there is a $\delta > 0$ such that for all $x, \ y \in X$ with $|x - y| < \delta$, we have $|f(x) - f(y)| < \varepsilon$.

Proposition 0.7.3. If a real-valued function f is continuous on a closed and bounded interval I, then f is uniformly continuous on I.

0.8. Derivative and Riemann Integral

A function $f\colon [a, b] \to \mathbf{R}$ is differentiable at a point $c \in [a, b]$ if

$$\lim_{h \to 0} \frac{f(c+h) - f(c)}{h}$$

exists, in which case, the limit is called the *derivative* of f at c and is denoted by $f'(c)$.

Let $[a, b]$ be a closed and bounded interval of $\mathbf{R}$ and f a real-valued function defined on $[a, b]$. As is well known, the (Riemann) integral

$$\int_a^b f(x)\,dx$$

is defined as the limit of Riemann sums

$$\sum_{j=1}^n f(\tilde{x}_j)(x_j - x_{j-1}),$$

where $x_j, j = 0,\ 1,\ 2, \ldots, n$, form a partition of $[a, b]$,

$$a = x_0 < x_1 < \ldots < x_n = b,$$

and $\tilde{x}_j \in [x_{j-1}, x_j]$, $j = 1,\ 2, \ldots, n$ are arbitrary. Recall that the integral exists, for instance, if f is continuous or monotone.

If f and g are Riemann integrable on $[a, b]$ and $f(x) \le g(x)$ for each $x \in [a, b]$, then $\int_a^b f(t)\,dt \le \int_a^b g(t)\,dt$.

If f is Riemann integrable on $[a, b], f \ge 0$ and $[\alpha, \beta] \subseteq [a, b]$, then f is Riemann integrable on $[\alpha, \beta]$ and $\int_\alpha^\beta f(t)\,dt \le \int_a^b f(t)\,dt$.

The following is known as the *fundamental theorem of integral calculus*.

Proposition 0.8.1. Let $f\colon [a, b] \to \mathbf{R}$ be integrable and let

$$F(x) = \int_a^x f(t)\,dt, \ a < x < b.$$

Then F is continuous on $[a, b]$. Moreover, if f is continuous at a point $c \in [a, b]$, then F is differentiable at c and

$$F'(c) = f(c).$$

Proposition 0.8.2. Suppose φ has a continuous derivative on $[a, b]$ and f is continuous on the image of the interval $[a, b]$. Then

$$\int_{\varphi(a)}^{\varphi(b)} f(t)\,dt = \int_a^b f(\varphi(u))\varphi'(u)\,du.$$

Proposition 0.8.3. Suppose $\{f_n\}_{n \geq 1}$ is a sequence of Riemann integrable functions on $[a, b]$ with uniform limit f. Then f is Riemann integrable on $[a, b]$ and

$$\lim \int_a^b f_n(t)\,dt = \int_a^b f(t)\,dt.$$

We end this section with a sketch of the proof of the Weierstrass approximation theorem.

Proposition 0.8.4. Suppose f is a real-valued continuous function defined on $[0, 1]$. Then there exists a sequence $\{P_n\}_{n \geq 1}$ of polynomials with real coefficients that converges uniformly to f on $[0, 1]$, that is,

$$\forall \varepsilon > 0, \ \exists n_0 \ni n \geq n_0 \Rightarrow |P_n(t) - f(t)| < \varepsilon \ \forall t \in [0, 1].$$

Proof. Without loss of generality, we may assume that $f(0) = f(1) = 0$. This is because $g(t) = f(t) - f(0) - (f(1) - f(0))t$ is a continuous function satisfying $g(0) = g(1) = 0$, and if $\{Q_n\}_{n \geq 1}$ is a sequence of polynomials converging uniformly to g, then the sequence of polynomials $\{P_n\}_{n \geq 1}$, where $P_n(t) = Q_n(t) + (f(1) - f(0))t + f(0)$, converges uniformly to f.

Extend the function f to the whole of $\mathbf{R}$ by setting $f(t) = 0$ for $t \in \mathbf{R} \setminus [0, 1]$. The extended function is clearly continuous on $\mathbf{R}$. For $n = 1, 2, \ldots$, let

$$Q_n(t) = \begin{cases} \alpha_n(1 - t^2)^n & |t| \leq 1 \\ 0 & |t| > 1, \end{cases}$$

where

$$\frac{1}{\alpha_n} = \int_{-1}^1 (1 - t^2)^n\,dt,$$

and let

$$P_n(t) = \int_{-1}^1 f(t - s)Q_n(s)\,ds. \tag{0.1}$$

Since $Q_n(t) = Q_n(-t)$ and $(1 - t^2)^n \geq 1 - nt^2$ for $-1 \leq t \leq 1$,

$$\int_{-1}^1 (1 - t^2)^n\,dt = 2\int_0^1 (1 - t^2)^n\,dt \geq 2\int_0^{1/\sqrt{n}} (1 - t^2)^n\,dt \geq 2\int_0^{1/\sqrt{n}} (1 - nt^2)\,dt \geq \frac{4}{3\sqrt{n}} \geq \frac{1}{\sqrt{n}}.$$

So, $\alpha_n \leq \sqrt{n}$. For any $\delta > 0$, this implies

$$Q_n(t) \leq \sqrt{n}(1 - \delta^2)^n, \quad \text{where } \delta \leq |t| \leq 1. \tag{0.2}$$

It is obvious that $\int_{-1}^{1} Q_n(t)dt = 1$. The function f, being continuous on $[0,1]$, is uniformly continuous; given $\varepsilon > 0$, there exists a $\delta > 0$ such that s, $t \in [0,1]$ and $|s - t| < \delta$ imply $|f(s) - f(t)| < \varepsilon/2$. As f vanishes outside $[0,1]$, we have $|f(s) - f(t)| < \varepsilon/2$ for all s, $t \in \mathbf{R}$ with $|s - t| < \delta$.

Since $f(t) = 0$ for $t \in \mathbf{R}\backslash[0,1]$, it follows by a simple change of variable that

$$P_n(t) = \int_0^1 f(s)Q_n(t - s)ds. \qquad (0.3)$$

Now, the integral on the right of (3) is a polynomial in t. Thus, $\{P_n\}_{n \geq 1}$ is a sequence of polynomials.

Let $M = \sup|f(t)| : t \in \mathbf{R}$. For $0 \leq t \leq 1$ and any positive integer n, using (0.1), (0.2) and (0.3), we have

$$|P_n(t) - f(t)| = \left| \int_{-1}^{1} [f(t - s) - f(t)]Q_n(s)ds \right|$$

$$\leq \int_{-1}^{1} |f(t - s) - f(t)|Q_n(s)ds$$

$$\leq 2M \int_{-1}^{-\delta} Q_n(s)ds + \left(\frac{\varepsilon}{2}\right) \int_{-\delta}^{\delta} Q_n(s)ds + 2M \int_{\delta}^{1} Q_n(s)ds$$

$$\leq 4M\sqrt{n}(1 - \delta^2)^n + \frac{\varepsilon}{2}.$$

This is less than ε for sufficiently large n, because when $0 < \delta < 1$, $\lim \sqrt{n}(1 - \delta^2)^n$ can be shown to be 0 as follows:

$$0 \leq \sqrt{n}(1 - \delta^2)^n = \sqrt{n}(1 + \beta)^{-n} \leq \beta^{-1}/\sqrt{n}, \text{ where } (1 - \delta^2) = 1/\beta, \ \beta > 0. \quad \square$$

0.9. Cantor's Construction

In this section we sketch one way of constructing $\mathbf{R}$ from $\mathbf{Q}$ (the axioms for $\mathbf{Q}$ will be assumed). The reasons for the inclusion of this approach are twofold; firstly, it is one of the quickest ways of obtaining $\mathbf{R}$, and secondly, it has a close connection with completion of metric spaces (see Section 1.5).

Definition 0.9.1. Let $\mathbf{Q}$ denote the field of rational numbers. A sequence $\{x_n\}_{n \geq 1}$ in $\mathbf{Q}$ is said to be **bounded** if there exists a rational number K such that

$$|x_n| \leq K \qquad \text{for all } n.$$

Definition 0.9.2. A sequence $\{x_n\}_{n \geq 1}$ in $\mathbf{Q}$ is said to be **Cauchy** if for each rational number $\varepsilon > 0$ there exists an integer n_0 such that

$$|x_n - x_m| < \varepsilon \qquad \text{for all } n, m \geq n_0.$$

Definition 0.9.3. A sequence $\{x_n\}_{n \geq 1}$ in $\mathbf{Q}$ is said to **converge to** a rational number x if for each rational number $\varepsilon > 0$ there exists an integer n_0 such that

$$|x_n - x| < \varepsilon \qquad \text{for all } n \geq n_0.$$

In symbols, $\lim_{n \to \infty} x_n = x$. The rational number x is called the **limit** of the sequence.

It may be easily verified that

(i) A convergent sequence in $\mathbf{Q}$ is a Cauchy sequence in $\mathbf{Q}$;

(ii) a Cauchy sequence in $\mathbf{Q}$ is bounded; in particular, every convergent sequence in $\mathbf{Q}$ is bounded;

(iii) if a sequence $\{x_n\}_{n \geq 1}$ converges to x as well as y, then $x = y$. Thus, the symbol $\lim_{n \to \infty} x_n$ is unambiguously defined when the sequence $\{x_n\}_{n \geq 1}$ converges.

Let $F_\mathbf{Q}$ denote the set of all Cauchy sequences in $\mathbf{Q}$.

Definition 0.9.4. A sequence $\{x_n\}_{n \geq 1}$ in $F_\mathbf{Q}$ is said to be **equivalent** to a sequence $\{y_n\}_{n \geq 1}$ in $F_\mathbf{Q}$ if and only if $\lim_{n \to \infty} |x_n - y_n| = 0$. In symbols, $\{\mathbf{x_n}\}_{\mathbf{n} \geq \mathbf{1}} \sim \{\mathbf{y_n}\}_{\mathbf{n} \geq \mathbf{1}}$.

The relation $\sim$ defined in $F_\mathbf{Q}$ is an equivalence relation, as is shown below:

(i) Reflexivity: $\{x_n\}_{n \geq 1} \sim \{x_n\}_{n \geq 1}$, since $|x_n - x_n| = 0$ for every n, so that $\lim_{n \to \infty} |x_n - x_n| = 0$.

(ii) Symmetry: If $\{x_n\}_{n \geq 1} \sim \{y_n\}_{n \geq 1}$, then $\lim_{n \to \infty} |x_n - y_n| = 0$; but $|x_n - y_n| = |y_n - x_n|$ for every n and therefore $\lim_{n \to \infty} |y_n - x_n| = 0$, so that $\{y_n\}_{n \geq 1} \sim \{x_n\}_{n \geq 1}$.

(iii) Transitivity: Suppose $\{x_n\}_{n \geq 1} \sim \{y_n\}_{n \geq 1}$ and $\{y_n\}_{n \geq 1} \sim \{z_n\}_{n \geq 1}$. Then $\lim_{n \to \infty} |x_n - y_n| = 0 = \lim_{n \to \infty} |y_n - z_n|$. Since $0 \leq |x_n - z_n| \leq |x_n - y_n| + |y_n - z_n|$ for all n, it follows that $\{x_n\}_{n \geq 1} \sim \{z_n\}_{n \geq 1}$.

Thus, the relation splits $F_\mathbf{Q}$ into equivalence classes. Any two members of the same equivalence class are equivalent, while no member of an equivalence class is equivalent to a member of any other equivalence class. The equivalence class containing the sequence $\{x_n\}_{n \geq 1}$ will be denoted by $[\{x_n\}_{n \geq 1}]$ or simply $[x_n]$ for short, i.e.,

$$[x_n] = \{\{y_n\}_{n \geq 1} \in F_\mathbf{Q} : \{y_n\}_{n \geq 1} \sim \{x_n\}_{n \geq 1}\}.$$

Henceforth we shall abbreviate $\{x_n\}_{n \geq 1}$ as simply $\{x_n\}$ whenever convenient.

Proposition 0.9.5. If $\{x_n\} \in F_\mathbf{Q}$ then $\lim_{n \to \infty} x_n = x$ if and only if $\{x_n\} \sim \{x\}$, where $\{x\}$ denotes the constant sequence with each term equal to x.

Proof. If $\{x_n\} \sim \{x\}$, then by definition of $\sim$, it follows that $\lim_{n \to \infty} |x_n - x| = 0$. On the other hand, if $\lim_{n \to \infty} |x_n - x| = 0$, then $\{x_n\} \sim \{x\}$, since the sequence $\{x\}$ has limit x. $\qquad\qquad\square$

Proposition 0.9.6. If $\{x_n\}$ and $\{y_n\}$ are in $F_{\mathbf{Q}}$, then so are the sequences $\{x_n + y_n\}$ and $\{x_n y_n\}$.

Proof. Let $\varepsilon > 0$ be a rational number. There exist integers n_1 and n_2 such that

$$|x_n - x_m| < \frac{\varepsilon}{2} \qquad \text{for all } n, m \geq n_1$$

and

$$|y_n - y_m| < \frac{\varepsilon}{2} \qquad \text{for all } n, m \geq n_2.$$

Let $n_0 \geq \max\{n_1, \ n_2\}$. Then for $n, \ m \geq n_0$ we have

$$|(x_n + y_n) - (x_m + y_m)| \leq |x_n - x_m| + |y_n - y_m| < \varepsilon.$$

So $\{x_n + y_n\}$ is a Cauchy sequence of rational numbers.

Since $\{x_n\}$ and $\{y_n\}$ are Cauchy sequences of rational numbers, there exist rational numbers K_1 and K_2 such that

$$|x_n| \leq K_1 \text{ and } |y_n| \leq K_2 \qquad \text{for all } n.$$

Now, for $n, \ m \geq n_0$, we have

$$
\begin{aligned}
|x_n y_n - x_m y_m| &= |y_n(x_n - x_m) + x_m(y_n - y_m)| \\
&\leq |y_n| \cdot |x_n - x_m| + |x_m| \cdot |y_n - y_m| \\
&< \left(\frac{\varepsilon}{2}\right) K_2 + \left(\frac{\varepsilon}{2}\right) K_1 = (K_1 + K_2)\left(\frac{\varepsilon}{2}\right).
\end{aligned}
$$

Since K_1 and K_2 are fixed, it follows that $\{x_n y_n\}$ is a Cauchy sequence of rational numbers. $\qquad\square$

Proposition 0.9.7. If $\{x_n\}$, $\{y_n\}$, $\{x'_n\}$ and $\{y'_n\}$ are in $F_{\mathbf{Q}}$ and $\{x_n\} \sim \{x'_n\}$, $\{y_n\} \sim \{y'_n\}$, then $\{x_n + y_n\} \sim \{x'_n + y'_n\}$ and $\{x_n y_n\} \sim \{x'_n y'_n\}$.

Proof. The fact that $\{x_n + y_n\} \sim \{x'_n + y'_n\}$ is easy to prove. We proceed to prove the other part. Since every Cauchy sequence is bounded, there exist rational constants K_1 and K_2 such that

$$|x_n| \leq K_1 \text{ and } |y'_n| \leq K_2 \qquad \text{for all } n.$$

Let $\varepsilon > 0$ be a given rational number. Since $\{x_n\} \sim \{x'_n\}$, and $\{y_n\} \sim \{y'_n\}$, there exists n_0 such that for $n \geq n_0$ we have

$$|x_n - x'_n| < \left(\frac{\varepsilon}{2K_2}\right) \text{ and } |y_n - y'_n| < \left(\frac{\varepsilon}{2K_1}\right).$$

For $n \geq n_0$ we get

$$\begin{aligned} \left| x_n y_n - x_n' y_n' \right| &\leq \left| x_n(y_n - y_n') \right| + \left| y_n'(x_n - x_n') \right| \\ &\leq |x_n| \cdot |y_n - y_n'| + |y_n'| \cdot |x_n - x_n'| \\ &< \frac{\varepsilon}{2} + \frac{\varepsilon}{2} = \varepsilon, \end{aligned}$$

which, in turn, implies $\lim_{n\to\infty} \left| x_n y_n - x_n' y_n' \right| = 0$. □

Proposition 0.9.8. If $\{x_n\}$ is a Cauchy sequence in $\mathbf{Q}$ that does not have limit 0, then there exists a Cauchy sequence $\{y_n\}$ in $\mathbf{Q}$ such that $\lim_{n\to\infty} |x_n y_n - 1| = 0$.

Proof. Since $\{x_n\}$ does not have limit 0, there is a positive number α in $\mathbf{Q}$ such that for every $n \in \mathbf{N}$, there exists some $k \in \mathbf{N}$ such that $k \geq n$ and

$$|x_k| \geq \alpha.$$

Since $\{x_n\}$ is a Cauchy sequence, there exists n_0 in $\mathbf{N}$ such that

$$|x_n - x_m| < \frac{\alpha}{2} \qquad \text{for } m, n \geq n_0.$$

If $|x_{k_0}| \geq \alpha$, where $k_0 > n_0$, then

$$|x_n| = |x_{k_0} - (x_{k_0} - x_n)| \geq |x_{k_0}| - |x_{k_0} - x_n| > \alpha - \frac{\alpha}{2} = \frac{\alpha}{2}$$

for all $n \geq n_0$. Hence, $x_n \neq 0$ for all $n \geq n_0$.
 Let

$$y_n = \begin{cases} 1 & \text{if } n < n_0, \\ \dfrac{1}{x_n} & \text{if } n \geq n_0. \end{cases}$$

Then $\{y_n\}$ is a sequence in $\mathbf{Q}$. If ε is any positive rational number, there exists n_ε such that

$$|x_m - x_n| < \frac{\alpha^2 \varepsilon}{4} \quad \text{for all } m, n \geq n_\varepsilon.$$

Hence,

$$|y_m - y_n| = \frac{|x_m - x_n|}{|x_m||x_n|} < \left(\frac{\alpha^2 \varepsilon}{4}\right)\left(\frac{4}{\alpha^2}\right) = \varepsilon$$

for all $m, n \geq \max\{n_0, n_\varepsilon\}$. Thus $\{y_n\}$ is a Cauchy sequence in $\mathbf{Q}$. Since $x_n y_n = 1$ for all $n \geq n_0$, $\lim_{n\to\infty} |x_n y_n - 1| = 0$. □

0.10. Addition, Multiplication and Order in R

Definition 0.10.1. A **real number** is an equivalence class $[x_n]$ with respect to the equivalence relation $\sim$ defined in F_Q.

Let $\mathbf{R}$ denote the set of all real numbers. If $\xi \in \mathbf{R}$ is $[x_n]$, then

$$\xi = \{\, \{y_n\}_{n \geq 1} \in F_Q \colon \{y_n\}_{n \geq 1} \sim \{x_n\}_{n \geq 1}\}.$$

Definition 0.10.2. If $\xi = [x_n]$ and $\eta = [y_n]$ are in $\mathbf{R}$, we define the **sum $\xi + \eta$** as

$$\xi + \eta = [x_n] + [y_n] = [x_n + y_n]$$

and the **product $\xi\eta$** as

$$\xi\eta = [x_n] \cdot [y_n] = [x_n y_n].$$

Proposition 0.10.3. The operations of addition and multiplication are well defined. With these operations of addition and multiplication, $\mathbf{R}$ is a field.

Proof. It follows from Proposition 0.9.7 that addition and multiplication are well defined. It can be easily verified that $(\mathbf{R}, +, \cdot)$ is a field with $[\{1\}]$ and $[\{0\}]$ serving as the multiplicative and additive identities, respectively, and the additive inverse of $[x_n]$ being $[-x_n]$. When $[x_n] \neq [\{0\}]$, it is possible that $x_n = 0$ for some n; consequently, the proof of the existence of its multiplicative inverse requires some care: If $[x_n] \neq [\{0\}]$, then $\{x_n\}$ is not equivalent to $\{0\}$, so that $\lim_{n\to\infty} x_n \neq 0$ in $\mathbf{Q}$. Hence, by Proposition 0.9.8 there is a sequence $\{y_n\}$ in $\mathbf{Q}$ such that $\lim_{n\to\infty} x_n y_n = 1$, i.e., $\{x_n y_n\} \sim \{1\}$. It follows that $[y_n]$ is a multiplicative inverse of $[x_n]$. $\qquad\square$

Definition 0.10.4. A sequence $\{x_n\}$ in $\mathbf{Q}$ is said to be a **positive sequence** if there exists a positive rational number α and a positive integer m such that

$$x_n > \alpha \quad \text{for all } n \geq m.$$

The first of the following two facts can now be easily verified by the reader:

I. If $\{x_n\}$ and $\{y_n\}$ are positive sequences of rational numbers, then so are

$$\{x_n + y_n\} \text{ and } \{x_n y_n\}.$$

II. If $\{x_n\}$ is a positive sequence of rational numbers and $\{x_n\} \sim \{x'_n\}$, then $\{x'_n\}$ is also a positive sequence of rational numbers.

As regards II, there exists a positive rational number α and a positive integer m_1 such that

$$x_n > \alpha \qquad \text{for all } n \geq m_1.$$

Since $\{x_n\} \sim \{x'_n\}$, there exists a positive integer m_2 such that

$$\left| x_n - x'_n \right| < \frac{\alpha}{2} \qquad \text{for } n \geq m_2.$$

Let $m_0 = \max\{m_1, m_2\}$. For $n \geq m_0$, we have

$$x_n - \frac{\alpha}{2} < x'_n < x_n + \frac{\alpha}{2}.$$

So,

$$x'_n > \alpha - \frac{\alpha}{2} = \frac{\alpha}{2} \qquad \text{for } n \geq m_0.$$

Definition 0.10.5. A number $\xi \in \mathbf{R}$ is said to be a **positive** real number if it contains a positive sequence. The set of all positive real numbers will be denoted by $\mathbf{R}^+$.

We note that by II above, number $\xi \in \mathbf{R}$ is positive if and only if *every* sequence belonging to it is positive. Also,

$$\mathbf{R}^+ = \{\xi \in \mathbf{R} : \xi \text{ is positive}\}$$
$$= \{\xi \in \mathbf{R} : \text{some } \{x_n\} \in \xi \text{ is a positive sequence of rational numbers}\}.$$

It is clear from I above that $\mathbf{R}^+$ is closed under addition and multiplication, i.e., if ξ, $\eta \in \mathbf{R}^+$, then so are $\xi + \eta$ and $\xi\eta$.

Proposition 0.10.6. If $\xi \in \mathbf{R}$, then one and only one of the following must hold:

(i) $\xi = 0$,
(ii) $\xi \in \mathbf{R}^+$,
(iii) $-\xi \in \mathbf{R}^+$.

Definition 0.10.7. For ξ, $\eta \in \mathbf{R}$, we define $\xi > \boldsymbol{\eta}$ if $\xi - \eta \in \mathbf{R}^+$. Also, we define the **absolute value** $|\xi|$ of ξ in the usual manner to be 0 if $\xi = 0$, ξ if $\xi \in \mathbf{R}^+$ and to be $-\xi$ if $-\xi \in \mathbf{R}^+$.

With the above definition of order (the relation $>$), the field $\mathbf{R}$ can be shown to be an ordered field in the sense that the following statements are true:

Transitivity: $\xi > \eta > \zeta$ implies that $\xi > \zeta$.
Compatibility of order with addition: $\xi > \eta$ and $\zeta \in \mathbf{R}$ implies $\xi + \zeta > \eta + \zeta$.
Compatibility of order with multiplication: $\xi > \eta$ and $\zeta > 0$ implies $\xi\zeta > \eta\zeta$.

Proposition 0.10.8. The mapping $i : \mathbf{Q} \to \mathbf{R}$ defined by $i(x) = \{x\}$ is an isomorphism of $\mathbf{Q}$ into $\mathbf{R}$. Moreover, i preserves order and, hence, also absolute values.

Proof. If $x, y \in \mathbf{Q}$, then

$$i(x + y) = \{x + y\} = \{x\} \underset{\mathbf{R}}{+} \{y\} = i(x) \underset{\mathbf{R}}{+} i(y)$$

and

$$i(xy) = \{xy\} = \{x\} \underset{\mathbf{R}}{\cdot} \{y\} = i(x) \underset{\mathbf{R}}{\cdot} i(y).$$

Moreover, $x < y$ in $\mathbf{Q}$ if and only if $y - x > 0$ in $\mathbf{Q}$, so that $\{y - x\}$ is a positive sequence in $F_{\mathbf{Q}}$. $\qquad\qquad\square$

Proposition 0.10.9. For any sequence $\{x_n\} \in F_{\mathbf{Q}}$, we have $|[x_n]| = [|x_n|]$.

Proof. We begin by showing that $|[x_n]| = 0$ if and only if $[|x_n|] = 0$. To this end, $|[x_n]| = [\{0\}]$ if and only if $\{x_n\} \sim \{0\}$, which, in turn, is equivalent to $\lim_{n\to\infty} x_n = 0$. Similarly, $[|x_n|] = [\{0\}]$ if and only if $\lim_{n\to\infty} |x_n| = 0$. But $\lim_{n\to\infty} x_n = 0$ if and only if $\lim_{n\to\infty} |x_n| = 0$.

Now consider the case when $[x_n] > [\{0\}]$. As noted after Definition 0.10.5, $\{x_n\}$ must be a positive sequence, i.e., there exists a positive rational number α and a positive integer m such that $x_n > \alpha$ for $n \geq m$ (see Definition 0.10.4). By definition of absolute value in $\mathbf{Q}$, it follows that $|x_n| = x_n$ for $n \geq m$. This implies that $\{|x_n|\} \sim \{x_n\}$, so that $[|x_n|] = [x_n]$. On the other hand, by Definition 0.10.7, we also have $|[x_n]| = [x_n]$. Thus, $|[x_n]|$ and $[|x_n|]$ are both equal to $[x_n]$.

The case when $[x_n] < [\{0\}]$ is similar. $\qquad\qquad\square$

0.11. Completeness of R

Convergence in $\mathbf{R}$ is defined exactly as in Definition 0.9.3 but with 'rational' replaced by 'real'. We first prove that every Cauchy sequence of rationals converges in $\mathbf{R}$, or more precisely that its image under the order-preserving isomorphism $i: \mathbf{Q} \to \mathbf{R}$ has a limit in $\mathbf{R}$.

Proposition 0.11.1. If $\{x_n\}$ is a Cauchy sequence of rationals and $\{x_n\} \in \xi$, then $\lim_{n\to\infty} x_n = \xi$ in $\mathbf{R}$, i.e., $\lim_{n\to\infty} i(x_n) = \xi$, where i is as in Proposition 0.10.8.

Proof. Let ε be a positive real number. By Definition 0.10.5, ε contains a positive sequence $\{z_n\}$ of rational numbers. Therefore (see Definition 0.10.4) there exists a positive rational number 2α and some positive integer n_1 such that $z_n > 2\alpha$ for $n \geq n_1$. So, $z_n - \alpha > \alpha$ for $n \geq n_1$ and hence $\{z_n - \alpha\}$ is a positive rational sequence. It follows that $[z_n] > i(\alpha)$, i.e.,

$$\varepsilon > i(\alpha). \tag{0.3}$$

Since $\{x_n\}$ is a Cauchy sequence, there exists a positive integer n_2 such that $|x_n - x_m| < 2\alpha$ for $n, m \geq n_2$. Hence $\alpha - |x_n - x_m| > \alpha$ for $n, m \geq n_2$. Consequently, for each $n \geq n_2$, the sequence $\{|x_n - x_m|\}_{m \geq 1}$ has the property that

$$[|x_n - x_m|] < i(\alpha). \tag{0.4}$$

Therefore, $n \geq n_2$ implies

$$|i(x_n) - \xi| = |i(x_n) - [x_m]| = |[\{x_n - x_m\}_{m \geq 1}]| = [|x_n - x_m|] < i(\alpha) < \varepsilon,$$

using Proposition 0.10.9, (0.4) and (0.3) in that order. Thus, $\lim_{n \to \infty} x_n = \xi$. □

In order to avoid cluttered notation, we shall henceforth denote $i(x)$ by x. Thus, for example, in the next Corollary, $|\xi - x|$ means $|\xi - i(x)|$.

Corollary 0.11.2. If $\xi \in \mathbf{R}$ and $\varepsilon > 0$ in $\mathbf{R}$, then there is an $x \in \mathbf{Q}$ such that $|\xi - x| < \varepsilon$ is in $\mathbf{R}$.

Proof. Let $\{x_n\} \in \xi$. Then by Proposition 0.11.1, $\lim_{n \to \infty} x_n = \xi$. Therefore, there exists $n_0 \in \mathbf{N}$ such that

$$|\xi - x_n| < \varepsilon \text{ in } \mathbf{R} \qquad \text{for } n \geq n_0.$$

In particular, the number $x = x_{n_0}$ has the property that $x \in \mathbf{Q}$ and $|\xi - x| < \varepsilon$ in $\mathbf{R}$. □

Corollary 0.11.3. If $\xi < \eta$ in $\mathbf{R}$, there is a $z \in \mathbf{Q}$ such that $\xi < z < \eta$.

Proof. $\xi < \eta$ implies $2\xi = \xi + \xi < \xi + \eta < \eta + \eta = 2\eta$ and $2 = [\{1\}] + [\{1\}] > [\{0\}]$. It follows that $1/2 > [\{0\}]$ in $\mathbf{R}$ and hence that $\xi < (\xi + \eta)/2 < \eta$. Let $\zeta = (\xi + \eta)/2$. If

$$\varepsilon = \min\{\zeta - \xi, \ \eta - \zeta\},$$

then $\varepsilon > 0$ and by Corollary 0.11.2, there is a rational number z such that $\xi \leq \zeta - \varepsilon < z < \zeta + \varepsilon \leq \eta$. □

Corollary 0.11.4. $\mathbf{R}$ is Archimedean.

Proof. For $0 < \xi < \eta$ in $\mathbf{R}$, let x and y be rational numbers (see Corollary 0.11.3) such that

$$0 < x < \xi < \eta < y < \xi + \eta.$$

Since the field $\mathbf{Q}$ is Archimedean, there exists a positive integer n such that $nx > y$. Therefore,

$$n\xi > nx > y > \eta.$$ □

Theorem 0.11.5. Every Cauchy sequence of real numbers converges in $\mathbf{R}$.

Proof. Let $\{\xi_n\}_{n \geq 1}$ be a Cauchy sequence in $\mathbf{R}$. By Corollary 0.11.2, for each $n \in \mathbf{N}$, there exists a rational number x_n such that

$$|\xi_n - x_n| < 1/n.$$

We shall first show that $\{x_n\}_{n\geq 1}$ is a Cauchy sequence in $\mathbf{Q}$. For $\varepsilon > 0$ in $\mathbf{R}$, there exists $m_1 \in \mathbf{N}$ such that

$$|\xi_n - \xi_m| < \frac{\varepsilon}{3} \qquad \text{for } m, n \geq m_1.$$

Choose m_2 such that

$$\frac{1}{n} < \frac{\varepsilon}{3} \qquad \text{for all } n \geq m_2,$$

so that

$$|\xi_n - x_n| < \frac{\varepsilon}{3} \qquad \text{for all } n \geq m_2.$$

Let $m_0 = \max\{m_1, m_2\}$. Then $n,\ m \geq m_0$ implies

$$|x_n - x_m| = |x_n - \xi_n + \xi_n - \xi_m + \xi_m - x_m| \leq |x_n - \xi_n| + |\xi_n - \xi_m| + |\xi_m - x_m| < \varepsilon.$$

Therefore, $\{x_n\}_{n\geq 1}$ is a Cauchy sequence in $\mathbf{Q}$.

Now $[x_n]$ is a real number, say ξ. We shall show that $\lim_{n\to\infty} \xi_n = \xi$ in $\mathbf{R}$. For this purpose, consider any $\varepsilon > 0$ in $\mathbf{R}$. Let m_1 and m_2 be as above. By Proposition 0.11.1, $\lim_{n\to\infty} i(x_n) = \xi$ in $\mathbf{R}$, and therefore there exists $m_3 \in \mathbf{N}$ such that

$$|\xi - x_n| < \frac{2\varepsilon}{3} \text{ in } \mathbf{R} \qquad \text{for } n \geq m_3.$$

Therefore, for $n \geq m = \max\{m_2, m_3\}$, we have

$$|\xi_n - \xi| = |\xi_n - x_n + x_n - \xi| < |\xi_n - x_n| + |x_n - \xi| < \varepsilon.$$

This completes the proof that $\lim_{n\to\infty} \xi_n = \xi$ in $\mathbf{R}$. $\qquad\qquad\square$

Finally, the following result holds:

Theorem 0.11.6. Every nonempty subset of the real numbers that is bounded above has a supremum.

Proof. Suppose A is a subset containing an element α and having an upper bound β, so that $\alpha \leq \beta$. Since $\beta - \alpha \geq 0$, for any $n \in \mathbf{N}$, there exists an $m \in \mathbf{N}$ such that $m/n \geq \beta - \alpha$, i.e., $\alpha + m/n \geq \beta \in A$ and therefore $\alpha + m/n$ is an upper bound for A. Hence, for each $n \in \mathbf{N}$, the set

$$B_n = \{m \in \mathbf{N} : \alpha + \frac{m}{n} \text{is an upper bound for } A\}$$

is nonempty. Being a nonempty set of natural numbers, B_n must have a least element, say m_n. Therefore for each $n \in \mathbf{N}$,

$$y_n = \alpha + \frac{m_n}{n}$$

is an upper bound of A, i.e.,

$$x \leq y_n \qquad \text{for every } n \in \mathbf{N} \text{ and every } x \in A,$$

and also,

$$x_n = y_n - \frac{1}{n} = \alpha + \frac{m_n - 1}{n} \leq x \qquad \text{for some } x \in A.$$

Hence,

$$x_m \leq y_n \qquad \text{for all } m, n \in \mathbf{N}.$$

It follows that, for any $m, n \in \mathbf{N}$, we have

$$x_n - x_m \leq y_m - x_m = \frac{1}{m} \text{ and } x_m - x_n \leq y_n - x_n = \frac{1}{n}.$$

Therefore,

$$|x_n - x_m| \leq \max\left\{\frac{1}{n}, \frac{1}{m}\right\} \qquad \text{for all } m, n \in \mathbf{N}.$$

We use this inequality to argue that $\{x_n\}$ is a Cauchy sequence. Consider any $\varepsilon > 0$. Since $\mathbf{R}$ is Archimedean, there exists a positive integer $n_0 > 1/\varepsilon$. For $m, n \geq n_0$, it follows from the above inequality that $|x_n - x_m| < \varepsilon$. This shows that $\{x_n\}$ is a Cauchy sequence. Since $\mathbf{R}$ is complete (see Theorem 0.11.5) $\{x_n\}$ converges; let $\lim_{n \to \infty} x_n = \xi$ in $\mathbf{R}$.

We next show that ξ is an upper bound of A. If not, then there exists an $x \in A$ such that $\xi < x$. Since $\lim_{n \to \infty} x_n = \xi$ and $\mathbf{R}$ is Archimedean, there exists some $n \in \mathbf{N}$ such that

$$x_n - \xi \leq |x_n - \xi| < \frac{x - \xi}{2} \text{ and } \frac{1}{n} < \frac{x - \xi}{2}.$$

Then $y_n = x_n + 1/n < (x + \xi)/2 + (x - \xi)/2 = x$. But this is impossible because $x \in A$ and y_n is an upper bound of A. This contradiction shows that ξ is an upper bound of A.

Finally, we show that ξ is less than or equal to every upper bound of A. Suppose not. Consider any real number $\eta < \xi$. Let $\delta = \xi - \eta > 0$. Since $\lim_{n \to \infty} x_n = \xi$, therefore there exists $n \in \mathbf{N}$ (corresponding to the positive number δ) such that

$$\xi - x_n \leq |\xi - x_n| < \delta = \xi - \eta \text{ and hence } \eta < x_n.$$

But, as observed earlier, $x_n < x$ for some $x \in A$, whence we have $\eta < x$ for some $x \in A$. This implies that η is not an upper bound of A. Thus no real number less than ξ can be an upper bound of A. In other words, an upper bound of A cannot be less than ξ and must therefore, by Proposition 0.10.6, be greater than ξ or equal. $\qquad \square$

Remark 0.11.7. The above proof makes no explicit reference to real numbers being equivalence classes of Cauchy sequences. A close examination of the argument shows that it works in any ordered field having the Archimedean property and in which every Cauchy sequence converges.

1 Basic Concepts

In many branches of mathematics, it is convenient to have available a notion of distance between elements of an abstract set. For example, the proofs of some of the theorems in real analysis or analytic function theory depend only on a few properties of the distance between points and not on the fact that the points are in **R** or **C**. When these properties of distance are abstracted, they lead to the concept of a metric space. The notion of distance between points of an abstract set leads naturally to the discussion of convergence of sequences and Cauchy sequences in the set. Unlike the situation of real numbers, where each Cauchy sequence is convergent, there are metric spaces in which Cauchy sequences fail to converge. A metric space in which every Cauchy sequence converges is called a "complete" metric space. This property plays a vital role in analysis when one wishes to make an existence statement.

Our objective in this chapter is to define a metric space and list a large number of examples to emphasise the usefulness and the unifying force of the concept. We also define complete metric spaces, give several examples and describe their elementary properties. In Section 1.5, we shall prove that every metric space can be 'completed' in an appropriate sense.

1.1. Inequalities

The subject of inequalities has applications in every part of mathematics, and the study of metric spaces is no exception. In fact, the definition of a metric space involves an inequality which is a generalisation of the familiar triangle inequality, satisfied by the distance function in **R**. ($|x - y| \le |x - z| + |z - y|$ for all x, y, z in **R** or **C**.)

In this section, we establish some inequalities that will be required for confirming that some of the examples we list are indeed metric spaces. Theses examples will be invoked repeatedly.

Proposition 1.1.1. The function $f(x) = \dfrac{x}{1 + x}$, $x \ge 0$, is monotonically increasing.

Proof. Let $y > x \geq 0$. Then

$$\frac{1}{1+y} < \frac{1}{1+x} \text{ and so } 1 - \frac{1}{1+y} > 1 - \frac{1}{1+x}, \text{ i.e., } \frac{y}{1+y} > \frac{x}{1+x}. \qquad \square$$

Theorem 1.1.2. For any two real numbers x and y, the following inequality holds:

$$\frac{|x+y|}{1+|x+y|} \leq \frac{|x|}{1+|x|} + \frac{|y|}{1+|y|}.$$

Proof. Let x and y have the same sign. Without loss of generality, we may assume that $x \geq 0$ and $y \geq 0$, and so

$$\frac{|x+y|}{1+|x+y|} = \frac{x+y}{1+x+y} = \frac{x}{1+x+y} + \frac{y}{1+x+y}$$

$$\leq \frac{x}{1+x} + \frac{y}{1+y} = \frac{|x|}{1+|x|} + \frac{|y|}{1+|y|}.$$

Suppose x and y have different signs. We may assume that $|x| > |y|$. Then $|x+y| \leq |x|$.

It follows from Proposition 1.1.1 that

$$\frac{|x+y|}{1+|x+y|} \leq \frac{|x|}{1+|x|} \leq \frac{|x|}{1+|x|} + \frac{|y|}{1+|y|}.$$

This completes the proof. $\qquad \square$

The next proposition is the well known **arithmetic mean-geometric mean inequality,** or AM-GM inequality, for short.

Proposition 1.1.3. (AM-GM Inequality) If $a > 0$ and $b > 0$ and if $0 < \lambda < 1$ is fixed, then

$$a^\lambda b^{1-\lambda} \leq \lambda a + (1-\lambda)b. \qquad (1.1)$$

Proof. Since $y = \ln x$, $x > 0$, is concave, we have

$$\ln(\lambda a + (1-\lambda)b) \geq \lambda \ln a + (1-\lambda)\ln b,$$

i.e.,

$$\ln a^\lambda b^{1-\lambda} \leq \ln(\lambda a + (1-\lambda)b).$$

As $y = \exp x$ is a strictly increasing function, it follows from the above inequality that

$$a^\lambda b^{1-\lambda} \leq \lambda a + (1-\lambda)b. \qquad \square$$

Remark. When $x \geq 0$, $y \geq 0$, $p > 1$ and $1/p + 1/q = 1$, we have

$$xy \leq \frac{1}{p}x^p + \frac{1}{q}y^q.$$

The inequality is obvious when either x or y is 0. If $x \neq 0 \neq y$, then it follows from Proposition 1.1.3 upon writing $\lambda = 1/p$, $a^\lambda = x$ and $b^{1-\lambda} = y$.

Theorem 1.1.4. (Hölder's Inequality) Let $x_i \geq 0$ and $y_i \geq 0$ for $i = 1, 2, \ldots, n$, and suppose that $p > 1$ and $q > 1$ are such that $1/p + 1/q = 1$. Then

$$\sum_{i=1}^{n} x_i y_i \leq \left(\sum_{i=1}^{n} x_i^p \right)^{1/p} \left(\sum_{i=1}^{n} y_i^q \right)^{1/q}. \tag{1.2}$$

In the special case when $p = q = 2$, the above inequality reduces to

$$\sum_{i=1}^{n} x_i y_i \leq \left(\sum_{i=1}^{n} x_i^2 \right)^{1/2} \left(\sum_{i=1}^{n} y_i^2 \right)^{1/2}. \tag{1.3}$$

This is known as the **Cauchy-Schwarz inequality.**

Proof. We need consider only the case when $\sum_{i=1}^{n} x_i^p \neq 0 \neq \sum_{i=1}^{n} y_i^q$. To begin with, we assume that

$$\sum_{i=1}^{n} x_i^p = 1 = \sum_{i=1}^{n} y_i^q \tag{1.4}$$

In this case, the inequality (1.2) reduces to the form

$$\sum_{i=1}^{n} x_i y_i \leq 1 \tag{1.5}$$

To obtain (1.5), we put successively $x = x_i$ and $y = y_i$ for $i = 1, 2, \ldots, n$ in the inequality of the preceding Remark and then add up the inequalities so obtained. The general case can be reduced to the foregoing special case if we take in place of the numbers x_i, y_i the numbers

$$x_i' = \frac{x_i}{\left(\sum_{i=1}^{n} x_i^p \right)^{1/p}}, \quad y_i' = \frac{y_i}{\left(\sum_{i=1}^{n} y_i^q \right)^{1/q}},$$

for which the condition (1.4) is satisfied. It follows by what we have proved in the paragraph above that

$$\sum_{i=1}^{n} x_i' y_i' \leq 1.$$

This is equivalent to (1.2). This completes the proof. $\qquad \square$

Theorem 1.1.5. (Minkowski's Inequality) Let $x_i \geq 0$ and $y_i \geq 0$ for $i = 1, 2, \ldots, n$ and suppose that $p \geq 1$. Then

$$\left(\sum_{i=1}^{n} (x_i + y_i)^p \right)^{1/p} \leq \left(\sum_{i=1}^{n} x_i^p \right)^{1/p} + \left(\sum_{i=1}^{n} y_i^p \right)^{1/p}. \tag{1.6}$$

Proof. If $p = 1$, the inequality (1.6) is self-evident. So, assume that $p > 1$. We write $\sum_{i=1}^{n} (x_i + y_i)^p$ in the form

$$\sum_{i=1}^{n} (x_i + y_i)^p = \sum_{i=1}^{n} x_i(x_i + y_i)^{p-1} + \sum_{i=1}^{n} y_i(x_i + y_i)^{p-1}. \tag{1.7}$$

Let $q > 1$ be such that $1/p + 1/q = 1$. Apply Hölder's inequality to the two sums on the right side of (1.7) and obtain

$$\sum_{i=1}^{n} (x_i + y_i)^p \leq \left(\sum_{i=1}^{n} x_i^p \right)^{1/p} \left(\sum_{i=1}^{n} (x_i + y_i)^{(p-1)q} \right)^{1/q}$$
$$+ \left(\sum_{i=1}^{n} y_i^p \right)^{1/p} \left(\sum_{i=1}^{n} (x_i + y_i)^{(p-1)q} \right)^{1/q}$$
$$= \left[\left(\sum_{i=1}^{n} x_i^p \right)^{1/p} + \left(\sum_{i=1}^{n} y_i^p \right)^{1/p} \right] \left(\sum_{i=1}^{n} (x_i + y_i)^p \right)^{1/q}.$$

Dividing both sides of this inequality by $\left(\sum_{i=1}^{n} (x_i + y_i)^p \right)^{1/q}$, we obtain (1.6) in the case $\sum_{i=1}^{n} (x_i + y_i)^p \neq 0$. In the contrary case, (1.6) is self-evident. □

Theorem 1.1.6. (Minkowski's Inequality for Infinite Sums) Suppose that $p \geq 1$ and let $\{x_n\}_{n \geq 1}$, $\{y_n\}_{n \geq 1}$ be sequences of nonnegative terms such that $\sum_{n=1}^{\infty} x_n^p$ and $\sum_{n=1}^{\infty} y_n^p$ are convergent. Then $\sum_{n=1}^{\infty} (x_n + y_n)^p$ is convergent. Moreover,

$$\left(\sum_{n=1}^{\infty} (x_n + y_n)^p \right)^{1/p} \leq \left(\sum_{n=1}^{\infty} x_n^p \right)^{1/p} + \left(\sum_{n=1}^{\infty} y_n^p \right)^{1/p}.$$

Proof. For any positive integer m, we have from Theorem 1.1.5,

$$\left(\sum_{n=1}^{m} (x_n + y_n)^p \right)^{1/p} \leq \left(\sum_{n=1}^{m} x_n^p \right)^{1/p} + \left(\sum_{n=1}^{m} y_n^p \right)^{1/p}$$
$$\leq \left(\sum_{n=1}^{\infty} x_n^p \right)^{1/p} + \left(\sum_{n=1}^{\infty} y_n^p \right)^{1/p}.$$

Thus, $\{(\sum_{n=1}^{m}(x_n + y_n)^p)^{1/p}\}$, which is an increasing sequence of nonnegative real numbers, is bounded above by the sum

$$\left(\sum_{n=1}^{\infty} x_n^p\right)^{1/p} + \left(\sum_{n=1}^{\infty} y_n^p\right)^{1/p}.$$

It follows that $\sum_{n=1}^{\infty}(x_n + y_n)^p$ is convergent and that the desired inequality holds. □

Theorem 1.1.7. Let $p > 1$. For $a \geq 0$ and $b \geq 0$, we have

$$(a + b)^p \leq 2^{p-1}(a^p + b^p).$$

Proof. If either a or b is 0, the result is obvious. So assume $a > 0$ and $b > 0$. Now the function that maps every positive number x into x^p is convex when $p > 1$. So,

$$\left(\frac{a + b}{2}\right)^p \leq \frac{a^p + b^p}{2},$$

i.e.,

$$(a + b)^p \leq 2^{p-1}(a^p + b^p). \qquad \qquad \square$$

1.2. Metric Spaces

The notion of function, the concept of limit and the related concept of continuity play an important role in the study of mathematical analysis. The notion of limit can be formulated entirely in terms of distance. For example, a sequence $\{x_n\}_{n \geq 1}$ of real numbers converges to x if and only if for all $\varepsilon > 0$ there exists a positive integer n_0 such that $|x_n - x| < \varepsilon$ whenever $n \geq n_0$. A discerning reader will note that the above definition of convergence depends only on the properties of the distance $|\alpha - \beta|$ between pairs α, β of real numbers, and that the algebraic properties of real numbers have no bearing on it, except insofar as they determine properties of the distance such as,

$$|\alpha - \beta| > 0 \text{ when } \alpha \neq \beta, \quad |\alpha - \beta| = |\beta - \alpha| \text{ and } |\alpha - \gamma| \leq |\alpha - \beta| + |\beta - \gamma|.$$

There are many other sets of elements for which "distance between pairs of elements" can be defined, and doing so provides a general setting in which the notions of convergence and continuity can be studied. Such a setting is called a *metric space*. The approach through metric spaces illuminates many of the concepts of classical analysis and economises the intellectual effort involved in learning them.

We begin with the definition of a metric space.

Definition 1.2.1. A nonempty set X with a map $d : X \times X \to \mathbf{R}$ is called a **metric space** if the map d has the following properties:

(MS1) $d(x, y) \geq 0$ $x, \ y \in X$;
(MS2) $d(x, y) = 0$ if and only if $x = y$;
(MS3) $d(x, y) = d(y, x)$ $x, y \in X$;
(MS4) $d(x, y) \leq d(x, z) + d(z, y)$ $x, y, z \in X$.

The map d is called the **metric** on X or sometimes the **distance function** on X. The phrase "(X, d) is a metric space" means that d is a metric on the set X. Property (MS4) is often called the **triangle inequality**.

The four properties (MS1)–(MS4) are abstracted from the familiar properties of distance between points in physical space. It is customary to refer to elements of any metric space as **points** and $d(x, y)$ as the **distance between the points** x and y.

We shall often omit all mention of the metric d and write "the metric space X" instead of "the metric space (X, d)". This abuse of language is unlikely to cause any confusion. Different choices of metrics on the same set X give rise to different metric spaces. In such a situation, careful distinction between them must be maintained.

Suppose that (X, d) is a metric space and Y is a nonempty subset of X. The restriction d_Y of d to $Y \times Y$ will serve as a metric for Y, as it clearly satisfies the metric space axioms (MS1)–(MS4); so (Y, d_Y) is a metric space. By abuse of language, we shall often write (Y, d) instead of (Y, d_Y). This metric space is called a **subspace** of X or of (X, d) and the restriction d_Y is called the metric **induced** by d on Y.

Examples 1.2.2. (i) The function $d: \mathbf{R} \times \mathbf{R} \to \mathbf{R}^+$ defined by $d(x, y) = |x - y|$ is a metric on $\mathbf{R}$, the set of real numbers. To prove that d is a metric on $\mathbf{R}$, we need verify only (MS4), as the other axioms are obviously satisfied. For any $x, y, z \in \mathbf{R}$,

$$d(x, z) = |x - z| = |(x - y) + (y - z)| \leq |x - y| + |y - z| = d(x, y) + d(y, z).$$

It is known as the **usual** or **standard** metric on $\mathbf{R}$.

(ii) Let $X = \mathbf{R}^n = \{x = (x_1, x_2, \ldots, x_n): x_i \in \mathbf{R}, \ 1 \leq i \leq n\}$ be the set of real n-tuples. For $x = (x_1, x_2, \ldots, x_n)$ and $y = (y_1, y_2, \ldots, y_n)$ in $\mathbf{R}^n$, define

$$d(x, y) = \left(\sum_{i=1}^{n} (x_i - y_i)^2 \right)^{1/2}.$$

(For $n = 2$, $d(x, y) = ((x_1 - y_1)^2 + (x_2 - y_2)^2)^{1/2}$ is the usual distance in the Cartesian plane.) To verify that d is a metric on $\mathbf{R}^n$, we need only check (MS4), i.e., if $z = (z_1, z_2, \ldots, z_n)$, we must show that $d(x, y) \leq d(x, z) + d(z, y)$. For $k = 1, 2, \ldots, n$, set

$$a_k = x_k - z_k, \ b_k = z_k - y_k.$$

Then

$$d(x, z) = \left(\sum_{k=1}^{n} a_k^2 \right)^{1/2}, \ d(z, y) = \left(\sum_{k=1}^{n} b_k^2 \right)^{1/2},$$

and

1.2. Metric Spaces

29

$$d(x, y) = \left(\sum_{k=1}^{n} (a_k + b_k)^2 \right)^{1/2}.$$

We must show that

$$\left(\sum_{k=1}^{n} (a_k + b_k)^2 \right)^{1/2} \le \left(\sum_{k=1}^{n} a_k^2 \right)^{1/2} + \left(\sum_{k=1}^{n} b_k^2 \right)^{1/2}. \qquad (*)$$

Squaring both sides of (*), and using the equality

$$(a + b)^2 = a^2 + 2ab + b^2,$$

we see that (*) is equivalent to

$$\sum_{k=1}^{n} a_k b_k \le \left(\sum_{k=1}^{n} a_k^2 \right)^{1/2} \left(\sum_{k=1}^{n} b_k^2 \right)^{1/2},$$

which is just the Cauchy-Schwarz inequality (see Theorem 1.1.4). This metric is known as the **Euclidean metric** on $\mathbf{R}^n$.

When $n > 1$, $\mathbf{R}^{n-1}$ can be regarded as a subset of $\mathbf{R}^n$ in the usual way. The metric induced on $\mathbf{R}^{n-1}$ by the Euclidean metric of $\mathbf{R}^n$ is the Euclidean metric of $\mathbf{R}^{n-1}$.

(iii) Let $X = \mathbf{R}^n$. For $x = (x_1, x_2, \ldots, x_n)$ and $y = (y_1, y_2, \ldots, y_n)$ in $\mathbf{R}^n$, define

$$d_p(x, y) = \left(\sum_{i=1}^{n} |x_i - y_i|^p \right)^{1/p},$$

where $p \ge 1$. Note that when $p = 2$ this agrees with the Euclidean metric. To verify that d_p is a metric, we need only check that $d_p(x, y) \le d_p(x, z) + d_p(z, y)$ when $z = (z_1, z_2, \ldots, z_n) \in \mathbf{R}^n$. For $k = 1, 2, \ldots, n$, set $a_k = x_k - z_k$, $b_k = z_k - y_k$. Then

$$d_p(x, z) = \left(\sum_{k=1}^{n} |a_k|^p \right)^{1/p}, \quad d_p(z, y) = \left(\sum_{k=1}^{n} |b_k|^p \right)^{1/p}$$

and

$$d_p(x, y) = \left(\sum_{k=1}^{n} |a_k + b_k|^p \right)^{1/p}.$$

We need to show that

$$\left(\sum_{k=1}^{n} |a_k + b_k|^p \right)^{1/p} \le \left(\sum_{k=1}^{n} |a_k|^p \right)^{1/p} + \left(\sum_{k=1}^{n} |b_k|^p \right)^{1/p}.$$

However, this is just Minkowski's inequality (see Theorem 1.1.5).

Here again, when $n > 1$, the metric induced by d_p on the subset $\mathbf{R}^{n-1}$ is the corresponding metric of the same kind, i.e.,

$$d_p(x, y) = \left(\sum_{i=1}^{n-1} |x_i - y_i|^p \right)^{1/p} \quad \text{when } x, y \in \mathbf{R}^{n-1}.$$

(iv) Let $X = \mathbf{R}^n$. For $x = (x_1, x_2, \ldots, x_n)$ and $y = (y_1, y_2, \ldots, y_n)$ in $\mathbf{R}^n$, define

$$d_\infty(x, y) = \max_{1 \le i \le n} |x_i - y_i|.$$

Then d_∞ is a metric. Indeed, for $z = (z_1, z_2, \ldots, z_n) \in \mathbf{R}^n$, we have

$$|x_i - y_i| \le |x_i - z_i| + |z_i - y_i|$$
$$\le \max_{1 \le i \le n} |x_i - z_i| + \max_{1 \le i \le n} |z_i - y_i|,$$

from which it follows that

$$\max_{1 \le i \le n} |x_i - y_i| \le \max_{1 \le i \le n} |x_i - z_i| + \max_{1 \le i \le n} |z_i - y_i|.$$

Remark. Examples (ii), (iii) and (iv) show that we can define more than one distance function, i.e., different metrics in the set of ordered n-tuples.

(v) Let X be any nonempty set whatsoever and let

$$d(x, y) = \begin{cases} 0 & \text{if } x = y \\ 1 & \text{if } x \ne y. \end{cases}$$

It may be easily verified that d is a metric on X. It is called the **discrete metric** on X. If Y is a nonempty subset of X, the metric induced on Y by d is the discrete metric on Y.

Sequence spaces provide natural extensions of Examples (i), (iii) and (iv).

(vi) **The space of all bounded sequences.** Let X be the set of all bounded sequences of numbers, i.e., all infinite sequences

$$x = (x_1, x_2, \ldots) = \{x_i\}_{i \ge 1}$$

such that

$$\sup_i |x_i| < \infty.$$

If $x = \{x_i\}_{i \ge 1}$ and $y = \{y_i\}_{i \ge 1}$ belong to X, we introduce the distance

$$d(x, y) = \sup_i |x_i - y_i|.$$

It is clear that d is a metric on X. Indeed, if $z = \{z_i\}_{i \ge 1}$ is in X, then

$$|x_i - y_i| \le |x_i - z_i| + |z_i - y_i|$$
$$\le \sup_i |x_i - z_i| + \sup_i |z_i - y_i|$$
$$= d(x, z) + d(z, y).$$

Therefore,

$$d(x, y) = \sup_i |x_i - y_i| \le d(x, z) + d(z, y).$$

(vii) **The space** ℓ_p. Let X be the set of all sequences $x = \{x_i\}_{i \ge 1}$ such that

$$\left(\sum_{i=1}^{\infty} |x_i|^p \right)^{1/p} < \infty, \qquad p \ge 1.$$

If $x = \{x_i\}_{i \ge 1}$ and $y = \{y_i\}_{i \ge 1}$ belong to X, we introduce the distance

$$d(x, y) = \left(\sum_{i=1}^{\infty} |x_i - y_i|^p \right)^{1/p}.$$

It is a consequence of the Minkowski inequality for infinite sums (Theorem 1.1.6) that $d(x, y) \in \mathbf{R}$. Evidently, $d(x, y) \ge 0$, $d(x, y) = 0$ if and only if $x = y$, and $d(x, y) = d(y, x)$. The triangle inequality follows from the Minkowski inequality for infinite sums. In the special case when $p = 2$, the space ℓ_p is called the **space of square summable sequences**.

(viii) **The space of bounded functions.** Let S be any nonempty set and $\mathcal{B}(S)$ denote the set of all real- or complex-valued functions on S, each of which is bounded, i.e.,

$$\sup_{x \in S} |f(x)| < \infty.$$

If f and g belong to $\mathcal{B}(S)$, there exist $M > 0$ and $N > 0$ such that

$$\sup_{x \in S} |f(x)| \le M \text{ and } \sup_{x \in S} |g(x)| \le N.$$

It follows that $\sup_{x \in S} |f(x) - g(x)| < \infty$. Indeed,

$$|f(x) - g(x)| \le |f(x)| + |g(x)| \le \sup_{x \in S} |f(x)| + \sup_{x \in S} |g(x)|,$$

and so

$$0 \le \sup_{x \in S} |f(x) - g(x)| \le M + N.$$

Define

$$d(f, g) = \sup_{x \in S} |f(x) - g(x)|, \qquad f, g \in \mathcal{B}(S).$$

Evidently, $d(f,g) \geq 0$, $d(f,g) = 0$ if and only if $f(x) = g(x)$ for all $x \in S$ and $d(f,g) = d(g,f)$. It remains to verify the triangle inequality for $\mathcal{B}(S)$. By the triangle inequality for $\mathbf{R}$, we have

$$
\begin{aligned}
|f(x) - g(x)| &\leq |f(x) - h(x)| + |h(x) - g(x)| \\
&\leq \sup_{x \in S} |f(x) - h(x)| + \sup_{x \in S} |h(x) - g(x)| \\
&= d(f,h) + d(h,g)
\end{aligned}
$$

and so

$$
d(f,g) \leq d(f,h) + d(h,g),
$$

for all $f,g,h \in \mathcal{B}(S)$. The metric d is called the **uniform metric** (or **supremum metric**).

(ix) **The space of continuous functions**. Let X be the set of all continuous functions defined on $[a,b]$, an interval in $\mathbf{R}$. For $f,g \in X$, define

$$
d(f,g) = \sup_{x \in [a,b]} |f(x) - g(x)|.
$$

The measure of distance between the functions f and g is the largest vertical distance between their graphs (see Figure 1.1). Since the difference of two continuous functions is continuous, the composition of two continuous functions is continuous, and a continuous function defined on the closed and bounded interval $[a,b]$ is bounded, it follows that $d(f,g) \in \mathbf{R}$ for all $f,g \in X$. It may be verified as in Example (viii) that d is a metric on X. The space X with metric d defined as above is denoted by $C[a,b]$. All that we have said is valid whether all complex-valued continuous functions are taken into consideration or only real-valued ones are. When it is necessary to specify which, we write $C_{\mathbf{C}}[a,b]$ or $C_{\mathbf{R}}[a,b]$. Note that $C[a,b] \subseteq \mathcal{B}[a,b]$ and the metric described here is the one induced by the metric in Example (viii) and is also called the **uniform metric** (or **supremum metric**).

(x) The set of all continuous functions on $[a,b]$ can also be equipped with the metric

$$
d(f,g) = \int_a^b |f(x) - g(x)| dx.
$$

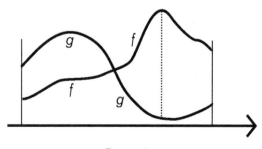

FIGURE 1.1

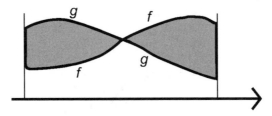

FIGURE 1.2

The measure of distance between the functions f and g represents the area between their graphs, indicated by shading in Figure 1.2. If $f, g \in C[a, b]$, then $|f - g| \in C[a, b]$, and the integral defining $d(f, g)$ is finite. It may be easily verified that d is a metric on $C[a, b]$. We note that the continuity of the functions enters into the verification of the "only if" part of (MS2).

(xi) Let $X = \mathbf{R} \cup \{\infty\} \cup \{-\infty\}$. Define $f : X \to \mathbf{R}$ by the rule

$$f(x) = \begin{cases} \dfrac{x}{1 + |x|} & \text{if } -\infty < x < \infty \\ 1 & \text{if } x = \infty \\ -1 & \text{if } x = -\infty. \end{cases}$$

Evidently, f is one-to-one and $-1 \le f(x) \le 1$. Define d on X as follows:

$$d(x, y) = |f(x) - f(y)|, x, y \in X.$$

If $x = y$ then $f(x) = f(y)$ and so $d(x, y) = 0$. On the other hand, if $d(x, y) = 0$, then $|f(x) - f(y)| = 0$, i.e., $f(x) = f(y)$. Since f is one-to-one, it follows that $x = y$. That d satisfies (MS3) and (MS4) is a consequence of the properties of the modulus of real numbers.

(xii) On $X = \mathbf{R} \cup \{\infty\} \cup \{-\infty\}$, define

$$d(x, y) = |\tan^{-1} x - \tan^{-1} y|, x, y \in X.$$

(Note that $\tan^{-1}(\infty) = \pi/2$, $\tan^{-1}(-\infty) = -\pi/2$ and that the function $\tan^{-1}$ is one-to-one.) It may now be verified that X is a metric space with metric d defined as above.

(xiii) **The extended complex plane.** Let $X = \mathbf{C} \cup \{\infty\}$. We represent X as the unit sphere in $\mathbf{R}^3$,

$$S = \{(x_1, x_2, x_3) : x_1, x_2, x_3 \in \mathbf{R}, x_1^2 + x_2^2 + x_3^2 = 1\},$$

and identify $\mathbf{C}$ with $\{(x_1, x_2, 0) : x_1, x_2 \in \mathbf{R}\}$. The line joining the north pole $(0,0,1)$ of S with the point $z = (x_1, x_2) \in \mathbf{C}$ intersects the sphere S at $(\xi, \eta, \zeta) \ne (0, 0, 1)$, say. We map $\mathbf{C}$ to S by sending $z = (x_1, x_2)$ to (ξ, η, ζ). The mapping is clearly one-to-one and each point of S other than $(0,0,1)$ is the image of some point in $\mathbf{C}$. The correspondence is completed by letting the point ∞ correspond to $(0, 0, 1)$. (See Figure 1.3.)

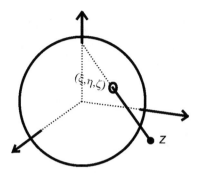

FIGURE 1.3

Analytically, the above representation is described by the formulae

$$\xi = \frac{2\Re z}{1 + |z|^2}, \quad \eta = \frac{2\Im z}{1 + |z|^2}, \quad \zeta = \frac{|z|^2 - 1}{|z|^2 + 1}.$$

Corresponding to the point at infinity, we have the point $(0, 0, 1)$. Also,

$$z = \frac{\xi + i\eta}{1 - \zeta}.$$

We define the distance between the points of X by

$$d(z_1, z_2) = \begin{cases} \dfrac{2|z_1 - z_2|}{\sqrt{1 + |z_1|^2}\sqrt{1 + |z_2|^2}} & \text{when } z_1, z_2 \in \mathbf{C} \\[3ex] \dfrac{2}{\sqrt{1 + |z_1|^2}} & \text{when } z_1 \in \mathbf{C} \text{ and } z_2 = \infty. \end{cases}$$

This is actually the chordal distance between those points on the sphere corresponding to the points $z_1, z_2 \in \mathbf{C} \cup \{\infty\}$. Evidently, $d(z_1, z_2) \geq 0$ and $d(z_1, z_2) = 0$ if and only if $z_1 = z_2$. Also, $d(z_1, z_2) = d(z_2, z_1)$. For the triangle inequality, the following elementary inequality will be needed:

$$\begin{aligned} (1 + zw)(1 + \bar{z}\bar{w}) &= 1 + zw + \bar{z}\bar{w} + |z|^2|w|^2 \\ &= 1 + 2\Re(zw) + |z|^2|w|^2 \\ &\leq 1 + 2|z||w| + |z|^2|w|^2 \\ &\leq 1 + |z|^2 + |w|^2 + |z|^2|w|^2 \\ &= (1 + |z|^2)(1 + |w|^2). \end{aligned} \tag{1.8}$$

From the identity

$$(z_1 - z_2)(1 + \bar{z}_3 z_3) = (z_1 - z_3)(1 + \bar{z}_3 z_2) + (z_3 - z_2)(1 + \bar{z}_3 z_1),$$

we have

$$|z_1 - z_2|(1 + |z_3|^2) \le |z_1 - z_3||1 + \bar{z}_3 z_2| + |z_3 - z_2||1 + \bar{z}_3 z_1|$$
$$\le |z_1 - z_3|(1 + |z_3|^2)^{1/2}(1 + |z_2|^2)^{1/2}$$
$$+ |z_3 - z_2|(1 + |z_3|^2)^{1/2}(1 + |z_1|^2)^{1/2}, \quad (1.9)$$

using inequality (1.8). Dividing both sides of (1.9) by

$$(1 + |z_1|^2)^{1/2}(1 + |z_2|^2)^{1/2}(1 + |z_3|^2),$$

we get the triangle inequality when none among z_1, z_2, z_3 is ∞. Other cases are left to the reader.

(xiv) Let d be a metric on a nonempty set X. We shall show that the function e defined by

$$e(x, y) = \min\{1, d(x, y)\},$$

where $x, y \in X$ is also a metric on X.

Since $d(x,y) \ge 0$, so $e(x,y) = \min\{1, d(x,y)\} \ge 0$. If $x = y$, then $d(x,y) = 0$, and hence, $e(x,y) = \min\{1, d(x,y)\} = 0$. On the other hand, if $e(x,y) = 0$, i.e., $\min\{1, d(x,y)\} = 0$, it follows that $d(x,y) = 0$, and hence that $x = y$. This proves (MS1) and (MS2). Since $d(x,y) = d(y,x)$, it is immediate from the definition of e that $e(x,y) = e(y,x)$. Thus, (MS3) also holds. It remains to verify only the triangle inequality (MS4).

Let $x, y, z \in X$. Observe that $e(x,y) \le 1$ for all $x, y \in X$; therefore, $e(x,y) \le e(x,z) + e(z,y)$ if either $e(x,z)$ or $e(z,y)$ is 1. But if both are less than 1, then $e(x,z) = d(x,z)$, and $e(z,y) = d(z,y)$. Accordingly,

$$e(x, y) = \min\{1, d(x,y)\} \le d(x,y) \le d(x,z) + d(z,y) = e(x,z) + e(z,y).$$

Proposition 1.2.3. Let (X, d) be a metric space. Define $d': X \times X \to \mathbf{R}$ by

$$d'(x, y) = \frac{d(x,y)}{1 + d(x,y)}.$$

Then d' is a metric on X. Besides, $d'(x,y) < 1$ for all $x, y \in X$.

Proof. We need prove only (MS4). Suppose $x, y, z \in X$. Then $d(x,y) \ge 0, d(x,z) \ge 0, d(z,y) \ge 0$ and $d(x,y) \le d(x,z) + d(z,y)$ because d is a metric on X. Now,

$$d'(x, z) + d'(z, y) = \frac{d(x,z)}{1 + d(x,z)} + \frac{d(z,y)}{1 + d(z,y)}$$
$$\ge \frac{d(x,z) + d(z,y)}{1 + d(x,z) + d(z,y)} \quad \text{by Theorem 1.1.2}$$
$$= \frac{1}{1 + \dfrac{1}{d(x,z) + d(z,y)}}$$
$$\ge \frac{1}{1 + \dfrac{1}{d(x,y)}} = \frac{d(x,y)}{1 + d(x,y)}. \qquad \square$$

Examples 1.2.4. (i) Let $X = \mathbf{R}$. For $x, y \in \mathbf{R}$, define

$$d'(x, y) = \frac{|x - y|}{1 + |x - y|}.$$

Then d' is a metric on $\mathbf{R}$, different from the usual metric described by $d(x, y) = |x - y|$ in Example 1.2.1(i).

(ii) **The space of all sequences of numbers.** Let X be the space of all sequences of numbers. Let $x = \{x_i\}_{i \geq 1}$ and $y = \{y_i\}_{i \geq 1}$ be elements of X. Define

$$d(x, y) = \sum_{i=1}^{\infty} \frac{1}{2^i} \frac{|x_i - y_i|}{1 + |x_i - y_i|}.$$

We prove that the axioms (MS1)–(MS4) are fulfilled by this distance. It is immediate that

$$d(x, y) \geq 0, \ d(x, y) = 0 \text{ if and only if } x = y \text{ and that } d(x, y) = d(y, x).$$

For the triangle inequality, we have

$$\begin{aligned} d(x, y) &= \sum_{i=1}^{\infty} \frac{1}{2^i} \frac{|x_i - y_i|}{1 + |x_i - y_i|} \\ &= \sum_{i=1}^{\infty} \frac{1}{2^i} \frac{|x_i - z_i + z_i - y_i|}{1 + |x_i - z_i + z_i - y_i|} \\ &\leq \sum_{i=1}^{\infty} \frac{1}{2^i} \frac{|x_i - z_i|}{1 + |x_i - z_i|} + \sum_{i=1}^{\infty} \frac{1}{2^i} \frac{|z_i - y_i|}{1 + |z_i - y_i|} \\ &= d(x, z) + d(z, y). \end{aligned}$$

In $\mathbf{R}^2$, define $d(x, y) = |x_2 - y_2|$, where $x = (x_1, x_2)$ and $y = (y_1, y_2)$. For $x = (0, 0)$ and $y = (1, 0)$, we have $d(x, y) = 0$ although $x \neq y$. Therefore, the "only if" part of (MS2) does not hold and so, d is not a metric on $\mathbf{R}^2$. It may be noted, however, that all other properties of a metric, including the "if" part of (MS2) do hold. In such a situation, d is called a **pseudometric**. More precisely, we have the following definition.

Definition 1.2.5. Let X be a nonempty set. A **pseudometric** on X is a mapping of $X \times X$ into $\mathbf{R}$ that satisfies the conditions:

(PMS1) $d(x, y) \geq 0$ $x, y \in X$;
(PMS2) $d(x, y) = 0$ if $x = y$;
(PMS3) $d(x, y) = d(y, x)$ $x, y \in X$;
(PMS4) $d(x, y) \leq d(x, z) + d(z, y)$ $x, y, z \in X$.

Another example of a pseudometric space is the following:

Example 1.2.6. Let X denote the set of all Riemann integrable functions on $[a, b]$. For $f, g \in X$, define

$$d(f, g) = \int_a^b |f(x) - g(x)| dx.$$

It is easily verified that d is a pseudometric on X, but it is not a metric on X. Indeed, if

$$f(x) = \begin{cases} 0 & \text{if } x = a \\ 1 & \text{if } a < x \leq b \end{cases}$$

and $g(x) = 1$ for all $x \in [a, b]$, then $d(f, g) = 0$. However, $f \neq g$.

Let (X, d) be a pseudometric space. Define a relation $\sim$ on X by

$$x \sim y \text{ if and only if } d(x, y) = 0.$$

Then $\sim$ is an equivalence relation on X. Since $d(x, x) = 0$, it follows that $x \sim x$. As $d(x, y) = d(y, x)$, the relation is also symmetric. We shall show that it is transitive: If $x \sim y$ and $y \sim z$, then $d(x, y) = 0$ and $d(y, z) = 0$; it follows from the triangle inequality (PMS4) that $d(x, z) = 0$, so that $x \sim z$. The set of all equivalence classes of X corresponding to the equivalence relation $\sim$ is denoted by $X/\sim$. As is customary, the equivalence class to which x belongs will be denoted by $[x]$. Moreover, $[x] = [y]$ if and only if $x \sim y$.

We now define a metric $\tilde{d}$ on $X/\sim$ as follows:

$$\tilde{d}([x], [y]) = d(x, y),$$

where $x \in [x]$ and $y \in [y]$. It is readily verified that the value of $\tilde{d}([x], [y])$ is independent of the choice of representatives x and y of the classes $[x]$ and $[y]$. We next show that $\tilde{d}$ is a metric on the set $X/\sim$.

$\tilde{d}([x], [y]) = 0$ implies $d(x, y) = 0$, where $x \in [x]$ and $y \in [y]$, which in turn implies that $x \sim y$, so that $[x] = [y]$. On the other hand, if $[x] = [y]$, then $x \sim y$ and so $d(x, y) = 0$, i.e., $\tilde{d}([x], [y]) = 0$. Since $d(x, y) = d(y, x)$, we have $\tilde{d}([x], [y]) = \tilde{d}([y], [x])$. Finally,

$$\tilde{d}([x], [y]) = d(x, y) \leq d(x, z) + d(z, y)$$
$$= \tilde{d}([x], [z]) + \tilde{d}([z], [y]) \qquad \text{if } z \not\sim x \text{ and } z \not\sim y.$$

The case where either $z \sim x$ or $z \sim y$ is trivial.

1.3. Sequences in Metric Spaces

As pointed out in Chapter 0, analysis is primarily concerned with matters involving limit processes. It is no wonder that mathematicians thinking about such matters studied and generalised the concept of convergence of sequences of real numbers and of continuous functions of a real variable. The reader will note that the basic facts about convergence are just as easily expressed in this setting.

Definition 1.3.1. Let (X,d) be a metric space. A **sequence** of points **in X** is a function f from $\mathbf{N}$ into X.

In other words, a sequence assigns to each $n \in \mathbf{N}$ a uniquely determined element of X. If $f(n) = x_n$, it is customary to denote the sequence by the symbol $\{x_n\}_{n \geq 1}$ or $\{x_n\}$ or by $x_1, x_2, \ldots, x_n, \ldots$.

Definition 1.3.2. Let d be a metric on a set X and $\{x_n\}$ be a sequence in the set X. An element $x \in X$ is said to be a **limit** of $\{x_n\}$ if, for every $\varepsilon > 0$, there exists a natural number n_0 such that

$$d(x_n, x) < \varepsilon \text{ whenever } n \geq n_0.$$

In this case, we also say that $\{x_n\}$ **converges to** x, and write it in symbols as $x_n \to x$. If there is no such x, we say that the sequence **diverges**. A sequence is said to be **convergent** if it converges to some limit, and **divergent** otherwise.

Remark 1. By comparing the above with the definition of convergence in $\mathbf{R}$ (or $\mathbf{C}$), we find that $x_n \to x$ if and only if $\lim_{n\to\infty} d(x_n, x) = 0$, where d denotes the usual metric in $\mathbf{R}$ (or $\mathbf{C}$).

Remark 2. In case there are two or more metrics on the set X, then it is necessary to specify which metric is intended to be used in applying the definition of convergence.

Remark 3. If $x \in X$ and $x' \in X$ are such that $x_n \to x$ and $x_n \to x'$, then $x = x'$. Indeed, for $\varepsilon > 0$, there exist natural numbers n_1 and n_2 such that $d(x_n, x) < \varepsilon/2$ whenever $n \geq n_1$ and $d(x_n, x') < \varepsilon/2$ whenever $n \geq n_2$. Consequently, $d(x, x') \leq d(x_n, x) + d(x_n, x') < \varepsilon$ provided that $n \geq \max\{n_1, n_2\}$. It follows that $d(x, x') \leq 0$ and hence that $d(x, x') = 0$ by (MS1) and $x = x'$ by (MS2). This means that a sequence can have at most one limit and we can speak of *the* (one and only) limit of a sequence, provided that the sequence under reference has a limit in the first place. We can now unambiguously denote the limit by $\lim x_n$ or $\lim_n x_n$ or by $\lim_{n\to\infty} x_n$.

We next consider the notion of convergence in specific metric spaces.

Examples 1.3.3. (i) Let $X = \mathbf{R}$ with $d(x,y) = |x - y|$, $x, y \in \mathbf{R}$. Let $\{x_n\}$ be a sequence of real numbers. It converges to $x \in \mathbf{R}$ in the sense of the metric space (X, d) if and only if

$$\lim_{n\to\infty} d(x_n, x) = \lim_{n\to\infty} |x_n - x| = 0.$$

So convergence in $\mathbf{R}$ with the usual metric is the same as convergence of a sequence of real numbers in the familiar sense.

(ii) Let $X = \mathbf{R}^n$ with

$$d(x, y) = d_p(x, y) = \left(\sum_{j=1}^{n} |x_j - y_j|^p \right)^{1/p},$$

where $x = (x_1, x_2, \ldots, x_n)$ and $y = (y_1, y_2, \ldots, y_n)$ are in $\mathbf{R}^n$ and $p \geq 1$. Let $x^{(k)} = (x_1^{(k)}, x_2^{(k)}, \ldots, x_n^{(k)})$, $k = 1, 2, \ldots$, be a sequence of elements in $\mathbf{R}^n$ that converges to $x = (x_1, x_2, \ldots, x_n)$, say, i.e.,

$$\lim_{k \to \infty} d(x^{(k)}, x) = \lim_{k \to \infty} \left(\sum_{j=1}^{n} |x_j^{(k)} - x_j|^p \right)^{1/p} = 0.$$

For every $\varepsilon > 0$, there exists an integer $k_0(\varepsilon)$ such that $|x_j^{(k)} - x_j| < \varepsilon$ for all $k \geq k_0(\varepsilon)$ and $j = 1, \ldots, n$. This says that $\lim_{k \to \infty} x_j^{(k)} = x_j$ for each j, i.e., that the sequence $\{x^{(k)}\}_{k \geq 1}$ converges to x coordinatewise. On the other hand, suppose that $x_j^{(k)} \to x_j$ as $k \to \infty$ for each $j = 1, \ldots, n$. Let $\varepsilon > 0$ be given. For each j there exists a positive integer k_j such that

$$|x_j^{(k)} - x_j| < \frac{\varepsilon}{n^{1/p}} \qquad \text{for } k \geq k_j.$$

Consequently,

$$d(x^{(k)}, x) = \left(\sum_{j=1}^{n} |x_j^{(k)} - x_j|^p \right)^{1/p} < \varepsilon \qquad \text{for } k \geq k' = \max \{k_1, k_2, \ldots, k_n\}.$$

Thus, convergence of sequences in $(\mathbf{R}^n, d_p)$ is equivalent to coordinatewise convergence.

(iii) Let $X = \mathbf{R}^n$. For $x = (x_1, x_2, \ldots, x_n)$ and $y = (y_1, y_2, \ldots, y_n)$ in $\mathbf{R}^n$, let

$$d_\infty(x, y) = \max \{|x_j - y_j| : 1 \leq j \leq n\}.$$

Let $\{x^{(k)}\}_{k \geq 1}$ be a sequence in $\mathbf{R}^n$ that converges to $x \in \mathbf{R}^n$, say, i.e.,

$$d_\infty(x^{(k)}, x) = \max \{|x_j^{(k)} - x_j| : 1 \leq j \leq n\} \to 0 \quad \text{as } k \to \infty.$$

Then, for every $\varepsilon > 0$, there exists an integer k_0 such that

$$|x_j^{(k)} - x_j| < \varepsilon \qquad \text{for } k \geq k_0 \text{ and each } j = 1, \ldots, n.$$

So the sequence $\{x^{(k)}\}_{k \geq 1}$ converges coordinatewise to x. On the other hand, if $\{x^{(k)}\}_{k \geq 1}$ converges coordinatewise to x, then for any $\varepsilon > 0$ and each $j = 1, \ldots, n$, there exists an integer k_j such that

$$|x_j^{(k)} - x_j| < \varepsilon \qquad \text{for } k \geq k_j.$$

If $k' = \max \{k_1, k_2, \ldots, k_n\}$, then we have

$$\max \{|x_j^{(k)} - x_j| : 1 \leq j \leq n\} < \varepsilon \qquad \text{for } k \geq k'.$$

Thus, the convergence of sequences in $(\mathbf{R}^n, d_\infty)$ is equivalent to coordinatewise convergence.

(iv) Let X be a nonempty set and d denote the discrete metric. For $x, y \in X$,

$$d(x, y) = \begin{cases} 1 & \text{if } x \neq y \\ 0 & \text{if } x = y \end{cases}.$$

Let $\{x_n\}_{n \geq 1}$ be a sequence in X that converges to x, say, $\lim_{n \to \infty} d(x_n, x) = 0$. For $\varepsilon = 1/2$, there exists an integer $n_0(1/2)$ such that

$$d(x_n, x) < \frac{1}{2} \text{ whenever } n \geq n_0\left(\frac{1}{2}\right).$$

But in the metric space under consideration, $d(x_n, x) < 1/2$ if and only if $d(x_n, x) = 0$ if and only if $x_n = x$. Thus, $x_n \to x$ in (X, d) if and only if all the x_n, except possibly finitely many, are equal to x.

(v) Let X denote the space of all sequences of numbers with metric d defined by

$$d(x, y) = \sum_{j=1}^{\infty} \frac{1}{2^j} \frac{|x_j - y_j|}{1 + |x_j - y_j|},$$

where $x = \{x_j\}_{j \geq 1}$ and $y = \{y_j\}_{j \geq 1}$ are in X. Let $\{x^{(n)}\}_{n \geq 1} = \{\{x_j^{(n)}\}_{j \geq 1}\}_{n \geq 1}$ be a sequence in X which converges to, say, $x \in X$. In other words,

$$d(x^{(n)}, x) = \sum_{j=1}^{\infty} \frac{1}{2^j} \frac{|x_j^{(n)} - x_j|}{1 + |x_j^{(n)} - x_j|} \to 0 \text{ as } n \to \infty.$$

This means that, for $\varepsilon > 0$, there exists an integer $n_0(\varepsilon)$ such that

$$\sum_{j=1}^{\infty} \frac{1}{2^j} \frac{|x_j^{(n)} - x_j|}{1 + |x_j^{(n)} - x_j|} < \varepsilon \text{ whenever } n \geq n_0(\varepsilon),$$

and so

$$\frac{1}{2^j} \frac{|x_j^{(n)} - x_j|}{1 + |x_j^{(n)} - x_j|} < \varepsilon,$$

for $n \geq n_0(\varepsilon)$ and for all $j = 1, 2, \ldots$. Since $\varepsilon > 0$ is arbitrary, it follows that $\lim_{n \to \infty} x_j^{(n)} = x_j$ for each j. On the other hand, let $\lim_{n \to \infty} x_j^{(n)} = x_j$ for each j. Consider any $\varepsilon > 0$. There exists an integer m such that

$$\sum_{j=m+1}^{\infty} \frac{1}{2^j} < \frac{\varepsilon}{2}.$$

Consequently,

$$d(x^{(n)}, x) = \sum_{j=1}^{\infty} \frac{1}{2^j} \frac{|x_j^{(n)} - x_j|}{1 + |x_j^{(n)} - x_j|}$$

$$= \sum_{j=1}^{m} \frac{1}{2^j} \frac{|x_j^{(n)} - x_j|}{1 + |x_j^{(n)} - x_j|} + \sum_{j=m+1}^{\infty} \frac{1}{2^j} \frac{|x_j^{(n)} - x_j|}{1 + |x_j^{(n)} - x_j|}$$

$$< \sum_{j=1}^{m} \frac{1}{2^j} \frac{|x_j^{(n)} - x_j|}{1 + |x_j^{(n)} - x_j|} + \frac{\varepsilon}{2}.$$

As the first sum contains only finitely many terms and $\lim_{n\to\infty} x_j^{(n)} = x_j$ for each j, there exists $n_0(\varepsilon)$ such that

$$\sum_{j=1}^{m} \frac{1}{2^j} \frac{|x_j^{(n)} - x_j|}{1 + |x_j^{(n)} - x_j|} < \frac{\varepsilon}{2} \quad \text{whenever } n \geq n_0(\varepsilon).$$

Hence, $d(x^{(n)}, x) < \varepsilon$ whenever $n \geq n_0(\varepsilon)$. Thus, convergence in the space of all sequences is coordinatewise covergence.

(vii) Let $X = \ell_p (p \geq 1)$ and let $d(x, y) = \left(\sum_{k=1}^{\infty} |x_k - y_k|^p\right)^{1/p}$, where $x = \{x_k\}_{k\geq 1}$ and $y = \{y_k\}_{k\geq 1}$ are in ℓ_p. Let $\{x^{(n)}\}_{n\geq 1} = \{\{x_k^{(n)}\}_{k\geq 1}\}_{n\geq 1}$ be a sequence in ℓ_p that converges to, say, $x \in \ell_p$, i.e.,

$$\lim_{n\to\infty} d(x^{(n)}, x) = \lim_{n\to\infty} \left(\sum_{k=1}^{\infty} |x_k^{(n)} - x_k|^p\right)^{1/p} = 0.$$

Then, for $\varepsilon > 0$ there exists a positive integer $n_0(\varepsilon)$ such that

$$\left(\sum_{k=1}^{\infty} |x_k^{(n)} - x_k|^p\right)^{1/p} < \varepsilon$$

for all $n \geq n_0(\varepsilon)$. Since $\varepsilon > 0$ is arbitrary, it follows that $\lim_{n\to\infty} x_k^{(n)} = x_k$ for each k. The converse is, however, not true, as the following example shows.

$$\text{Let } x_k^{(n)} = x_k + \delta_{kn}, \text{ where}$$

$$\delta_{kn} = \begin{cases} 0 & \text{if } k \neq n \\ 1 & \text{if } k = n. \end{cases}$$

Clearly,

$$|x_k^{(n)} - x_k| = \delta_{kn} = 0 \quad \text{if } n > k.$$

Consequently, $\lim_{n\to\infty} x_k^{(n)} = x_k$ for each k. However,

$$d(x^{(n)}, x) = \left(\sum_{k=1}^{\infty} |x_k^{(n)} - x_k|^p\right)^{1/p} = 1$$

for all n. Thus, $x^{(n)} \not\to x$ in the ℓ_p-metric.

The following does, however, hold.

Theorem 1.3.4. Let $\{x^{(n)}\}_{n\geq 1}$ be a sequence in ℓ_p such that $\lim_{n\to\infty} x_k^{(n)} = x_k$ for each k, where $x = \{x_k\}_{k\geq 1}$ is an element of ℓ_p. Suppose also that for every $\varepsilon > 0$ there exists an integer $m_0(\varepsilon)$ such that

$$\left(\sum_{k=m+1}^{\infty} |x_k^{(n)}|^p\right)^{1/p} < \varepsilon \quad \text{for } m \geq m_0(\varepsilon) \text{ and for all } n.$$

Then $\lim_{n\to\infty} d(x^{(n)}, x) = 0$.

Proof. Consider any $\varepsilon > 0$. Then for $m \geq m_0(\varepsilon/4)$, we have

$$\left(\sum_{k=m+1}^{\infty} |x_k^{(n)}|^p \right)^{1/p} < \frac{\varepsilon}{4} \qquad \text{for all } n, \tag{1.10}$$

and hence,

$$\left(\sum_{k=m+1}^{N} |x_k^{(n)}|^p \right)^{1/p} < \frac{\varepsilon}{4} \qquad \text{for all } N \geq m+1 \text{ and all } n.$$

Taking the limit as $n \to \infty$ in this finite sum, we get

$$\left(\sum_{k=m+1}^{N} |x_k|^p \right)^{1/p} \leq \frac{\varepsilon}{4} \qquad \text{for all } N \geq m+1.$$

Taking the limit as $N \to \infty$, we get

$$\left(\sum_{k=m+1}^{\infty} |x_k|^p \right)^{1/p} \leq \frac{\varepsilon}{4}. \tag{1.11}$$

Select any $m \geq m_0(\varepsilon/4)$. Then, for any $n \in \mathbf{N}$, we have

$$\left(d(x^{(n)}, x) \right)^p = \sum_{k=1}^{\infty} |x_k^{(n)} - x_k|^p$$

$$= \sum_{k=1}^{m} |x_k^{(n)} - x_k|^p + \sum_{k=m+1}^{\infty} |x_k^{(n)} - x_k|^p$$

$$\leq \sum_{k=1}^{m} |x_k^{(n)} - x_k|^p + 2^{p-1} \left(\sum_{k=m+1}^{\infty} |x_k^{(n)}|^p + \sum_{k=m+1}^{\infty} |x_k|^p \right) \text{ by Theorem 1.1.7}$$

$$\leq \sum_{k=1}^{m} |x_k^{(n)} - x_k|^p + 2^p \left(\frac{\varepsilon}{4} \right)^p \text{ by (1.10) and (1.11).}$$

By hypothesis, $\lim_{n \to \infty} x_k^{(n)} = x_k$ for each k. Therefore, for the finite sum above we know that there exists $n_0(\varepsilon)$ such that

$$\sum_{k=1}^{m} |x_k^{(n)} - x_k|^p < \left(\frac{\varepsilon}{4} \right)^p \qquad \text{whenever } n \geq n_0(\varepsilon).$$

Together with the preceding inequality this implies that

$$\left(d(x^{(n)}, x) \right)^p < (1 + 2^p) \left(\frac{\varepsilon}{4} \right)^p,$$

or

$$d(x^{(n)}, x) < (1 + 2^p)^{1/p} \left(\frac{\varepsilon}{4}\right) \quad \text{whenever } n \geq n_0(\varepsilon).$$

But $(1 + 2^p)^{1/p} \leq (2^p + 2^p)^{1/p} = (2^{p+1})^{1/p} \leq (2^{2p})^{1/p} = (4^p)^{1/p} = 4.$ Therefore, $d(x^{(n)}, x) < \varepsilon$ whenever $n \geq n_0(\varepsilon)$. Since the existence of such $n_0(\varepsilon)$ has been established for an arbitrary $\varepsilon > 0$, it has been shown that $\lim_{n \to \infty} d(x^{(n)}, x) = 0.$ $\square$

Examples 1.3.5. (i) Let $X = C[a, b]$, the space of all continuous functions on $[a, b]$. It does not matter whether we consider only real-valued functions or also allow complex-valued functions to be in $C[a, b]$. Define

$$d(f, g) = \int_a^b |f(x) - g(x)| dx.$$

A sequence $\{f_n\}_{n \geq 1}$ in X converges to $f \in X$ if and only if

$$\lim_{n \to \infty} d(f_n, f) = \lim_{n \to \infty} \int_a^b |f_n(x) - f(x)| dx = 0.$$

Convergence in this metric does not imply pointwise convergence, as the following example shows: Let $f(x) = 1$ for all $x \in [0, 1]$ and

$$f_n(x) = \begin{cases} nx & \text{when } 0 \leq x \leq \dfrac{1}{n} \\ 1 & \text{when } \dfrac{1}{n} < x \leq 1. \end{cases}$$

The functions f_n are illustrated in Fig. 1.4. Now,

$$\int_0^1 |f_n(x) - f(x)| dx = \int_0^{1/n} (1 - nx) dx = \frac{1}{2n} \to 0 \quad \text{as } n \to \infty.$$

So $d(f_n, f) \to 0$ as $n \to \infty$, i.e., $f_n \to f$ in the metric space. But $f_n(0) = 0$ for all n while $f(0) = 1$, so that $f_n \not\to f$ pointwise on $[0, 1]$.

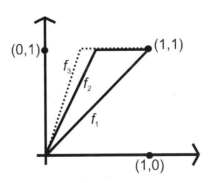

FIGURE 1.4

Indeed, the notion of pointwise convergence of sequences of functions on $[a, b]$, i.e.,

$$\lim_{n\to\infty} f_n = f \quad \text{if and only if} \quad \lim_{n\to\infty} f_n(x) = f(x)\ \forall x \in [a, b],$$

cannot be described by any metric on a nontrivial class of functions (Dugundji [8; p. 273]).

(ii) Let $X = C[a, b]$, and $d(f, g) = \sup\{|f(x) - g(x)|: x \in [a, b]\}$. Let $\{f_n\}_{n\geq 1}$ be a sequence of elements of $C[a, b]$ that converges to, say, $f \in C[a, b]$; i.e.,

$$\lim_{n\to\infty} d(f_n, f) = \lim_{n\to\infty} \sup\{|f_n(x) - f(x)|: x \in [a, b]\} = 0.$$

This means that for $\varepsilon > 0$ there exists an integer $n_0(\varepsilon)$ such that

$$\sup\{|f_n(x) - f(x)| : x \in [a, b]\} < \varepsilon \quad \text{whenever } n \geq n_0(\varepsilon),$$

i.e., the sequence $\{f_n\}_{n\geq 1}$ converges uniformly to f on $[a,b]$. The converse is also true. Indeed, if $f_n \to f$ uniformly on $[a,b]$, then for any $\varepsilon > 0$ there exists an integer $n_0(\varepsilon)$ such that

$$\sup\{|f_n(x) - f(x)| : x \in [a, b]\} < \varepsilon \quad \text{whenever } n \geq n_0(\varepsilon),$$

i.e., $\lim_{n\to\infty} d(f_n, f) = 0$. Therefore, convergence of a sequence of functions in $C[a, b]$ in the uniform metric is the same as uniform convergence.

(iii) Let X denote the space $\mathcal{B}(S)$ of all bounded functions defined on a nonempty set S and

$$d(f, g) = \sup\{|f(x) - g(x)| : x \in S\},$$

where $f, g \in X$. As in (ii) above, a sequence $\{f_n\}_{n\geq 1}$ in X converges to f if and only if $f_n \to f$ uniformly on S. Therefore, convergence of a sequence of functions in $\mathcal{B}(S)$ in the uniform metric is the same as uniform convergence. The reader may fill in the necessary details.

1.4. Cauchy Sequences

In real analysis (function theory), we have encountered Cauchy's principle of convergence. (Recall that a sequence $\{x_n\}_{n\geq 1}$ of numbers is said to be *Cauchy*, or to *satisfy the Cauchy criterion*, if and only if, for all $\varepsilon > 0$, there exists an integer $n_0(\varepsilon)$ such that $|x_n - x_m| < \varepsilon$ whenever $m \geq n_0(\varepsilon)$ and $n \geq n_0(\varepsilon)$. The Cauchy principle states that a sequence in $\mathbf{R}$ or $\mathbf{C}$ is convergent if and only if it is Cauchy.) The principle enables us to prove the convergence of a sequence without prior knowledge of its limit.

The real sequence

$$x_1 = \frac{1}{2},\ x_2 = \frac{3}{4},\ x_3 = \frac{7}{8},\ x_4 = \frac{15}{16}, \ldots$$

is such that for $m \geq n$ the distance between the terms is given by

$$|x_n - x_m| = \left| \left(1 - \frac{1}{2^n} \right) - \left(1 - \frac{1}{2^m} \right) \right| = \frac{1}{2^n} - \frac{1}{2^m},$$

which tends to zero as m, n tend to infinity. In other words, the real sequence $\{x_n\}_{n \geq 1}$, where $x_n = 1 - 1/2^n$, satisfies the Cauchy criterion and hence converges by Cauchy's principle of convergence.

A similar situation arises with sequences of functions; in fact, it comes up more often than with real or complex sequences. An extension of the idea of Cauchy sequences to metric spaces turns out to be useful.

Definition 1.4.1. Let d be a metric on a set X. A sequence $\{x_n\}_{n \geq 1}$ in the set X is said to be a **Cauchy sequence** if, for every $\varepsilon > 0$, there exists a natural number n_0 such that

$$d(x_n, x_m) < \varepsilon \quad \text{whenever } n \geq n_0 \text{ and } m \geq n_0.$$

Remark 1. A sequence $\{x_n\}$ in **R** or **C** is a Cauchy sequence in the sense familiar from elementary analysis if and only if it is a Cauchy sequence according to Definition 1.4.1 in the sense of the usual metric on **R** or **C**.

Remark 2. It is cumbersome to keep referring to a "sequence in a set X with metric d", especially if it is understood which metric is intended and no symbol such as d has been introduced to denote it. We shall therefore adopt the standard phrase "sequence in a metric space X".

Examples 1.4.2. (i) The sequence $\{x_n\}_{n \geq 1}$, where $x_n = 1 + 1/2 + \ldots + 1/n$, does not satisfy Cauchy's criterion of convergence. Indeed,

$$|x_{2n} - x_n| = \frac{1}{n+1} + \frac{1}{n+2} + \ldots + \frac{1}{2n}$$
$$\geq \frac{1}{2n} + \frac{1}{2n} + \ldots + \frac{1}{2n} = \frac{1}{2}.$$

So, it is not the case here that $|x_n - x_m| \to 0$ for large m and n.

(ii) In $C[0,1]$, the sequence f_1, f_2, f_3, $\ldots$ given by

$$f_n(x) = \frac{nx}{n+x}, \qquad x \in [0,1]$$

is Cauchy in the uniform metric. For $m \geq n$ the function

$$\frac{mx}{m+x} - \frac{nx}{n+x} = \frac{(m-n)x^2}{(m+x)(n+x)},$$

being continuous on $[0,1]$, assumes its maximum at some point $x_0 \in [0,1]$. So,

$$d(f_m, f_n) = \sup \{|f_m(x) - f_n(x)| : x \in [0, 1]\}$$
$$= \frac{(m-n)x_0^2}{(m+x_0)(n+x_0)} \leq \frac{x_0^2}{n+x_0} \leq \frac{1}{n} \to 0$$

for large m and n. Moreover, the sequence $\{f_n\}_{n \geq 1}$ converges to some limit. For $f(x) = x$,

$$|f_n(x) - f(x)| = \left| \frac{nx}{n+x} - x \right| = \frac{x^2}{n+x} \leq \frac{1}{n} \to 0 \quad \text{as } n \to \infty.$$

Therefore, $\{f_n\}_{n \geq 1}$ converges to the limit f, where $f(x) = x$ for all $x \in [0,1]$.

Proposition 1.4.3.   A convergent sequence in a metric space is a Cauchy sequence.

Proof. Let $\{x_n\}$ be a sequence in a set X with metric d, and let x be an element of X such that $\lim_{n \to \infty} x_n = x$. Given any $\varepsilon > 0$, there exists some natural number n_0 such that $d(x_n, x) < \varepsilon/2$ whenever $n \geq n_0$. Consider any natural numbers n and m such that $n \geq n_0$ and $m \geq n_0$. Then $d(x_n, x) < \varepsilon/2$ and $d(x_m, x) < \varepsilon/2$. Therefore $d(x_n, x_m) \leq d(x_n, x) + d(x_m, x) < \varepsilon/2 + \varepsilon/2 = \varepsilon$. $\qquad\qquad \square$

Does the converse of Proposition 1.4.3 hold? If a sequence $\{x_n\}_{n \geq 1}$ in a metric space (X, d) fulfills the Cauchy condition of Definition 1.4.1, does it follow that the sequence converges?

Examples 1.4.4. (i) Let X denote the set of all rational numbers with the usual metric, namely, $d(x, y) = |x - y|$ for $x, y \in X$. It is well known that the sequence

$$1.4, \ 1.41, \ 1.414, \ldots$$

converges to $\sqrt{2}$. It is therefore Cauchy. However, it does not converge to a point of X. So, a Cauchy sequence need not converge to a point of the space.

(ii) Another example of a Cauchy sequence that does not converge to a point of the space is the following: Let $X = C[0, 1]$ with metric d defined by

$$d(f, g) = \int_0^1 |f(x) - g(x)| \, dx, \quad f, g \in C[0, 1].$$

Let $\{f_n\}_{n \geq 2}$ be the sequence in $C[0, 1]$ defined by

$$f_n(x) = \begin{cases} 0, & 0 \leq x \leq \frac{1}{2} - \frac{1}{n}, \\[2mm] n\left(x - \frac{1}{2}\right) + 1, & \frac{1}{2} - \frac{1}{n} < x \leq \frac{1}{2}, \\[2mm] 1, & \frac{1}{2} < x \leq 1. \end{cases}$$

Clearly, $\{f_n\}_{n \geq 2}$ is a sequence in $C[0, 1]$. Moreover, it is Cauchy because

$$d(f_m, \, f_n) = \int_0^1 |f_n(x) - f_m(x)| \, dx$$

$$\leq \int_{1/2 - 1/m}^{1/2} f_m(x) \, dx + \int_{1/2 - 1/n}^{1/2} f_n(x) \, dx = (1/2)\left(\frac{1}{m} + \frac{1}{n}\right).$$

The measure $d(f_m, \, f_n)$ is the shaded area in Fig. 1.5.

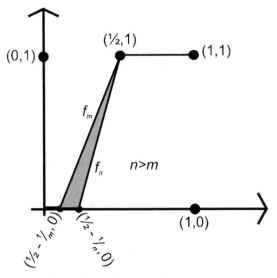

FIGURE 1.5

Suppose now that there is a continuous function f such that $d(f_n, f) \to 0$. It will be shown that this leads to a contradiction. Since

$$d(f_n, f) = \int_0^{1/2-1/n} |f(x)| dx + \int_{1/2-1/n}^{1/2} |f_n(x) - f(x)| dx + \int_{1/2}^1 |1 - f(x)| dx$$

and $d(f_n, f) \to 0$ as $n \to \infty$, it follows that

$$\int_0^{1/2} |f(x)| dx = 0 \quad \text{and} \quad \int_{1/2}^1 |1 - f(x)| dx = 0.$$

Since f is continuous, we see that $f(x) = 0$ for $0 \le x \le 1/2$ and $f(x) = 1$ for $1/2 < x \le 1$, which is impossible.

Thus, the metric spaces in which Cauchy sequences are guaranteed to converge are special and we need a name for them.

Definition 1.4.5. A metric space (X,d) is said to be **complete** if every Cauchy sequence in X is convergent.

It follows from Cauchy's principle of convergence that $\mathbf{R}$, $\mathbf{C}$ and $\mathbf{R}^n$ equipped with their standard metrics (namely, $d(x, y) = |x - y|$ for $x, y \in \mathbf{R}$; $d(z, w) = |z - w|$ for $z, w \in \mathbf{C}$ and $d(x, y) = \left(\sum_{i=1}^n (x_i - y_i)^2\right)^{1/2}$ for $x = (x_1, x_2, \ldots, x_n)$ and $y = (y_1, y_2, \ldots, y_n)$ in $\mathbf{R}^n$) are complete metric spaces. The metric space (X, d), where X denotes the set of rationals and $d(x, y) = |x - y|$ for $x, y \in X$, has been observed to be an incomplete metric space (see Example 1.4.4(i)). That the metric space (X, d) of rationals is incomplete also follows on considering the sequence $\{x_n\}_{n \ge 1}$, where

$$x_n = 1 + \frac{1}{1!} + \frac{1}{2!} + \ldots + \frac{1}{n!},$$

as this is a Cauchy sequence but it converges to the irrational number e.

In our next proposition, we need the following definition.

Definition 1.4.6. Let $\{x_n\}_{n \geq 1}$ be a given sequence in a metric space (X,d) and let $\{n_k\}_{k \geq 1}$ be a sequence of positive integers such that $n_1 < n_2 < n_3 < \ldots$. Then the sequence $\{x_{n_k}\}_{k \geq 1}$ is called a **subsequence** of $\{x_n\}_{n \geq 1}$. If $\{x_{n_k}\}_{k \geq 1}$ converges, its limit is called a **subsequential limit** of $\{x_n\}_{n \geq 1}$.

It is clear that a sequence $\{x_n\}_{n \geq 1}$ in X converges to x if and only if every subsequence of it converges to x.

Proposition 1.4.7. If a Cauchy sequence of points in a metric space (X,d) contains a convergent subsequence, then the sequence converges to the same limit as the subsequence.

Proof. Let $\{x_n\}_{n \geq 1}$ be a Cauchy sequence in (X,d). Then for every positive number ε there exists an integer $n_0(\varepsilon)$ such that

$$d(x_m, x_n) < \varepsilon \quad \text{whenever } m, n \geq n_0(\varepsilon).$$

Denote by $\{x_{n_k}\}$ a convergent subsequence of $\{x_n\}_{n \geq 1}$ and its limit by x. It follows that

$$d(x_{n_m}, x_n) < \varepsilon \quad \text{whenever } m, n \geq n_0(\varepsilon),$$

since $\{n_k\}$ is a strictly increasing sequence of positive integers. Now,

$$d(x, x_n) \leq d(x, x_{n_m}) + d(x_{n_m}, x_n) < d(x, x_{n_m}) + \varepsilon \quad \text{whenever } m, \ n \geq n_0(\varepsilon).$$

Letting $m \to \infty$, we have

$$d(x, x_n) \leq \varepsilon \quad \text{whenever } n \geq n_0(\varepsilon).$$

So, the sequence $\{x_n\}_{n \geq 1}$ converges to x. $\quad\square$

We next show that the spaces $(\mathbf{R}^n, d_p)$, ℓ_p, $\mathcal{B}(S)$ and $C[a, b]$ are complete.

Proposition 1.4.8. The metric space $X = \mathbf{R}^n$ with the metric given by

$$d_p(x, y) = \left(\sum_{i=1}^n |x_i - y_i|^p \right)^{1/p}, \ p \geq 1,$$

where $x = (x_1, x_2, \ldots, x_n)$ and $y = (y_1, y_2, \ldots, y_n)$ are in $\mathbf{R}^n$, is a complete metric space.

Proof. Let $\{x^{(m)}\}_{m \geq 1}$, $x^{(m)} = (x_1^{(m)}, x_2^{(m)}, \ldots, x_n^{(m)})$, denote a Cauchy sequence in (X,d), i.e., $d_p(x^{(m)}, x^{(m')}) \to 0$ as $m, m' \to \infty$. Then, for a given $\varepsilon > 0$ there exists an integer $n_0(\varepsilon)$ such that

$$\left(\sum_{k=1}^{n} |x_k^{(m)} - x_k^{(m')}|^p\right)^{1/p} < \varepsilon \qquad \text{for all } m, m' \geq n_0(\varepsilon). \qquad (1.12)$$

Hence, $|x_k^{(m)} - x_k^{(m')}| < \varepsilon$ for all $m, m' \geq n_0(\varepsilon)$ and all $k = 1, 2, \ldots, n$. Upon fixing k and using Cauchy's principle of convergence, it follows that $\{x_k^{(m)}\}_{m \geq 1}$ converges to a limit x_k, say, i.e., $\lim_{m \to \infty} x_k^{(m)} = x_k$. Let $x = (x_1, x_2, \ldots, x_n)$ and $m \geq n_0(\varepsilon)$. It follows from (1.12) above that

$$\sum_{k=1}^{n} |x_k^{(m)} - x_k^{(m')}|^p < \varepsilon^p \qquad (1.13)$$

for all $m' \geq n_0(\varepsilon)$. Letting $m' \to \infty$ in (1.13), we have

$$\sum_{k=1}^{n} |x_k^{(m)} - x_k|^p \leq \varepsilon^p$$

for all $m \geq n_0(\varepsilon)$, i.e., $x^{(m)} \to x$ in (X, d). $\square$

Corollary 1.4.9. Let $X = \mathbf{R}^n$ and $d_\infty(x, y) = \max\{|x_k - y_k|: 1 \leq k \leq n\}$. Then (X, d_∞) is a complete metric space.

Proof. Observe that for any real numbers $u_1, u_2, \ldots, u_n$

$$\left(\sum_{k=1}^{n} u_k^2\right)^{1/2} \leq \sum_{k=1}^{n} |u_k| \leq n \max\{|u_k|: 1 \leq k \leq n\} \leq n\left(\sum_{k=1}^{n} u_k^2\right)^{1/2}. \qquad (*)$$

Indeed, the first inequality follows on squaring both sides. The second one is obvious. The third inequality is a consequence of the fact that

$$\left(\sum_{k=1}^{n} u_k^2\right)^{1/2} \geq |u_j|, \quad j = 1, 2, \ldots, n.$$

Let $x = (x_1, x_2, \ldots, x_n)$ and $y = (y_1, y_2, \ldots, y_n)$ belong to $\mathbf{R}^n$. It follows from (*) that

$$\left(\sum_{k=1}^{n} (x_k - y_k)^2\right)^{1/2} \leq n \max\{|x_k - y_k|: 1 \leq k \leq n\} \leq n\left(\sum_{k=1}^{n} (x_k - y_k)^2\right)^{1/2}. \qquad (1.14)$$

The inequality (1.14) shows that a sequence in $\mathbf{R}^n$ is Cauchy (respectively, convergent) in the sense of one of these metrics if and only if it is Cauchy (respectively, convergent) in the sense of the other metric. Since $\mathbf{R}^n$ with $d_2(x, y) = \left(\sum_{k=1}^{n} (x_k - y_k)^2\right)^{1/2}$ as metric is complete (see Proposition 1.4.8), it is also complete with the metric d_∞. $\square$

Proposition 1.4.10. The metric space (X, d), where X denotes the space of all sequences $x = (x_1, x_2, \ldots)$ of real numbers for which $\left(\sum_{k=1}^{\infty} |x_k|^p\right)^{1/p} < \infty (p \geq 1)$ and d is the metric given by

$$d_p(x, y) = \left(\sum_{k=1}^{\infty} |x_k - y_k|^p \right)^{1/p} \qquad x, y \in X,$$

is a complete metric space, i.e., the space ℓ_p is complete.

Proof. Let $\{x^{(m)}\}_{m \geq 1}$, $x^{(m)} = (x_1^{(m)}, x_2^{(m)}, \ldots)$, denote a Cauchy sequence in (X,d). Then, for a given $\varepsilon > 0$, there exists a positive integer $n_0(\varepsilon)$ such that

$$d_p(x^{(n)}, x^{(m)}) = \left(\sum_{k=1}^{\infty} |x_k^{(n)} - x_k^{(m)}|^p \right)^{1/p} < \varepsilon \qquad (1.15)$$

for all $n, m \geq n_0(\varepsilon)$. This implies $|x_k^{(n)} - x_k^{(m)}| < \varepsilon$ for all $n, m \geq n_0(\varepsilon)$; i.e., for each k the sequence $\{x_k^{(m)}\}_{m \geq 1}$ is a Cauchy sequence of real numbers. So by Cauchy's principle of convergence, $\lim_{m \to \infty} x_k^{(m)} = x_k$, say. Let x be the sequence $(x_1, x_2, \ldots)$. It will be shown that $x \in X$ and $\lim_{m \to \infty} x^{(m)} = x$. From (1.15) we have

$$\sum_{k=1}^{N} |x_k^{(n)} - x_k^{(m)}|^p < \varepsilon^p \qquad (1.16)$$

for any positive integer N, provided that $n, m \geq n_0(\varepsilon)$. Letting $m \to \infty$ in (1.16), we obtain

$$\sum_{k=1}^{N} |x_k^{(n)} - x_k|^p \leq \varepsilon^p$$

for any positive integer N and all $n \geq n_0(\varepsilon)$. The sequence $\left\{ \sum_{k=1}^{N} |x_k^{(n)} - x_k|^p \right\}_{N \geq 1}$ is a monotonically increasing sequence which is bounded above and therefore has a finite limit $\sum_{k=1}^{\infty} |x_k^{(n)} - x_k|^p$, which is less than or equal to ε^p. Hence,

$$\left(\sum_{k=1}^{\infty} |x_k^{(n)} - x_k|^p \right)^{1/p} \leq \varepsilon \text{ for all } n \geq n_0(\varepsilon). \qquad (1.17)$$

Observe that

$$\left(\sum_{k=1}^{\infty} |x_k|^p \right)^{1/p} \leq \left(\sum_{k=1}^{\infty} |x_k^{(n)} - x_k|^p \right)^{1/p} + \left(\sum_{k=1}^{\infty} |x_k^{(n)}|^p \right)^{1/p}$$

by Minkowski's inequality, Proposition 1.1.6, and consequently, $x \in X$. Moreover, $x^{(m)} \to x$ in X in view of (1.17) above. $\square$

Proposition 1.4.11. Let X denote the space of all bounded sequences and let

$$d(x, y) = \sup\{|x_k - y_k| : 1 \leq k\}$$

be the associated metric. Then (X,d) is a complete metric space.

Proof. As the argument is simpler than that of Proposition 1.4.10, the reader is invited to fill in the details.   □

Proposition 1.4.12. The space $\mathcal{B}(S)$ of all real- or complex-valued functions f on S, each of which is bounded, with the uniform metric $d(f,g) = \sup\{|f(x) - g(x)|: x \in S\}$, is complete.

Proof. Let $\{f_n\}$ be a Cauchy sequence in $\mathcal{B}(S)$. For each $s \in S$, we have $|f_n(s) - f_m(s)| \leq d(f_n, f_m)$, so that the sequence $\{f_n(s)\}$ in $\mathbf{C}$ is a Cauchy sequence and therefore convergent. Define $f: S \to \mathbf{C}$ by $f(s) = \lim_{n \to \infty} f_n(s)$. We shall prove first that $f \in \mathcal{B}(S)$ and then prove that $\lim_{n \to \infty} d(f_n, f) = 0$.

Since $1 > 0$, therefore by the Cauchy property of $\{f_n\}$, there exists some n_0 such that

$$d(f_n, f_m) < 1 \quad \text{whenever } n \geq n_0 \text{ and } m \geq n_0.$$

In particular, $d(f_n, f_{n_0}) < 1$, and hence $|f_n(s) - f_{n_0}(s)| < 1$ for all $s \in S$, whenever $n \geq n_0$. Since $f_{n_0} \in \mathcal{B}(S)$, there exists some $M > 0$ such that $|f_{n_0}(s)| \leq M$ for all $s \in S$. Therefore,

$$|f_n(s)| \leq |f_n(s) - f_{n_0}(s)| + |f_{n_0}(s)| \leq 1 + M \qquad \forall s \in S \text{ whenever } n \geq n_0.$$

It follows that $|f(s)| = \lim_{n \to \infty} |f_n(s)| \leq 1 + M$ for every $s \in S$. Thus, $f \in \mathcal{B}(S)$, as claimed.

Now consider any $\varepsilon > 0$. By the Cauchy property of $\{f_n\}$, there exists some n_0 such that

$$d(f_n, f_m) < \varepsilon \qquad \text{whenever } n \geq n_0 \text{ and } m \geq n_0.$$

Therefore,

$$|f_n(s) - f_m(s)| < \varepsilon \qquad \forall s \in S \text{ whenever } n \geq n_0 \text{ and } m \geq n_0.$$

It follows upon letting $m \to \infty$ that

$$|f_n(s) - f(s)| \leq \varepsilon \qquad \text{whenever } s \in S \text{ and } n \geq n_0.$$

This says that $\lim_{n \to \infty} d(f_n, f) = 0$.   □

Proposition 1.4.13. Let $X = C[a, b]$ and $d(f,g) = \sup\{|f(x) - g(x)|: a \leq x \leq b\}$ be the associated metric. Then (X, d) is a complete metric space.

Proof. Let $\{f_n\}_{n \geq 1}$ be a Cauchy sequence in $C[a,b]$. Then for every $\varepsilon > 0$ there exists an integer $n_0(\varepsilon)$ such that $m, n \geq n_0(\varepsilon)$ implies $d(f_m, f_n) = \sup\{|f_m(x) - f_n(x)|: a \leq x \leq b\} < \varepsilon$. In particular, for every $x \in [a, b]$, the sequence $\{f_n(x)\}_{n \geq 1}$ is a Cauchy sequence of numbers. By Cauchy's principle of convergence, $f_n(x) \to f(x)$, say, as $n \to \infty$. We have thus defined a function f with domain $[a,b]$. It remains to show that $f \in C[a, b]$ and that $\lim_{n \to \infty} d(f_n, f) = 0$.

Since for every $x \in [a, b]$,

$$|f_m(x) - f_n(x)| < \varepsilon$$

provided that $m, n \geq n_0(\varepsilon)$, it follows upon letting $m \to \infty$ that

$$|f_n(x) - f(x)| \leq \varepsilon \qquad (1.18)$$

for all $n \geq n_0(\varepsilon)$ and all $x \in [a, b]$.

To see why f is continuous, consider any $x_0 \in [a, b]$ and any $\eta > 0$. According to what has been noted in the preceding paragraph, there exists an integer $n_1(\eta)$ such that, for every $x \in [a, b]$, we have $|f_n(x) - f(x)| < \eta/3$ provided that $n \geq n_1(\eta)$. Select $m \geq n_1(\eta)$. Then

$$|f_m(x) - f_m(x)| < \frac{\eta}{3} \qquad \text{for all } x \in [a, b]. \qquad (1.19)$$

Now use the continuity of f_m to obtain $\delta > 0$ such that

$$|f_m(x) - f_m(x_0)| < \frac{\eta}{3} \quad \text{for } |x - x_0| < \delta. \qquad (1.20)$$

Since

$$|f(x) - f(x_0)| \leq |f(x) - f_m(x)| + |f_m(x) - f_m(x_0)| + |f_m(x_0) - f(x_0)|,$$

it follows from (1.19) and (1.20) that $|f(x) - f(x_0)| < \eta$ whenever $|x - x_0| < \delta$.

Therefore, $f \in C[a, b]$. Moreover, (1.18) says that $\lim_{n \to \infty} d(f_n, f) = 0$. As already noted, this completes the proof. $\square$

Examples 1.4.14. (i) Let X be any nonempty set and let d be defined by

$$d(x, y) = \begin{cases} 0 & \text{if } x = y, \\ 1 & \text{if } x \neq y. \end{cases}$$

Then (X, d) is a complete metric space.

Indeed, if $\{x_n\}_{n \geq 1}$ is a Cauchy sequence, then for $0 < \varepsilon < 1$ there exists a positive integer $n_0(\varepsilon)$ such that $d(x_n, x_m) < \varepsilon$ for all $m, n \geq n_0(\varepsilon)$. So for $n \geq n_0(\varepsilon)$, we have $x_n = x_{n_0}$. Thus, any Cauchy sequence in (X, d) is of the form

$$(x_1, x_2, \ldots, x_{n_0}, x_{n_0}, \ldots),$$

which is clearly convergent to the limit x_{n_0}.

(ii) Let $\mathbf{N}$ denote the set of natural numbers. Define

$$d(m, n) = |\frac{1}{m} - \frac{1}{n}|, \qquad m, n \in \mathbf{N}.$$

Then $(\mathbf{N}, d)$ is an incomplete metric space.

That $(\mathbf{N}, d)$ is a metric space is clear. The sequence $\{n\}_{n \geq 1}$ can be shown to be Cauchy by arguing as follows. Let $\varepsilon > 0$ and let n_0 be the least integer greater than $1/\varepsilon$. If $m, \; n \geq n_0$ then

$$d(m, n) = \left|\frac{1}{m} - \frac{1}{n}\right| \le \max\left\{\frac{1}{m}, \frac{1}{n}\right\} \le \frac{1}{n_0} < \varepsilon.$$

Thus, the sequence is Cauchy. To see why it does not converge, suppose that it were to converge if possible to, say, $p \in \mathbf{N}$. Let n_1 be any integer greater than $2p$. Then $n \ge n_1$ implies that

$$d(p, n) = \left|\frac{1}{p} - \frac{1}{n}\right| = \frac{1}{p} - \frac{1}{n} \ge \frac{1}{p} - \frac{1}{n_1} > \frac{1}{p} - \frac{1}{2p} = \frac{1}{2p}.$$

This shows that the sequence cannot converge to p and therefore does not converge at all.

(iii) Let $X = \mathbf{C} \cup \{\infty\}$ and let d be given by

$$d(z_1, z_2) = \begin{cases} \dfrac{2|z_1 - z_2|}{\sqrt{1 + |z_1|^2}\sqrt{1 + |z_2|^2}} & \text{when } z_1, z_2 \in \mathbf{C}, \\[4mm] \dfrac{2}{\sqrt{1 + |z_1|^2}} & \text{when } z_1 \in \mathbf{C} \text{ and } z_2 = \infty. \end{cases}$$

Then (X, d) is a complete metric space.

Let $\{z_n\}_{n \ge 1}$ be a Cauchy sequence in (X, d). If the sequence $\{z_n\}_{n \ge 1}$ contains the point at infinity infinitely often, then it contains a convergent subsequence, namely the subsequence each of whose terms is ∞. In this case, the sequence $\{z_n\}_{n \ge 1}$ converges to ∞ by Proposition 1.4.7. On the other hand, if the point at infinity appears only finitely many times in the sequence, then we may assume without loss of generality that the sequence consists of points of $\mathbf{C}$ only, as the deletion (or insertion) of finitely many terms does not alter the convergence behaviour of a sequence.

Case I. If the sequence $\{|z_n|\}_{n \ge 1}$ is unbounded, then for every natural number k there exists a term z_{n_k} of the sequence such that $|z_{n_k}| > k$, where these terms can be chosen so that $n_{k+1} > n_k$, $k = 1, 2, \ldots$. Now,

$$d(z_{n_k}, \infty) = \frac{2}{\sqrt{1 + |z_{n_k}|^2}} < \frac{2}{\sqrt{1 + k^2}} \to 0 \quad \text{as } k \to \infty.$$

We thus have $\lim_{k \to \infty} z_{n_k} = \infty$ in (X, d). By Proposition 1.4.7, it follows that $\lim_{n \to \infty} z_n = \infty$.

Case II. The sequence $\{|z_n|\}_{n \ge 1}$ is bounded, say by $M > 0$. Let $\varepsilon > 0$ be given. There exists $n_0 \in \mathbf{N}$ such that $m, n \ge n_0$ implies

$$\frac{2|z_n - z_m|}{\sqrt{1 + |z_n|^2}\ \sqrt{1 + |z_m|^2}} < \varepsilon.$$

Since $|z_n| < M$ for all n, it follows that $|z_n - z_m| < (1/2)\varepsilon(1 + M^2)$. This shows that $\{z_n\}_{n \ge 1}$ is a Cauchy sequence in the usual metric in $\mathbf{C}$, and hence $\lim_{n \to \infty} |z_n - z| = 0$ for some $z \in \mathbf{C}$. Since $d(z_n, z_m) \le 2|z_n - z_m|$ always, it follows that $d(z_n, z) \to 0$ as $n \to \infty$. Thus, (X, d) is a complete metric space.

(iv) Let $X = \mathbf{R}$ and $d: \mathbf{R} \times \mathbf{R} \to \mathbf{R}$ be defined by

$$d(x, y) = \frac{|x - y|}{\sqrt{1 + x^2}\sqrt{1 + y^2}}. \tag{1}$$

We shall show that $(\mathbf{R}, d)$ is a metric space which is not complete.

As the proof that $(\mathbf{R}, d)$ is a metric space is no different from that of (xiii) of Example 1.2.2, it is not included. We shall only show that $(\mathbf{R}, d)$ is not complete.

Consider the sequence $\{n\}_{n \geq 1}$ of natural numbers. Observe that

$$d(n, m) = \frac{|n - m|}{\sqrt{1 + n^2}\sqrt{1 + m^2}} = \frac{|\frac{1}{n} - \frac{1}{m}|}{\sqrt{\frac{1}{n^2} + 1}\sqrt{\frac{1}{m^2} + 1}}$$

$$\leq |\frac{1}{n} - \frac{1}{m}| \leq \frac{1}{n} + \frac{1}{m}.$$

The right hand side of the above inequality can be made arbitrarily small by choosing m and n sufficiently large. Thus $\{n\}_{n \geq 1}$ is a Cauchy sequence in the metric space $(\mathbf{R}, d)$. We claim that it does not converge in $\mathbf{R}$: Suppose, if possible, that it were to converge to some limit $p \in \mathbf{R}$. Then

$$d(n, p) = \frac{|n - p|}{\sqrt{1 + n^2}\sqrt{1 + p^2}} = \frac{|1 - \frac{p}{n}|}{\sqrt{1 + \frac{1}{n^2}}\sqrt{1 + p^2}},$$

so that

$$\lim_{n \to \infty} d(n, p) = \frac{1}{\sqrt{1 + p^2}} \neq 0.$$

This shows that the Cauchy sequence $\{n\}_{n \geq 1}$ does not converge to $p \in \mathbf{R}$. Since p is arbitrary, the argument is complete.

1.5. Completion of a Metric Space

Let (X, d) be a metric space that is not complete. It is always possible to construct a larger space which is complete and contains *just* enough points so that every Cauchy sequence in X has a limit in the larger space. In fact, we need to adjoin new points to (X, d) and extend d to all these new points in such a way that the formerly nonconvergent Cauchy sequences find limits among these new points and the new points are limits of sequences in X.

Definition 1.5.1. Let (X, d) be an arbitrary metric space. A complete metric space (X^*, d^*) is said to be a **completion** of the metric space (X, d) if

(i) X is a subspace of X^*;
(ii) every point of X^* is the limit of some sequence in X.

For example, the space of all real numbers is the completion of the space of rationals. Also, the closed interval $[0,1]$ is the completion of $(0,1)$, $[0,1)$, $(0,1]$ and itself. In fact, any complete metric space is its own completion. We note that the Weierstrass approximation theorem (Proposition 0.8.4) shows that the metric space $C_R[a, b]$ of Example 1.2.2(ix) is the completion of its subset consisting of polynomials.

Definition 1.5.2. Let (X,d) and (X', d') be two metric spaces. A mapping f of X into X' is an **isometry** if

$$d'(f(x), f(y)) = d(x, y)$$

for all $x, y \in X$. The mapping f is also called an **isometric embedding** of X into X'. If, however, the mapping is onto, the spaces X and X' themselves, between which there exists an isometric mapping, are said to be **isometric**. It may be noted that an isometry is always one-to-one.

Theorem 1.5.3. Every metric space has a completion and any two completions are isometric to each other.

Proof. Let (X,d) be a metric space and let $\hat{X}$ denote the set of all Cauchy sequences in X. We define two Cauchy sequences $\{x_n\}$ and $\{y_n\}$ in X to be **equivalent** if $\lim_{n \to \infty} d(x_n, y_n) = 0$ and write this in symbols as $\{x_n\} \sim \{y_n\}$. We shall now show that this is an equivalence relation in $\hat{X}$, i.e., the relation $\sim$ is reflexive, symmetric and transitive.

Reflexivity: $\{x_n\} \sim \{x_n\}$, since $d(x_n, x_n) = 0$ for every n and so $\lim_{n \to \infty} d(x_n, x_n) = 0$. Symmetry: If $\{x_n\} \sim \{y_n\}$, then $\lim_{n \to \infty} d(x_n, y_n) = 0$; but $d(x_n, y_n) = d(y_n, x_n)$ for every n, and, therefore, $\lim_{n \to \infty} d(y_n, x_n) = 0$, so that $\{y_n\} \sim \{x_n\}$. Transitivity: If $\{x_n\} \sim \{y_n\}$ and $\{y_n\} \sim \{z_n\}$, then $\lim_{n \to \infty} d(x_n, y_n) = 0$ and $\lim_{n \to \infty} d(y_n, z_n) = 0$. We shall show that $\lim_{n \to \infty} d(x_n, z_n) = 0$. Since

$$0 \le d(x_n, z_n) \le d(x_n, y_n) + d(y_n, z_n)$$

for all n, it follows that

$$0 \le \lim_{n \to \infty} d(x_n, z_n) \le \lim_{n \to \infty} d(x_n, y_n) + \lim_{n \to \infty} d(y_n, z_n) = 0,$$

i.e., $\lim_{n \to \infty} d(x_n, z_n) = 0$.

Thus, $\sim$ is an equivalence relation and $\hat{X}$ splits into equivalence classes. Any two members of the same equivalence class are equivalent, and no member of an equivalence class is equivalent to a member of any other equivalence class. Let $\tilde{X}$ denote the set of all equivalence classes; the elements of $\tilde{X}$ will be denoted by $\tilde{x}, \tilde{y}$, etc. Observe that if a Cauchy sequence $\{x_n\}$ has a limit $x \in X$, and if $\{y_n\}$ is equivalent to $\{x_n\}$, then $\lim_{n\to\infty} y_n = x$. This follows immediately from the following inequality:

$$d(y_n, x) \leq d(y_n, x_n) + d(x_n, x).$$

Moreover, if $\{x_n\}$ and $\{y_n\}$ are two nonequivalent sequences, then $\lim_{n\to\infty} x_n \neq \lim_{n\to\infty} y_n$.

For, if $\lim_{n\to\infty} x_n = x = \lim_{n\to\infty} y_n$, then the inequality

$$0 \leq d(x_n, y_n) \leq d(x_n, x) + d(x, y_n)$$

leads to $\lim_{n\to\infty} d(x_n, y_n) = 0$, contradicting the fact that $\{x_n\}$ and $\{y_n\}$ are two nonequivalent sequences. The constant sequence $(x, x, \ldots, x, \ldots)$ is evidently Cauchy and has limit x.

Define a mapping $f: X \to \tilde{X}$ as follows: $f(x) = \tilde{x}$, where $\tilde{x}$ denotes the equivalence class each of whose members converges to x. Thus the constant sequence $(x, x, \ldots, x, \ldots)$ is a representative of $\tilde{x}$. In view of the observations made above, the mapping f is one-to-one. We next define a metric ρ in $\tilde{X}$. For $\tilde{x}, \tilde{y} \in \tilde{X}$, set

$$\rho(\tilde{x}, \tilde{y}) = \lim_{n\to\infty} d(x_n, y_n), \text{ where } \{x_n\} \in \tilde{x} \text{ and } \{y_n\} \in \tilde{y}.$$

Observe that

$$|d(x_n, y_n) - d(x_m, y_m)| \leq d(x_n, x_m) + d(y_m, y_n),$$

where $\{x_n\} \in \tilde{x}$ and $\{y_n\} \in \tilde{y}$ and so $\{d(x_n, y_n)\}_{n \geq 1}$ is a Cauchy sequence of real numbers. Hence, $\lim_{n\to\infty} d(x_n, y_n)$ exists, for $\mathbf{R}$ is a complete metric space. We first show that ρ is well defined. Indeed, if $\{x'_n\} \sim \{x_n\}$ and $\{y'_n\} \sim \{y_n\}$, then

$$\lim_{n\to\infty} d(x_n, x'_n) = 0 \text{ and } \lim_{n\to\infty} d(y_n, y'_n) = 0.$$

From the relation

$$d(x'_n, y'_n) \leq d(x'_n, x_n) + d(x_n, y_n) + d(y_n, y'_n)$$

and the relation

$$d(x_n, y_n) \leq d(x_n, x'_n) + d(x'_n, y'_n) + d(y'_n, y_n),$$

it follows that

$$\lim_{n\to\infty} d(x_n, y_n) = \lim_{n\to\infty} d(x'_n, y'_n).$$

We next show that ρ is a metric on $\tilde{X}$.

Since $d(x_n, y_n) \geq 0$ for all n, it follows that $\lim_{n\to\infty} d(x_n, y_n) \geq 0$. Thus, $\rho(\tilde{x}, \tilde{y}) \geq 0$. If $\tilde{x} = \tilde{y}$, then $\rho(\tilde{x}, \tilde{y}) = \lim_{n\to\infty} d(x_n, y_n)$, where $\{x_n\} \in \tilde{x}$, $\{y_n\} \in \tilde{y}$ and $\{x_n\} \sim \{y_n\}$. But this means (by definition of $\sim$) that $\lim_{n\to\infty} d(x_n, y_n) = 0$. Therefore $\rho(\tilde{x}, \tilde{y}) = 0$. Conversely, if $\rho(\tilde{x}, \tilde{y}) = 0$, then $\lim_{n\to\infty} d(x_n, y_n) = 0$ (by definition of ρ) and hence $\{x_n\} \sim \{y_n\}$, so that $\tilde{x} = \tilde{y}$. Since $d(x_n, y_n) = d(y_n, x_n)$ for all n, we have $\rho(\tilde{x}, \tilde{y}) = \rho(\tilde{y}, \tilde{x})$. Finally,

$$\rho(\tilde{x}, \tilde{z}) = \lim_{n \to \infty} d(x_n, z_n) \leq \lim_{n \to \infty} d(x_n, y_n) + \lim_{n \to \infty} d(y_n, z_n) = \rho(\tilde{x}, \tilde{y}) + \rho(\tilde{y}, \tilde{z}),$$

where $\{x_n\} \in \tilde{x}$, $\{y_n\} \in \tilde{y}$ and $\{z_n\} \in \tilde{z}$.

The metric ρ defined in $\tilde{X}$ has the property that $\rho(\tilde{x}, \tilde{y}) = \rho(f(x), f(y)) = d(x, y)$ for all $x, y \in X$, i.e., f is an isometric embedding of X into $\tilde{X}$. Next we prove that every $\tilde{x} \in \tilde{X}$ is the limit of a sequence in $f(X)$. For this purpose, suppose $\{x_n\}$ is a Cauchy sequence in the equivalence class $\tilde{x}$. For any $k \in \mathbf{N}$, there exists a positive integer n_k such that $d(x_n, x_{n_k}) < 1/k$ for $n \geq n_k$. Let $\tilde{y}_k$ be the equivalence class containing all Cauchy sequences converging to x_{n_k}, i.e., $\tilde{y}_k = f(x_{n_k})$. Then

$$\rho(\tilde{x}, f(x_{n_k})) = \rho(\tilde{x}, \tilde{y}_k) = \lim_{n \to \infty} d(x_n, x_{n_k}) \leq \frac{1}{k}.$$

Thus, $\tilde{x} = \lim_{k \to \infty} f(x_{n_k})$.

Finally, we show that $\tilde{X}$ is complete.

Let $\{\tilde{x}^{(n)}\}$ be a Cauchy sequence in $\tilde{X}$. Since each $\tilde{x}^{(n)}$ is the limit of a sequence in $f(X)$, we can find $\tilde{y}^{(n)} \in f(X)$ such that $\rho(\tilde{x}^{(n)}, \tilde{y}^{(n)}) < \frac{1}{n}$. Then the sequence $\{\tilde{y}^{(n)}\}$ in $\tilde{X}$ can be shown to be Cauchy by arguing as follows:

$$\rho(\tilde{y}^{(n)}, \tilde{y}^{(m)}) \leq \rho(\tilde{y}^{(n)}, \tilde{x}^{(n)}) + \rho(\tilde{x}^{(n)}, \tilde{x}^{(m)}) + \rho(\tilde{x}^{(m)}, \tilde{y}^{(m)})$$
$$\leq \frac{1}{n} + \rho(\tilde{x}^{(n)}, \tilde{x}^{(m)}) + \frac{1}{m}.$$

The right hand side can be made as small as desired by choosing m and n large enough, for $\{\tilde{x}^{(n)}\}$ is Cauchy. Since $\tilde{y}^{(n)}$ is in $f(X)$, there exists some $y_n \in X$ such that $f(y_n) = \tilde{y}^{(n)}$. The sequence $\{y_n\}$ in X must be Cauchy because $\{\tilde{y}^{(n)}\}$ is a Cauchy sequence in $\tilde{X}$ and f is an isometry. Therefore, $\{y_n\}$ belongs to some equivalence class $\tilde{x} \in \tilde{X}$. We shall show that $\lim_{n \to \infty} \rho(\tilde{x}^{(n)}, \tilde{x}) = 0$. To this end, we take any $\varepsilon > 0$ and observe that

$$\rho(\tilde{x}^{(n)}, \tilde{x}) \leq \rho(\tilde{x}^{(n)}, \tilde{y}^{(n)}) + \rho(\tilde{y}^{(n)}, \tilde{x}) < \frac{1}{n} + \rho(\tilde{y}^{(n)}, \tilde{x})$$

and

$$\rho(\tilde{y}^{(n)}, \tilde{x}) = \rho(f(y_n), \tilde{x}) = \lim_{n \to \infty} d(y_n, y_p) \leq \varepsilon$$

for sufficiently large n, since $\{y_n\}$ is a Cauchy sequence in X. This implies that $\lim_{n \to \infty} \rho(\tilde{x}^{(n)}, \tilde{x}) = 0$, thereby completing the proof that $\tilde{X}$ is complete.

Uniqueness: Let (X^*, d^*) and (X^{**}, d^{**}) be any two completions of (X, d). We shall show that (X^*, d^*) and (X^{**}, d^{**}) are isometric.

Let $x^* \in X^*$ be arbitrary. By the definition of completion, there exists a sequence $\{x_n\}_{n \geq 1}$ in X such that $\lim_{n \to \infty} x_n = x^*$. The sequence $\{x_n\}_{n \geq 1}$ may be assumed to belong to X^{**}. Since X^{**} is complete, $\{x_n\}_{n \geq 1}$ converges in X^{**} to x^{**}, say, i.e., $\lim_{n \to \infty} x_n = x^{**}$. Define $\varphi: X^* \to X^{**}$ by setting $\varphi(x^*) = x^{**}$. It is clear that the mapping φ is one-to-one and does not depend on the choice of the sequence $\{x_n\}_{n \geq 1}$ converging to x^*. Moreover, by construction, $\varphi(x) = x$ for $x \in X$ and $d^{**}(\varphi(x^*), \varphi(y^*)) = d^*(x^*, y^*)$ for all $x^*, y^* \in X^*$.

This completes the proof. □

Remarks. (i) The proof explicitly assumes the completeness of **R**. Hence, the above method of completion cannot be employed for constructing the real number system from the rational number system.

(ii) There exist other methods of completion of an incomplete space. One such method will be provided in Example. 17 of Chapter 3 (Section 3.8).

1.6. Exercises

1. For $a, b > 0$ and $0 < p < 1$, show that $(a + b)^p \leq a^p + b^p$.

   Hint: The function $g(x) = \frac{(1+x)^p}{1+x^p}, x \geq 1$, has derivative $\frac{p(1+x)^{p-1}(1-x^{p-1})}{(1+x^p)^2}$, which is positive when $x > 1$. So, $g(x)$ is increasing for $x \geq 1$. Moreover, $g(x) \to 1$ as $x \to \infty$, and hence, $g(x) \leq 1$ for $x \geq 1$. Put $x = a/b$ or b/a, whichever is ≥ 1.

2. Let $X = \mathbf{C}^n$ and $0 < p < 1$. Define d_p by

$$d_p(z, w) = \left(\sum_{k=1}^{n} |z_k - w_k|^p \right)^{1/p},$$

   where $z = (z_1, z_2, \ldots, z_n)$ and $w = (w_1, w_2, \ldots, w_n)$ are in $\mathbf{C}^n$. Does d_p define a metric on $\mathbf{C}^n$?
   Hint: No. For $z = (1, 1, 0, 0, \ldots, 0), \zeta = (0, 1, 0, 0, \ldots, 0)$ and $w = (0, 0, 0, 0, \ldots, 0)$, $d_p(z, w) = 2^{1/p}, d_p(z, \zeta) = 1 = d_p(\zeta, w)$, so that $d_p(z, w) > d_p(z, \zeta) + d_p(\zeta, w)$.

3. Let X be the set of all sequences $\{z_k\}_{k \geq 1}$ of numbers that are p-summable, i.e., $\sum_{k=1}^{\infty} |z_k|^p < \infty$, with $d: X \times X \to \mathbf{R}$ defined by

$$d(z, w) = \sum_{k=1}^{\infty} |z_k - w_k|^p, \qquad \text{where } 0 < p < 1$$

   and $z = \{z_k\}_{k \geq 1}, w = \{w_k\}_{k \geq 1}$ are in X. Then (X, d) is a metric space.
   Hint: For $z = \{z_k\}_{k \geq 1}, w = \{w_k\}_{k \geq 1}$ and $\zeta = \{\zeta_k\}_{k \geq 1}$,

$$|z_k - w_k|^p \leq (|z_k - \zeta_k| + |\zeta_k - w_k|)^p \leq |z_k - \zeta_k|^p + |\zeta_k - w_k|^p,$$

   using the monotonicity of the function $x \mapsto x^p$ for $x > 0$, $0 < p < 1$ and Exercise 1 above. So,

$$\sum_{k=1}^{n} |z_k - w_k|^p \leq \sum_{k=1}^{n} |z_k - \zeta_k|^p + \sum_{k=1}^{n} |\zeta_k - w_k|^p \leq \sum_{k=1}^{\infty} |z_k - \zeta_k|^p + \sum_{k=1}^{\infty} |\zeta_k - w_k|^p,$$

   which implies $d(z, w) \leq d(z, \zeta) + d(\zeta, w)$.

4. Show that, in the definition of a metric space, the axiom (MS4) can be replaced by the following 'weaker' axiom:

(MS4*) If $x, y, z \in X$ are distinct, then $d(x, z) \le d(x, y) + d(y, z)$.
Hint: For $x = y, d(x, z) = d(y, z) = d(y, y) + d(y, z) = d(x, y) + d(y, z)$. The argment for $y = z$ is similar. For $x = z, d(x, z) = 0 \le d(x, y) + d(y, z)$.

5. Show that $d(x, y) = \sqrt{|x - y|}$ defines a metric on the set of reals.
6. (a) Let $X = \mathbf{R}$ and for $x, y \in \mathbf{R}$, define $d(x,y)$ by

$$d(x, y) = \begin{cases} |x - y| + 1 & \text{if exactly one among } x \text{ and } y \text{ is strictly positive,} \\ |x - y| & \text{otherwise.} \end{cases}$$

   Prove that (X,d) is a metric space.
   (b) In any space X with metric d, let f be a self map that is one-to-one. Set $D(x, y) = d(f(x), f(y))$. Prove that D is a metric on X.
   (c) In any space X with metric d, let U and V be disjoint nonempty subsets with union X; define $D(x,y)$ to be $d(x, y) + 1$ if exactly one among x and y belongs to U, and to be $d(x,y)$, otherwise. Prove that D is a metric on X.
   Hint: (a) If $x > 0$ and $z \le 0$, then $d(x, z) = |x - z| + 1 \le |x - y| + |y - z| + 1$. If $y \le 0$, this equals $d(x, y) + d(y, z)$ because $|x - y| + 1 = d(x, y)$. If $y > 0$, this again equals $d(x, y) + d(y, z)$ because now $|y - z| + 1 = d(y, z)$.
   (b) Each of properties (MS1),(MS3) and (MS4) for D is an immediate consequence of the corresponding property for d. Property (MS2) follows for D from the corresponding property for d in conjunction with the fact that f is one-to-one. Note that for $\mathbf{R}$ with the usual metric and f defined as $f(x) = x$ for $x < 0$ and $f(x) = x + 1$ for $x \ge 0$, the metric D is precisely the one of Exercise 6(a).
   (c) Imitate the argument for (a).

7. (a) Let $X = \mathbf{R}^2$, and for $x, y \in \mathbf{R}^2$ define $d(x,y)$ by

$$d(x, y) = d((x_1, x_2), (y_1, y_2)) = \begin{cases} |x_1 - y_1| & \text{if } x_2 = y_2 , \\ |x_1| + |x_2 - y_2| + |y_1| & \text{if otherwise.} \end{cases}$$

   Show that (X, d) is a metric space.
   (b) Let $X = \mathbf{R}$, and for $x, y \in \mathbf{R}$ define $d(x, y) = |x| + |x - y| + |y|$ when $x \ne y$, and $d(x, y) = 0$ when $x = y$. Show that d is a metric on $\mathbf{R}$.
   Hint: (a) Let $x = (x_1, x_2), y = (y_1, y_2)$ and $z = (z_1, z_2)$ be in X. We note firstly that $|x_1 - y_1| \le d(x, y)$. If $x_2 = y_2$, then $d(x, y) = |x_1 - y_1| \le |x_1 - z_1| + |z_1 - y_1| \le d(x, z) + d(z, y)$. If $x_2 \ne y_2$, then z_2 cannot be equal to both x_2 and y_2; so assume $z_2 \ne x_2$. Then

$$d(x, y) = |x_1| + |x_2 - y_2| + |y_1| \le |x_1| + |x_2 - z_2| + |z_2 - y_2| + |y_1|$$
$$\le \begin{cases} (|x_1| + |x_2 - z_2| + |z_1|) + |z_1 - y_1| & \text{if } y_2 = z_2 , \\ (|x_1| + |x_2 - z_2| + |z_1|) + (|z_1| + |z_2 - y_2| + |y_1|) & \text{if } y_2 \ne z_2 \end{cases}$$
$$= d(x, z) + d(z, y).$$

   (b) Straightforward.

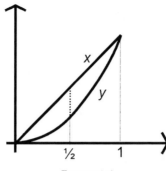

FIGURE 1.6

8. Let $X = C[0,1]$, and for $x, y \in X$ define $d(x,y) = \sup\{|x(t) - y(t)| : 0 \le t \le 1\}$.
   Calculate the distance between $x(t) = t$ and $y(t) = t^2$.
   Hint: $|x(t) - y(t)| = t - t^2 = -(1/2 - t)^2 + 1/4$; so $\sup\ \{|x(t) - y(t)| : 0 \le t \le 1\} = 1/4$.
   Alternate method: Consider the function $f(t) = t - t^2$. We have $f'(t) = 0$ if and only if $t = 1/2$, and $f''(1/2) = -2 < 0$. The function has a local maximum at $t = 1/2$, which is also an absolute maximum because $f(0) = 0 = f(1)$. The absolute maximum value is $f(1/2) = 1/4$. (See Fig. 1.6.)

9. Let X denote the set of all real sequences and let $x = \{x_i\}_{i \ge 1}, y = \{y_i\}_{i \ge 1}$ be arbitrary elements in X. Define

$$d(x,y) = \sum_{k=1}^{\infty} \frac{1}{k^2} \min\{|x_k - y_k|, \ 1\}.$$

Prove that (X,d) is a metric space. Let $\{x^{(n)}\}_{n \ge 1} = \{\{x_i^{(n)}\}_{i \ge 1}\}_{n \ge 1}$ be a sequence in X. Show that $x^{(n)} \to x$ in (X,d) if and only if $x_k^{(n)} \to x_k$ for all k.
Hint: $\min\{|x_n - z_n|, \ 1\} \le 1$. If $\min\{|x_n - y_n|, \ 1\} = 1$ or $\min\{|y_n - z_n|, \ 1\} = 1$, then $\min\{|x_n - z_n|, \ 1\} \le \min\{|x_n - y_n|, \ 1\} + \min\{|y_n - z_n|, \ 1\}$. So,

$$\sum_{k=1}^{n} \frac{1}{k^2} \min\{|x_k - z_k|, \ 1\} \le \sum_{k=1}^{n} \frac{1}{k^2} \min\{|x_k - y_k|, \ 1\} + \sum_{k=1}^{n} \frac{1}{k^2} \min\{|y_k - z_k|, \ 1\}$$

$$\le \sum_{k=1}^{\infty} \frac{1}{k^2} \min\{|x_k - y_k|, \ 1\} + \sum_{k=1}^{\infty} \frac{1}{k^2} \min\{|y_k - z_k|, \ 1\},$$

and hence, $d(x,z) \le d(x,y) + d(y,z)$. In the case $\min\{|x_n - y_n|, 1\} < 1$ and $\min\{|y_n - z_n|, 1\} < 1$, $\min\{|x_n - y_n|, 1\} + \min\{|y_n - z_n|, 1\} = |x_n - y_n| + |y_n - z_n| \ge |x_n - z_n| \ge \min\{|x_n - z_n|, 1\}$.
   Assume $x_k^{(n)} \to x_k$ for all k. For a fixed $\varepsilon > 0$, choose an integer p such that $\sum_{q=p+1}^{\infty} 1/q^2 < (1/2)\varepsilon$. There exists an integer n_0 such that $n \ge n_0$ implies

$$|x_q^{(n)} - x_q| < \frac{3}{\pi^2}\varepsilon \text{ for } q = 1, 2, \ldots, p.$$

Consequently, for $n \geq n_0$, we have

$$d(x^{(n)}, x) = \sum_{q=1}^{p} \frac{1}{q^2} \min\{|x_q^{(n)} - x_q|, 1\} + \sum_{q=p+1}^{\infty} \frac{1}{q^2} \min\{|x_q^{(n)} - x_q|, 1\}$$

$$\leq \sum_{q=1}^{p} \frac{1}{q^2}\left(\frac{3}{\pi^2}\varepsilon\right) + \sum_{q=p+1}^{\infty} \frac{1}{q^2}$$

$$< \frac{3\varepsilon}{\pi^2}\left(\frac{\pi^2}{6}\right) + \frac{\varepsilon}{2} = \varepsilon.$$

The converse is left to the reader.

10. Let $X = C(\mathbf{R})$, the set of real-valued continuous functions on $\mathbf{R}$, and let $x, y \in X$. Define

$$d_n(x, y) = \sup\{|x(t) - y(t)| : t \in [-n, n]\}$$

and

$$d(x, y) = \sum_{n=1}^{\infty} \frac{1}{2^n} \frac{d_n(x, y)}{1 + d_n(x, y)}.$$

Prove that d is a metric on X.

Hint: Clearly, $d_n, n = 1, 2, \ldots$, is a pseudometric on X. If $d_n(x, y) = 0$, then $x(t) = y(t)$ for $t \in [-n, n]$. Therefore, if $d_n(x, y) = 0$ for every n, then $x = y$. By the argument of Proposition 1.2.3, $\rho_n = \frac{d_n(x, y)}{1 + d_n(x, y)}, n = 1, 2, \ldots$, is also a pseudometric on X. Since for all $x, y \in X, \rho_n < 1, n = 1, 2, \ldots$, it follows by the comparison test that $d(x, y)$ is well defined. Now,

$$\rho_n(x, z) \leq \rho_n(x, y) + \rho_n(y, z), \qquad x, y, z \in C(\mathbf{R})$$

and so

$$\sum_{n=1}^{m} 2^{-n}\rho_n(x, z) \leq \sum_{n=1}^{m} 2^{-n}\rho_n(x, y) + \sum_{n=1}^{m} 2^{-n}\rho_n(y, z)$$

$$\leq \sum_{n=1}^{\infty} 2^{-n}\rho_n(x, y) + \sum_{n=1}^{\infty} 2^{-n}\rho_n(y, z).$$

Consequently, $d(x, z) \leq d(x, y) + d(y, z), x, y, z \in C(\mathbf{R})$. The other properties of a pseudometric are easy to establish. One can argue that d is actually a metric because $d(x, y) = 0$ implies that $\rho_n(x, y) = 0$ for every n, which further implies that $d_n(x, y) = 0$ for every n, and hence that $x = y$.

11. Let $C^1[0,1]$ be the class of all continuous functions $x: [0,1] \to \mathbf{R}$ having continuous derivatives x' on $[0,1]$. For any two such functions x and y, let

$$d(x,y) = \sup\{|x(t) - y(t)| : t \in [0,1]\} + \sup\{|x'(t) - y'(t)| : t \in [0,1]\}.$$

Show that $C^1[0,1]$ is a complete metric space.

Hint: Let $\{x_n\}_{n \geq 1}$ be a Cauchy sequence in $C^1[0,1]$, so that by Proposition 1.4.13, $x_n \to x$ uniformly and $x'_n \to y$ uniformly, where x and y are continuous functions on $[0,1]$. Then

$$x_n(t) - x_n(0) = \int_0^t x'_n(s)\,ds.$$

Since $x'_n \to y$ uniformly, we have $\int_0^t x'_n(s)\,ds \to \int_0^t y(s)\,ds$. Also, $x_n \to x$ uniformly and therefore

$$x(t) - x(0) = \int_0^t y(s)\,ds,$$

i.e., $x'(t) = y(t)$. Since each x'_n is continuous and $x'_n \to y$ uniformly, therefore y is continuous. Hence, $x \in C^1[0,1]$.

12. Show that the sequence $\{x_n\}_{n \geq 1}$, where $x_n = (1, 1/2, \ldots, 1/n, 0, 0, \ldots)$ converges to $x = (1, 1/2, 1/3, \ldots, 1/n, \ldots)$ in ℓ_2.

Hint: $d(x_n, x) = \left(\sum_{k=n+1}^{\infty} (1/k^2)\right)^{1/2} \to 0$ as $n \to \infty$.

13. Let $\{x_n\}_{n \geq 1}$ be a sequence in a metric space (X,d) such that the three subsequences $\{x_{2n}\}_{n \geq 1}$, $\{x_{2n+1}\}_{n \geq 1}$ and $\{x_{3n}\}_{n \geq 1}$ are all convergent. Show that $\{x_n\}_{n \geq 1}$ is convergent.

Hint: By hypothesis, there exists an integer n_0 such that $d(x_n, x_m) < \varepsilon/3$ for n and m even and $\geq n_0$; $d(x_n, x_m) < \varepsilon/3$ for n and m odd and $\geq n_0$; $d(x_n, x_m) < \varepsilon/3$ for n and m multiples of 3 and $\geq n_0$. For n even and m odd, choose l_1 even and l_2 odd, each a multiple of 3 and $l_1, l_2 \geq n_0$. Then $d(x_n, x_m) \leq d(x_n, x_{l_1}) + d(x_{l_1}, x_{l_2}) + d(x_{l_2}, x_m) < \varepsilon$ whenever $n, m \geq n_0$.

14. Let $\{x_n\}_{n \geq 1}$ be a sequence of real numbers defined by

$$x_1 = a, x_2 = b \text{ and } x_{n+2} = \frac{1}{2}(x_{n+1} + x_n) \text{ for } n = 1, 2, \ldots$$

Prove that $\{x_n\}_{n \geq 1}$ is Cauchy.

Hint:

$$x_{n+2} - x_{n+1} = \frac{1}{2}(x_n - x_{n+1}); \text{ so} |x_{n+2} - x_{n+1}| = \frac{1}{2}|x_{n+1} - x_n| = \frac{1}{2^2}|x_n - x_{n-1}|$$

$$= \ldots = \frac{1}{2^n}|x_2 - x_1| = \frac{1}{2^n}|b - a|.$$

Hence, if $n > m$,

$$|x_n - x_m| = |x_n - x_{n-1} + x_{n-1} - x_{n-2} + x_{n-2} + \ldots + x_{m+1} - x_m|$$

$$\leq |x_n - x_{n-1}| + |x_{n-1} - x_{n-2}| + \ldots + |x_{m+1} - x_m|$$

$$\leq \left\{ \frac{1}{2^{n-2}} + \frac{1}{2^{n-3}} + \ldots + \frac{1}{2^{m-1}} \right\} |b - a|$$

$$= \frac{1}{2^{m-1}} \frac{1 - \left(\frac{1}{2}\right)^{n-m}}{1 - \frac{1}{2}} |b - a| \leq \frac{1}{2^{m-2}} |b - a|.$$

15. Describe the convergent sequences in the metric spaces of Exercises 6 and 7.

16. Let (X,d) be a metric space and $C_b(X, \mathbf{R})$ denote the set of all continuous bounded real-valued functions defined on X, equipped with the uniform metric

$$d(f, g) = \sup\{|f(x) - g(x)| : x \in X\}.$$

Show that $C_b(X, \mathbf{R})$ is a complete metric space.

17. Let X be the set $\mathbf{C}$ of complex numbers and $d: X \times X \to \mathbf{R}$ be defined by $d(z_1, z_2) = 0$ when $z_1 = z_2$ but $|z_1| + |z_2|$ when $z_1 \neq z_2$. Show that (X,d) is a metric space which is complete.

Hint: If z_1, z_2 and z_3 are distinct, then $d(z_1, z_3) = |z_1| + |z_3| \leq |z_1| + |z_2| + |z_2| + |z_3| = d(z_1, z_2) + d(z_2, z_3)$. If $z_1 = z_2$, and z_3 is distinct from z_1 (and hence from z_2), then $d(z_1, z_3) = |z_1| + |z_3| = |z_2| + |z_3| = d(z_2, z_3) = d(z_1, z_2) + d(z_2, z_3)$, since $d(z_1, z_2) = 0$. The other cases are trivial. We note that $|z| = d(z,0)$ whether $z = 0$ or not. Now let $\{z_n\}$ be a Cauchy sequence. If there is some p such that $z_m = z_n$ whenever $m, n \geq p$, then $\{z_n\}$ converges to z_p. If there is no such p, then there is some subsequence $\{z_{n(k)}\}$, all points of which are distinct, so that $d(z_{n(k)}, z_{n(j)}) = |z_{n(k)}| + |z_{n(j)}|$ when $j \neq k$. We argue that the subsequence converges to zero: Consider any $\varepsilon > 0$; there exists n_0 such that $d(z_m, z_n) < \varepsilon$ whenever $m, n \geq n_0$, so that $|z_{n(k)}| + |z_{n(j)}| = d(z_{n(k)}, z_{n(j)}) < \varepsilon$ whenever $k, j \geq n_0$ and $j \neq k$. In particular, $|z_{n(k)}| < \varepsilon$ whenever $k \geq n_0$. However, as noted earlier, $|z_{n(k)}| = d(z_{n(k)}, 0)$. Therefore the subsequence converges to zero. The rest follows by Proposition 1.4.7.

2 Topology of a Metric Space

The real number system has two types of properties. The first type are algebraic properties, dealing with addition, multiplication and so on. The other type, called topological properties, have to do with the notion of distance between numbers and with the concept of limit. In this chapter, we study topological properties in the framework of metric spaces. We begin by looking at the notions of open and closed sets, limit points, closure and interior of a set and some elementary results involving them. The concept of base of a metric topology and related ideas are also discussed. In the final section, we deal with the important concept of category due to Baire and its usefulness in existence proofs. Also included are some theorems due to Baire.

2.1. Open and Closed Sets

There are special types of sets that play a distinguished role in analysis; these are the open and closed sets. To expedite the discussion, it is helpful to have the notion of a neighbourhood in metric spaces.

Definition 2.1.1. Let (X, d) be a metric space. The set

$$S(x_0, r) = \{x \in X : d(x_0, x) < r\}, \qquad \text{where } r > 0 \text{ and } x_0 \in X,$$

is called the **open ball** of radius r and centre x_0. The set

$$\bar{S}(x_0, r) = \{x \in X : d(x_0, x) \leq r\}, \qquad \text{where } r > 0 \text{ and } x_0 \in X,$$

is called the **closed ball** of radius r and centre x_0.

A few concrete examples are in order.

Examples 2.1.2. (i) The open ball $S(x_0, r)$ on the real line is the bounded open interval $(x_0 - r, \ x_0 + r)$ with midpoint x_0 and total length $2r$. Conversely, it is clear that any bounded open interval on the real line is an open ball. So the open balls on the real line are precisely the bounded open intervals. The closed balls $\bar{S}(x_0, r)$ on the real line are precisely the bounded closed intervals but containing more than one point.

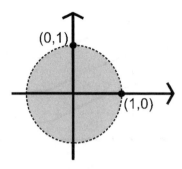

FIGURE 2.1

(ii) The open ball $S(x_0, r)$ in $\mathbf{R}^2$ with metric d_2 (see Example 1.2.2(iii)) is the inside of the circle with centre x_0 and radius r as in Fig. 2.1. Open balls of radius 1 and centre $(0,0)$, when the metric is d_1 or d_∞ (see Example 1.2.2(iv) for the latter) are illustrated in Figs. 2.2 and 2.3.

(iii) If (X, d) denotes the discrete metric space (see Example 1.2.2(v)), then $S(x, r) = \{x\}$ for all $x \in X$ and any positive $r \le 1$, whereas $S(x, r) = X$ for all $x \in X$ and any $r > 1$.

(iv) Consider the metric space $C_\mathbf{R}[a, b]$ of Example 1.2.2(ix). The open ball $S(x_0, r)$, where $x_0 \in C_\mathbf{R}[a, b]$ and $r > 0$, consists of all continuous functions $x \in C_\mathbf{R}[a, b]$ whose graphs lie within a band of vertical width $2r$ and is centred around the graph of x_0. (See Fig. 2.4.)

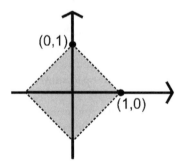

FIGURE 2.2

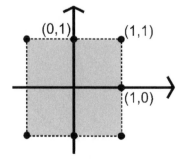

FIGURE 2.3

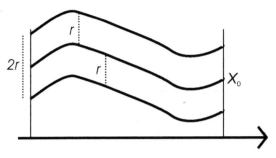

FIGURE 2.4

Definition 2.1.3. Let (X, d) be a metric space. A **neighbourhood** of the point $x_0 \in X$ is any open ball in (X, d) with centre x_0.

Definition 2.1.4. A subset G of a metric space (X, d) is said to be **open** if given any point $x \in G$, there exists $r > 0$ such that $S(x, r) \subseteq G$, i.e., each point of G is the centre of some open ball contained in G. Equivalently, every point of the set has a neighbourhood contained in the set.

Theorem 2.1.5. In any metric space (X, d), each open ball is an open set.

Proof. First observe that $S(x,r)$ is nonempty, since $x \in S(x, r)$. Let $y \in S(x, r)$, so that $d(y, x) < r$, and let $r' = r - d(y, x) > 0$. We shall show that $S(y, r') \subseteq S(x, r)$, as illustrated in Fig. 2.5. Consider any $z \in S(y, r')$. Then we have

$$d(z, x) \leq d(z, y) + d(y, x) < r' + d(y, x) = r,$$

which means $z \in S(x, r)$. Thus, for each $y \in S(x, r)$, there is an open ball $S(y, r') \subseteq S(x, r)$. Therefore $S(x, r)$ is an open subset of X. $\square$

Examples 2.1.6. (i) In **R**, any bounded open interval is an open subset because it is an open ball. It is easy to see that even an unbounded open interval is an open subset.

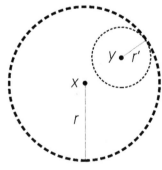

FIGURE 2.5

(ii) In ℓ_2, let $G = \{x = \{x_i\}_{i \geq 1} : \sum_1^\infty |x_i|^2 < 1\}$. Then G is an open subset of ℓ_2. Indeed, $G = S(0, 1)$ is the open ball with centre $0 = (0, 0, \dots)$ and radius 1. It is now a consequence of Theorem 2.1.5 that G is open.

(iii) In a discrete metric space X, any subset G is open, because any $x \in G$ is the centre of the open ball $S(x, 1/2)$ which is nothing but $\{x\}$.

The following are fundamental properties of open sets.

Theorem 2.1.7. Let (X, d) be a metric space. Then

  (i) $\emptyset$ and X are open sets in (X, d);

 (ii) the union of any finite, countable or uncountable family of open sets is open;

(iii) the intersection of any finite family of open sets is open.

Proof. (i) As the empty set contains no points, the requirement that each point in $\emptyset$ is the centre of an open ball contained in it is automatically satisfied. The whole space X is open, since every open ball centred at any of its points is contained in X.

(ii) Let $\{G_\alpha : \alpha \in \Lambda\}$ be an arbitrary family of open sets and $H = \cup_{\alpha \in \Lambda} G_\alpha$. If H is empty, then it is open by part (i). So assume H to be nonempty and consider any $x \in H$. Then $x \in G_\alpha$ for some $\alpha \in \Lambda$. Since G_α is open, there exists an $r > 0$ such that $S(x, r) \subseteq G_\alpha \subseteq H$. Thus, for each $x \in H$ there exists an $r > 0$ such that $S(x, r) \subseteq H$. Consequently, H is open.

(iii) Let $\{G_i : 1 \leq i \leq n\}$ be a finite family of open sets in X and let $G = \cap_{i=1}^n G_i$. If G is empty, then it is open by part (i). Suppose G is nonempty and let $x \in G$. Then $x \in G_j, j = 1, \dots, n$. Since G_j is open, there exists $r_j > 0$ such that $S(x, r_j) \subseteq G_j, j = 1, \dots, n$. Let $r = \min\{r_1, r_2, \dots, r_n\}$. Then $r > 0$ and $S(x, r) \subseteq S(x, r_j), j = 1, \dots, n$. Therefore the ball $S(x, r)$ centred at x satisfies

$$S(x, \ r) \subseteq \bigcap_{j=1}^n S(x, \ r_j) \subseteq G.$$

This completes the proof. $\qquad\qquad\qquad\qquad\qquad\qquad\qquad\qquad\qquad\qquad$ $\square$

Remark 2.1.8. The intersection of an infinite number of open sets need not be open. To see why, let $S_n = S(0, \frac{1}{n}) \subseteq \mathbf{C}$, $n = 1, 2, \dots$. Each S_n is an open ball in the complex plane and hence an open set in $\mathbf{C}$. However,

$$\bigcap_{n=1}^\infty S_n = \{0\},$$

which is not open, since there exists no open ball in the complex plane with centre 0 that is contained in $\{0\}$.

The following theorem characterises open subsets in a metric space.

Theorem 2.1.9. A subset G in a metric space (X, d) is open if and only if it is the union of all open balls contained in G.

Proof. Suppose that G is open. If G is empty, then there are no open balls contained in it. Thus, the union of all open balls contained in G is a union of an empty class, which is empty and therefore equal to G. If G is nonempty, then since G is open, each of its points is the centre of an open ball contained entirely in G. So, G is the union of all open balls contained in it. The converse follows immediately from Theorem 2.1.5 and Theorem 2.1.7.   □

Remark 2.1.10. The above Theorem 2.1.9 describes the structure of open sets in a metric space. This information is the best possible in an arbitrary metric space. For open subsets of **R**, Theorem 2.1.9 can be improved.

Theorem 2.1.11. Each nonempty open subset of **R** is the union of a countable family of disjoint open intervals. Moreover, the endpoints of any open interval in the family lie in the complement of the set and are no less than the infimum and no greater than the supremum of the set.

Proof. Let G be a nonempty open subset of **R** and let $x \in G$. Since G is open, there exists a bounded open interval with centre x and contained in G. So there exists some $y > x$ such that $(x, y) \subseteq G$ and some $z < x$ such that $(z, x) \subseteq G$. Let

$$a = \inf \{z : (z, x) \subseteq G\} \text{ and } b = \sup \{y : (x, y) \subseteq G\}. \tag{1}$$

Then $a < x < b$ and $I_x = (a, b)$ is an open interval containing x. We shall show that $a \notin G$, $b \notin G$ and $I_x \subseteq G$. This is obvious if $a = -\infty$ or if $b = \infty$. So, assume $-\infty < a$ and $\infty > b$. If a were to be in G, we would have $(a - \varepsilon, a + \varepsilon) \subseteq G$ for some $\varepsilon > 0$, whence we would also have $(a - \varepsilon, x) \subseteq G$, contradicting (1). The argument that $b \notin G$ is similar. Now suppose $w \in I_x$ we shall show that $w \in G$. If $w = x$, then of course $w \in G$. Let $w \neq x$, so that either $a < w < x$ or $x < w < b$. We need consider only the former case: $a < w < x$. Since $a < w$, it follows from (1) that there exists some $z < w$ such that $(z, x) \subseteq G$. Since $w < x$, this implies that $w \in G$.

Consider the collection of open intervals $\{I_x\}$, $x \in G$. Since each $x \in G$ is contained in I_x and each I_x is contained in G, it follows that $G = \bigcup \{I_x : x \in G\}$. We shall next show that any two intervals in the collection $\{I_x : x \in G\}$ are disjoint. Let (a, b) and (c, d) be two intervals in this collection with a point in common. Then we must have $c < b$ and $a < d$. Since c does not belong to G, it does not belong to (a, b) and so $c \leq a$. Since a does not belong to G, and hence also does not belong to (c, d), we also have $a \leq c$. Therefore, $c = a$. Similarly, $b = d$, which shows that (a,b) and (c, d) are actually the same interval. Thus, $\{I_x : x \in G\}$ is a collection of disjoint intervals.

Now we establish that the collection is countable. Each nonempty open interval contains a rational number. Since disjoint intervals cannot contain the same number and the rationals are countable, it follows that the collection $\{I_x : x \in G\}$ is countable.

Finally, we note that it follows from (1) that $a \geq \inf G$ and $b \leq \sup G$.   □

Definition 2.1.12. Let A be a subset of a metric space (X, d). A point $x \in X$ is called an **interior point** of A if there exists an open ball with centre x contained in A, i.e.,

$$x \in S(x, r) \subseteq A \text{ for some } r > 0,$$

or equivalently, if x has a neighbourhood contained in A. The set of all interior points of A is called the **interior of** A and is denoted either by $\text{Int}(A)$ or A°. Thus

$$\text{Int}(A) = A^\circ = \{x \in A : S(x, r) \subseteq A \text{ for some } r > 0\}.$$

Observe that $\text{Int}(A) \subseteq A$.

Example 2.1.13. The interior of the subset $[0, 1] \subseteq \mathbf{R}$ can be shown to be $(0,1)$. Let $x \in (0, 1)$. Since $(0,1)$ is open, there exists $r > 0$ such that $(x - r, x + r) \subseteq [0,1]$. So, x is an interior point of $[0,1]$. Also, 0 is not an interior point of $[0,1]$, because there exists no $r > 0$ such that $(-r, r) \subseteq [0, 1]$. Similarly, 1 is also not an interior point of $[0, 1]$.

The next theorem relates interiors to open sets and provides a characterisation of open subsets in terms of interiors.

Theorem 2.1.14. Let A be a subset of a metric space (X, d). Then

(i) A° is an open subset of A that contains every open subset of A;
(ii) A is open if and only if $A = A^\circ$.

Proof. (i) Let $x \in A^\circ$ be arbitrary. Then, by definition, there exists an open ball $S(x, r) \subseteq A$. But $S(x,r)$ being an open set (see Theorem 2.1.5), each point of it is the centre of some open ball contained in $S(x,r)$ and consequently also contained in A. Therefore each point of $S(x,r)$ is an interior point of A, i.e., $S(x, r) \subseteq A^\circ$. Thus, x is the centre of an open ball contained in A°. Since $x \in A^\circ$ is arbitrary, it follows that each $x \in A^\circ$ has the property of being the centre of an open ball contained in A°. Hence, A° is open.

It remains to show that A° contains every open subset $G \subseteq A$. Let $x \in G$. Since G is open, there exists an open ball $S(x, r) \subseteq G \subseteq A$. So $x \in A^\circ$. This shows that $x \in G \Rightarrow x \in A^\circ$. In other words, $G \subseteq A^\circ$.

(ii) is immediate from (i). $\square$

The following are basic properties of interiors.

Theorem 2.1.15. Let (X, d) be a metric space and A, B be subsets of X. Then

(i) $A \subseteq B \Rightarrow A^\circ \subseteq B^\circ$;
(ii) $(A \cap B)^\circ = A^\circ \cap B^\circ$;
(iii) $(A \cup B)^\circ \supseteq A^\circ \cup B^\circ$.

Proof. (i) Let $x \in A^\circ$. Then there exists an $r > 0$ such that $S(x, r) \subseteq A$. Since $A \subseteq B$, we have $S(x, r) \subseteq B$, i.e., $x \in B^\circ$.

(ii) $A \cap B \subseteq A$ as well as $A \cap B \subseteq B$. It follows from (i) that $(A \cap B)^\circ \subseteq A^\circ$ as well as $(A \cap B)^\circ \subseteq B^\circ$, which implies that $(A \cap B)^\circ \subseteq A^\circ \cap B^\circ$. On the other hand, let $x \in A^\circ \cap B^\circ$. Then $x \in A^\circ$ and $x \in B^\circ$. Therefore, there exist $r_1 > 0$ and $r_2 > 0$ such that $S(x, r_1) \subseteq A$ and $S(x, r_2) \subseteq B$. Let $r = \min\{r_1, r_2\}$. Clearly, $r > 0$ and $S(x, r) \subseteq A \cap B$, i.e., $x \in (A \cap B)^\circ$.

(iii) $A \subseteq A \cup B$ as well as $B \subseteq A \cup B$. Now apply (i).   □

Remark 2.1.16. The following example shows that $(A \cup B)^\circ$ need not be the same as $A^\circ \cup B^\circ$. Indeed, if $A = [0, 1]$ and $B = [1, 2]$, then $A \cup B = [0, 2]$. Since $A^\circ = (0, 1), B^\circ = (1, 2)$ and $(A \cup B)^\circ = (0, 2)$, we have $(A \cup B)^\circ \neq A^\circ \cup B^\circ$.

Definition 2.1.17. Let X be a metric space and F a subset of X. A point $x \in X$ is called a **limit point** of F if each open ball with centre x contains at least one point of F different from x, i.e.,

$$(S(x, r) - \{x\}) \cap F \neq \varnothing.$$

The set of all limit points of F is denoted by F' and is called the **derived set** of F.

Examples 2.1.18. (i) The subset $F = \{1, 1/2, 1/3, \ldots\}$ of the real line has 0 as a limit point; in fact, 0 is its only limit point. Thus the derived set of F is $\{0\}$, i.e., $F' = \{0\}$.

(ii) The subset $\mathbf{Z}$ of integers of the real line, consisting of all the integers, has no limit point. Its derived set $\mathbf{Z}'$ is $\varnothing$.

(iii) Each real number is a limit point of the subset of rationals: $\mathbf{Q}' = \mathbf{R}$.

(iv) If (X, d) is a discrete metric space (see Example 1.2.2(v)) and $F \subseteq X$, then F has no limit points, since every open ball of radius 1 consists only of the centre.

(v) Consider the subset $F = \{(x, y) \in \mathbf{C} : x > 0, y > 0\}$ of the complex plane. Each point of the subset $\{(x, y) \in \mathbf{C} : x \geq 0, y \geq 0\}$ is a limit point of F. In fact, the latter set is precisely F'.

(vi) For an interval $I \subseteq \mathbf{R}$, the set I' consists of not only all the points of I but also any endpoints I may have, even if they do not belong to I. Thus $(0, 1)' = (0, 1]' = [0, 1)' = [0, 1]' = [0, 1]$.

Proposition 2.1.19. Let (X, d) be a metric space and $F \subseteq X$. If x_0 is a limit point of F, then every open ball $S(x_0, r), r > 0$, contains an infinite number of points of F.

Proof. Suppose that the ball $S(x_0, r)$ contains only a finite number of points of F. Let $y_1, y_2, \ldots, y_n$ denote the points of $S(x_0, r) \cap F$ that are distinct from x_0. Let

$$\delta = \min\{d(y_1, x_0), d(y_2, x_0), \ldots, d(y_n, x_0)\}.$$

Then the ball $S(x_0, \delta)$ contains no point of F distinct from x_0, contradicting the assumption that x_0 is a limit point of F.   □

The following characterisation of the limit points of a set in a metric space is useful.

Proposition 2.1.20. Let (X, d) be a metric space and $F \subseteq X$. Then a point x_0 is a limit point of F if and only if it is possible to select from the set F a sequence of distinct points $x_1, x_2, \ldots, x_n, \ldots$ such that $\lim_n d(x_n, x_0) = 0$.

Proof. If $\lim_n d(x_n, x_0) = 0$, where $x_1, x_2, \ldots, x_n, \ldots$ is a sequence of distinct points of F, then every ball $S(x_0, r)$ with centre x_0 and radius r contains each of x_n, where $n \geq n_0$ for some suitably chosen n_0. As $x_1, x_2, \ldots, x_n, \ldots$ in F are distinct, it follows that $S(x_0, r)$ contains a point of F different from x_0. So, x_0 is a limit point of F.

On the other hand, assume that x_0 is a limit point of F. Choose a point $x_1 \in F$ in the open ball $S(x_0, 1)$ such that x_1 is different from x_0. Next, choose a point $x_2 \in F$ in the open ball $S(x_0, 1/2)$ different from x_0 as well as from x_1; this is possible by Proposition 2.1.19. Continuing this process in which, at the nth step of the process we choose a point $x_n \in F$ in $S(x_0, 1/n)$ different from $x_1, x_2, \ldots, x_{n-1}$, we have a sequence $\{x_n\}$ of distinct points of the set F such that $\lim_n d(x_n, x_0) = 0$.   □

Definition 2.1.21. A subset F of the metric space (X, d) is said to be **closed** if it contains each of its limit points, i.e., $F' \subseteq F$.

Examples 2.1.22. (i) The set $\mathbf{Z}$ of integers is a closed subset of the real line.

(ii) The set $F = \{1, 1/2, 1/3, \ldots, 1/n, \ldots\}$ is not closed in $\mathbf{R}$. In fact, $F' = \{0\}$, which is not contained in F.

(iii) The set $F = \{(x, y) \in \mathbf{C} : x \geq 0, y \geq 0\}$ is a closed subset of the complex plane $\mathbf{C}$. In this case, the derived set is $F' = F$.

(iv) Each subset of a discrete metric space is closed.

Proposition 2.1.23. Let F be a subset of the metric space (X, d). The set of limit points of F, namely, F' is a closed subset of (X, d), i.e., $(F')' \subseteq F'$.

Proof. If $F' = \varnothing$ or $(F')' = \varnothing$, then there is nothing to prove. Let $F' \neq \varnothing$ and let $x_0 \in (F')'$. Choose an arbitrary open ball $S(x_0, r)$ with centre x_0 and radius r. By the definition of limit point, there exists a point $y \in F'$ such that $y \in S(x_0, r)$. If $r' = r - d(y, x_0)$, then $S(y, r')$ contains infinitely many points of F by Proposition 2.1.19. But $S(y, r') \subseteq S(x_0, r)$ as in the proof of Theorem 2.1.5. So, infinitely many points of F lie in $S(x_0, r)$. Therefore, x_0 is a limit point of F, i.e., $x_0 \in F'$. Thus, F' contains all its limit points and hence F' is closed.   □

Theorem 2.1.24. Let (X, d) be a metric space and let F_1, F_2 be subsets of X.

(i) If $F_1 \subseteq F_2$, then $F_1' \subseteq F_2'$.
(ii) $(F_1 \cup F_2)' = F_1' \cup F_2'$.
(iii) $(F_1 \cap F_2)' \subseteq F_1' \cap F_2'$.

Proof. The proofs of (i) and (iii) are obvious. For the proof of (ii), observe that $F_1' \cup F_2' \subseteq (F_1 \cup F_2)'$, which follows from (i). It remains to show that

$(F_1 \cup F_2)' \subseteq F_1' \cup F_2'$. Let $x_0 \in (F_1 \cup F_2)'$. Then there exists a sequence $\{x_n\}_{n \geq 1}$ of distinct points in $F_1 \cup F_2$ such that $d(x_n, x_0) \to 0$ as $n \to \infty$ by Proposition 2.1.20. If an infinite number of points x_n lie in F_1, then $x_0 \in F_1'$, and, consequently, $x_0 \in F_1' \cup F_2'$. If only a finite number of points of $\{x_n\}_{n \geq 1}$ lie in F_1, then $x_0 \in F_2' \subseteq F_1' \cup F_2'$. We therefore have $x_0 \in F_1' \cup F_2'$ in either case. This completes the proof of (ii).   □

Definition 2.1.25. Let F be a subset of a metric space (X, d). The set $F \cup F'$ is called the **closure** of F and is denoted by $\bar{F}$.

Corollary 2.1.26. The closure $\bar{F}$ of $F \subseteq X$, where (X, d) is a metric space, is closed.

Proof. In fact, by Proposition 2.1.23 and Theorem 2.1.24(ii),

$$(\bar{F})' = (F \cup F')' = F' \cup (F')' \subseteq F' \cup F' = F' \subseteq \bar{F}.$$

Corollary 2.1.27. (i) Let F be a subset of a metric space (X, d). Then F is closed if and only if $F = \bar{F}$.
  (ii) If $A \subseteq B$, then $\bar{A} \subseteq \bar{B}$.
  (iii) If $A \subseteq F$ and F is closed, then $\bar{A} \subseteq F$.

Proof. (i) If $F = \bar{F}$, then it follows from Corollary 2.1.26 that F is closed. On the other hand, suppose that F is closed; then

$$\bar{F} = F \cup F' = F \subseteq \bar{F}.$$

It follows from the above relations that $F = \bar{F}$.
  (ii) This is an immediate consequence of Theorem 2.1.24(i).
  (iii) This is an immediate consequence of (ii) above.   □

Proposition 2.1.28. Let (X, d) be a metric space and $F \subseteq X$. Then the following statements are equivalent:

  (i) $x \in \bar{F}$;
  (ii) $S(x, \varepsilon) \cap F \neq \emptyset$ for every open ball $S(x, \varepsilon)$ centred at x;
  (iii) there exists an infinite sequence $\{x_n\}$ of points (not necessarily distinct) of F
       such that $x_n \to x$.

Proof. (i)$\Rightarrow$(ii). Let $x \in \bar{F}$. If $x \in F$, then obviously $S(x, \varepsilon) \cap F \neq \emptyset$. If $x \notin F$, then by the definition of closure, we have $x \in F'$. By definition of a limit point,

$$(S(x, \varepsilon) \backslash \{x\}) \cap F \neq \emptyset$$

and, a fortiori,

$$S(x, \varepsilon) \cap F \neq \emptyset.$$

(ii)$\Rightarrow$(iii). For each positive integer n, choose $x_n \in S(x, 1/n) \cap F$. Then the sequence $\{x_n\}$ of points in F converges to x. In fact, upon choosing $n_0 > 1/\varepsilon$, where $\varepsilon > 0$ is arbitrary, we have $d(x_n, x) < 1/n < 1/n_0 < \varepsilon$, i.e., $x_n \in S(x, \varepsilon)$ whenever $n \geq n_0$.

(iii)$\Rightarrow$(i) If the sequence $\{x_n\}_{n\geq 1}$ of points in F consists of finitely many distinct points, then there exists a subsequence $\{x_{n_k}\}$ such that $x_{n_k} = x$ for all k. So, $x \in F$. If however, $\{x_n\}_{n\geq 1}$ contains infinitely many distinct points, then there exists a subsequence $\{x_{n_k}\}$ consisting of distinct points and $\lim_k d(x_{n_k}, x) = 0$, for $\lim_n d(x_n, x) = 0$ by hypothesis. By Proposition 2.1.20, it follows that $x \in F' \subseteq \bar{F}$. $\square$

Condition (ii) of Definition 1.5.1 of a completion can be rephrased in view of condition (i) and Proposition 2.1.28 (iii) as saying that the closure of metric space X as a subset of its completion X^* must be the whole of X^*.

The following proposition is an easy consequence of Theorem 2.1.24.

Proposition 2.1.29. Let F_1, F_2 be subsets of a metric space (X, d). Then

(i) $\overline{(F_1 \cup F_2)} = \bar{F}_1 \cup \bar{F}_2$;
(ii) $\overline{(F_1 \cap F_2)} \subseteq \bar{F}_1 \cap \bar{F}_2$.

Proof. Using Theorem 2.1.24 (ii), we have

$$\overline{(F_1 \cup F_2)} = (F_1 \cup F_2) \cup (F_1 \cup F_2)' = (F_1 \cup F_2) \cup (F_1' \cup F_2')$$
$$= (F_1 \cup F_1') \cup (F_2 \cup F_2') = \bar{F}_1 \cup \bar{F}_2,$$

which establishes (i). The proof of (ii) is equally simple.

Remarks 2.1.30. (i) It is not necessarily the case that the closure of an arbitrary union is the union of the closures of the subsets in the union. If $\{A_\alpha\}_{\alpha\in\Lambda}$ is an infinite family of subsets of (X, d), it follows from Corollary 2.1.27 (ii) that

$$\bigcup_{\alpha\in\Lambda} \bar{A}_\alpha \subseteq \overline{\bigcup_{\alpha\in A} A_\alpha}.$$

Equality need not hold, as the following example shows: If $A_n = \{r_n\}$, $n = 1, 2, \ldots$ and $r_1, r_2, \ldots, r_n, \ldots$ is an enumeration of rationals, then $\bar{A}_n = \overline{\{r_n\}} = \{r_n\}$ and $\bigcup_{n=1}^\infty \bar{A}_n = \mathbf{Q}$, whereas $\overline{\bigcup_{n=1}^\infty A_n} = \bar{\mathbf{Q}} = \mathbf{R}$.

(ii) In Proposition 2.1.29 (ii), equality need not hold. For example, if F_1 denotes the set of rationals in $\mathbf{R}$ and F_2 the set of irrationals in $\mathbf{R}$, then $\overline{(F_1 \cap F_2)} = \bar{\varnothing} = \varnothing$ but $\bar{F}_1 = \bar{F}_2 = \mathbf{R}$.

Proposition 2.1.31. Let (X, d) be a metric space. The empty set $\varnothing$ and the whole space X are closed sets.

Proof. Since the empty set has no limit points, the requirement that a closed set contain all its limit points is automatically satisfied by the empty set.

Since the whole space contains all points, it certainly contains all its limit points (if any), and is thus closed. $\square$

The following is a useful characterisation of closed sets in terms of open sets.

Theorem 2.1.32. Let (X, d) be a metric space and F be a subset of X. Then F is closed in X if and only if F^c is open in X.

Proof. Suppose F is closed in X. We show that F^c is open in X. If $F = \varnothing$ (respectively, X), then $F^c = X$ (respectively, $\varnothing$) and it is open by Theorem 2.1.7(i); so we may suppose that $F \neq \varnothing \neq F^c$. Let x be a point in F^c. Since F is closed and $x \notin F$, x cannot be a limit point of F. So there exists an $r > 0$ such that $S(x, r) \subseteq F^c$. Thus, each point of F^c is contained in an open ball contained in F^c. This means F^c is open.

For the converse, suppose F^c is open. We show that F is closed. Let $x \in X$ be a limit point of F. Suppose, if possible, that $x \notin F$. Then $x \in F^c$, which is assumed to be open. Therefore, there exists $r > 0$ such that $S(x, r) \subseteq F^c$, i.e.,

$$S(x, r) \cap F = \varnothing.$$

Thus, x cannot be a limit point of F, which is a contradiction. Hence, x belongs to F. $\square$

Theorem 2.1.33. Let (X, d) be a metric space and $\bar{S}(x, r) = \{y \in X : d(y, x) \leq r\}$ be a closed ball in X. Then $\bar{S}(x, r)$ is closed.

Proof. We show that $(\bar{S}(x, r))^c$ is open in X (see Theorem 2.1.32). Let $y \in (\bar{S}(x, r))^c$. Then $d(y, x) > r$. If $r_1 = d(y, x) - r$, then $r_1 > 0$. Moreover, $S(y, r_1) \subseteq (\bar{S}(x, r))^c$. Indeed, if $z \in S(y, r_1)$, then

$$d(z, x) \geq d(y, x) - d(y, z) > d(y, x) - r_1 = r.$$

Thus, $z \notin \bar{S}(x, r)$, i.e., $z \in (\bar{S}(x, r))^c$. $\square$

The following fundamental properties of closed sets are analogues of the properties of open sets formulated in Theorem 2.1.7 and are easy consequences of it along with de Morgan's laws (see Chapter 0, p. 3) and Proposition 2.1.31.

Theorem 2.1.34. Let (X, d) be a metric space. Then

(i) $\varnothing$ and X are closed;
(ii) any intersection of closed sets is closed;
(iii) a finite union of closed sets is closed.

Proof. (i) This is a restatement of Proposition 2.1.31.

(ii) Let $\{F_\alpha\}$ be a family of closed sets in X and $F = \bigcap_\alpha F_\alpha$. Then by Theorem 2.1.32, F is closed if F^c is open. Since $F^c = \bigcup_\alpha F_\alpha^c$ by de Morgan's laws, and since each F_α^c is open (Theorem 2.1.32), $\bigcup_\alpha F_\alpha^c$ is open by Theorem 2.1.7, i.e., F^c is open.

(iii) This proof is similar to (ii). $\square$

Remark 2.1.35. An arbitrary union of closed sets need not be closed. Indeed, $\bar{S}(0, 1 - 1/n)$, $n \geq 2$, is a closed subset of the complex plane, but

$$\bigcup_{n=2}^{\infty} \bar{S}\left(0, 1 - \frac{1}{n}\right) = S(0, 1)$$

is not closed (because each point z satisfying $|z| = 1$ is a limit point of $S(0, 1)$ but is not contained in $S(0, 1)$).

An explicit characterisation of open sets on the real line is the content of Theorem 2.1.11. We now turn to the study of closed sets on the real line. Observe that closed intervals and finite unions of closed intervals are closed sets in **R**. Since a set consisting of a single point is a closed interval with identical endpoints, single point sets, and consequently finite sets, are closed sets as well.

Theorem 2.1.36. Let F be a nonempty bounded closed subset of **R** and let $\alpha = \inf F$ and $\beta = \sup F$. Then $\alpha \in F$ and $\beta \in F$.

Proof. We need only show that if $\alpha \notin F$, then α is a limit point of F. By the definition of infimum, for any $\varepsilon > 0$, there exists at least one member $x \in F$ such that $\alpha \leq x < \alpha + \varepsilon$. But $\alpha \notin F$, whereas $x \in F$. So,

$$\alpha < x < \alpha + \varepsilon.$$

Thus, every neighbourhood of α contains at least one member $x \in F$ which is different from α. Hence, α is a limit point of F. $\qquad \square$

Definition 2.1.37. Let F be a nonempty bounded subset of **R** and let $\alpha = \inf F$ and $\beta = \sup F$. The closed interval $[\alpha, \beta]$ is called the **smallest closed interval containing** F.

Theorem 2.1.38. If $[\alpha, \beta]$ is the smallest closed interval containing F, where F is a nonempty bounded closed subset of **R**, then

$$[\alpha, \beta] \backslash F = (\alpha, \beta) \cap F^c$$

and so is open in **R**.

Proof. Let $x_0 \in [\alpha, \beta] \backslash F$; this means that $x_0 \in [\alpha, \beta]$, $x_0 \notin F$. If $x_0 \notin F$, then $x_0 \neq \alpha$ and $x_0 \neq \beta$, because α and β do belong to F, by Theorem 2.1.36. It follows that $x_0 \in (\alpha, \beta)$. Moreover, it is obvious that $x_0 \in F^c$, so that

$$[\alpha, \beta] \backslash F \subseteq (\alpha, \beta) \cap F^c.$$

The reverse inclusion is obvious. $\qquad \square$

The following characterisation of closed subsets of **R** is a direct consequence of Theorems 2.1.11 and 2.1.38.

Theorem 2.1.39. Let F be a nonempty bounded closed subset of **R**. Then F is either a closed interval or is obtained from some closed interval by removing a countable family of pairwise disjoint open intervals whose endpoints belong to F.

Proof. Let $[\alpha, \beta]$ be the smallest closed interval containing F, where $\alpha = \inf F$ and $\beta = \sup F$. By Theorem 2.1.38,

$$[\alpha, \beta] \backslash F = (\alpha, \beta) \cap F^c$$

is open and hence is a countable union of disjoint open intervals by Theorem 2.1.11. Moreover, the endpoints of the open intervals do not belong to $[\alpha, \beta] \backslash F$ but do belong to $[\alpha, \beta]$. So they belong to F. The set F thus has the desired property. $\square$

This seemingly simple looking process of writing a nonempty bounded closed subset of **R** leads to some very interesting and useful examples. The following example, which is of particular importance, is due to Cantor.

Example 2.1.40. (Cantor) Divide the closed interval $I = [0, 1]$ into three equal parts by the points 1/3 and 2/3 and remove the open interval (1/3, 2/3) from I. Divide each of the remaining two closed intervals [0, 1/3] and [2/3, 1] into three equal parts by the points 1/9, 2/9 and by 7/9, 8/9, respectively, and remove the open intervals (1/9, 2/9) and (7/9, 8/9). Now divide each of the remaining four intervals [0, 1/9], [2/9, 1/3], [2/3, 7/9] and [8/9, 1] into three equal parts and remove the middle third open intervals. Continue this process indefinitely. The open set G removed in this way from $I = [0, 1]$ is the union of disjoint open intervals

$$G = \left(\frac{1}{3}, \frac{2}{3}\right) \cup \left(\frac{1}{9}, \frac{2}{9}\right) \cup \left(\frac{7}{9}, \frac{8}{9}\right) \cup \dots.$$

The complement of G in [0,1], denoted by P, is called the **Cantor set**. Important properties of this set are listed in the Exercise 16 and Section 6.4.

The completeness of **R** can also be characterised in terms of nested sequences of bounded closed intervals. An analogue of this result for metric spaces is proved in Theorem 2.1.44. We begin with some relevant definitions.

2.1.41. Definition. Let (X, d) be a metric space and let A be a nonempty subset of X. We say that A is **bounded** if there exists $M > 0$ such that

$$d(x, y) \leq M \qquad x, y \in A.$$

If A is bounded, we define the **diameter** of A as

$$\text{diam}(A) = d(A) = \sup\{d(x, y) : x, y \in A\}.$$

If A is not bounded, we write $d(A) = \infty$.

We define the distance between the point $x \in X$ and the subset B of X by

$$d(x, B) = \inf\{d(x, y) : y \in B\},$$

and, in an analogous manner, we define the distance between two nonempty subsets B and C of X by

$$d(B, C) = \inf\{d(x, y) : x \in B, y \in C\}.$$

2.1.42. Examples. (i) Recall that a subset A of $\mathbf{R}$ (respectively, $\mathbf{R}^2$) is bounded if and only if A is contained in an interval (respectively, square) of finite length (respectively, whose edge has finite length). Thus, our definition of bounded set in an arbitrary metric space is consistent with the definition of bounded set of real numbers (respectively, bounded set of pairs of real numbers).

(ii) The interval $(0, \infty)$ is not a bounded subset of $\mathbf{R}$. However, if $\mathbf{R}$ is equipped with the discrete metric, then every subset A of this discrete space (in particular, the set $(0, \infty)$) is bounded, since $d(x, y) \leq 1$ for $x, y \in A$. Indeed, $d(A) = 1$, provided A contains more than one point. Moreover, any subset of any discrete metric space has diameter 1 if it contains more than one point.

(iii) If $\mathbf{R}$ is equipped with the nondiscrete metric $d(x, y) = |x - y|/[1 + |x - y|]$, then every subset is bounded and $d(\mathbf{R}) = 1$.

(iv) In the space (ℓ_2, d) (see Example 1.2.2(vii)), consider the set

$$Y = \{e_1, e_2, \ldots, e_n, \ldots\},$$

where e_n denotes the sequence all of whose terms are equal to 0 except the nth term, which is equal to 1. If $j \neq k$, then $d(e_j, e_k) = \sqrt{2}$. Hence, Y is bounded and $d(Y) = \sqrt{2}$.

2.1.43. Proposition. If A is a subset of the metric space (X, d), then $d(A) = d(\overline{A})$.

Proof. If $x, y \in \overline{A}$, then there exist sequences $\{x_n\}_{n \geq 1}$ and $\{y_n\}_{n \geq 1}$ in A such that $d(x, x_n) < \varepsilon/2$ and $d(y, y_n) < \varepsilon/2$ for $n \geq n_0$, say, where $\varepsilon > 0$ is arbitrary. Now for $n \geq n_0$, we have

$$d(x, y) \leq d(x, x_n) + d(x_n, y_n) + d(y_n, y)$$
$$\leq \frac{\varepsilon}{2} + d(x_n, y_n) + \frac{\varepsilon}{2}$$
$$\leq d(A) + \varepsilon,$$

and so $d(\overline{A}) \leq d(A)$, since $\varepsilon > 0$ is arbitrary. Clearly, $d(A) \leq d(\overline{A})$. $\square$

Let $\{I_n\}_{n \geq 1}$ be a sequence of intervals in $\mathbf{R}$. The sequence $\{I_n\}_{n \geq 1}$ is said to be **nested** if $I_{n+1} \subseteq I_n$, $n = 1, 2, \ldots$. The sequence $I_n = (0, 1/n)$, $n \in \mathbf{N}$, is nested. However $\bigcap_{n=1}^{\infty} I_n$ is empty. Similarly, the sequence $J_n = [n, \infty), n \in \mathbf{N}$, is nested with $\bigcap_{n=1}^{\infty} J_n = \varnothing$. In the metric space of rationals, the nested sequence $K_n = \{x \in \mathbf{Q} : |x - \sqrt{2}| < 1/n\}$ is such that $\bigcap_{n=1}^{\infty} K_n = \varnothing$, since $\sqrt{2}$ belongs to K_n for no n. The reader will note that the sequence $\{I_n\}_{n \geq 1}$ consists of intervals that are not closed, the sequence $\{J_n\}_{n \geq 1}$ consists of intervals that are not bounded, whereas the sequence $\{K_n\}_{n \geq 1}$ is in $\mathbf{Q}$, which is not complete. It is a very important property of real numbers that every nested sequence of closed bounded intervals does have a nonempty intersection. An analogue of this result holds in metric spaces.

Theorem 2.1.44. (Cantor) Let (X, d) be a metric space. Then (X, d) is complete if and only if, for every nested sequence $\{F_n\}_{n \geq 1}$ of nonempty closed subsets of X, that is,
(a) $F_1 \supseteq F_2 \supseteq \ldots \supseteq F_n \supseteq \ldots$ such that (b) $d(F_n) \to 0$ as $n \to \infty$,
the intersection $\bigcap_{n=1}^{\infty} F_n$ contains one and only one point.

Proof. First suppose that (X, d) is complete. For each positive integer n, let x_n be any point in F_n. Then by (a),

$$x_n, x_{n+1}, x_{n+2}, \ldots$$

all lie in F_n. Given $\varepsilon > 0$, there exists by (b) some integer n_0 such that $d(F_{n_0}) < \varepsilon$. Now, $x_{n_0}, x_{n_0+1}, x_{n_0+2}, \ldots$ all lie in F_{n_0}. For $m, n \geq n_0$, we then have $d(x_m, x_n) \leq d(F_{n_0}) < \varepsilon$. This shows that the sequence $\{x_n\}_{n \geq 1}$ is a Cauchy sequence in the complete metric space X. So, it is convergent. Let $x \in X$ be such that $\lim_{n \to \infty} x_n = x$. Now for any given n, we have the sequence $x_n, x_{n+1}, \ldots \subseteq F_n$. In view of this,

$$x = \lim_{n \to \infty} x_n \in \bar{F}_n = F_n$$

since F_n is closed. Hence,

$$x \in \bigcap_{n=1}^{\infty} F_n.$$

If $y \in X$ and $y \neq x$, then $d(y, x) = \alpha > 0$. There exists n large enough so that $d(F_n) < \alpha = d(y, x)$, which ensures that $y \notin F_n$. Hence, y cannot be in $\bigcap_{n=1}^{\infty} F_n$.

To prove the converse, let $\{x_n\}_{n \geq 1}$ be any Cauchy sequence in X. For each natural number n, let

$$F_n = \overline{\{x_m : m \geq n\}}.$$

Then $\{F_n\}_{n \geq 1}$ is a nested sequence of closed sets and since $\{x_n\}_{n \geq 1}$ is a Cauchy sequence,

$$\lim_{n \to \infty} d(F_n) = 0,$$

using Proposition 2.1.43.
    Let

$$\bigcap_{n=1}^{\infty} F_n = \{x\}.$$

If $\varepsilon > 0$, then there exists a natural number n_0 such that $d(F_{n_0}) < \varepsilon$. But $x \in F_{n_0}$ and thus $n \geq n_0$ implies $d(x_n, x) < \varepsilon$. $\qquad \square$

2.2. Relativisation and Subspaces

Let (X, d) be a metric space and Y a nonempty subset of X. If d_Y denotes the restriction of the function d to the set $Y \times Y$, then d_Y is a metric for Y and (Y, d_Y) is a metric space (see Section 1.2). If $Z \subseteq Y \subseteq X$, we may speak of Z being open (respectively, closed) relative to Y as well as open (respectively, closed) relative to X. It may happen that Z is an open (respectively, closed) subset of Y but not of X. For example, let X be $\mathbf{R}^2$ with metric d_2 and $Y = \{(x, 0): x \in \mathbf{R}\}$ with the induced metric. Then Y is a closed subset of X (for $Y^c = \{(x, y) \in \mathbf{R}^2: y \neq 0\}$ is open in X). If $Z = \{(x, 0): 0 < x < 1\}$, then Z considered as a subset of Y is open in Y. However, Z considered as a subset of X is not open in X. In fact, no point $(x, 0) \in Z$ is an interior point of Z (Z considered as a subset of X) because any neighbourhood of $(x, 0)$ in X is the ball $S((x, 0), r), r > 0$, which is not contained in Z. Thus, $Z = \{(x, 0): 0 < x < 1\}$ is an open subset of $Y = \{(x, 0): x \in \mathbf{R}\}$ but not of $X = (\mathbf{R}^2, d_2)$.

The above examples illustrate that the property of a set being open (respectively closed) depends on the metric space of which it is regarded a subset. The following theorem characterises open (respectively closed) sets in a subspace Y in terms of open (respectively closed) subsets in the space X. First we shall need a lemma.

Lemma 2.2.1. Let (X, d) be a metric space and Y a subspace of X. Let $z \in Y$ and $r > 0$. Then

$$S_Y(z, r) = S_X(z, r) \cap Y,$$

where $S_Y(z, r)$ (respectively $S_X(z, r)$) denotes the ball with centre z and radius r in Y (respectively X).

Proof. We have

$$
\begin{aligned}
S_X(z, r) \cap Y &= \{x \in X : d(x, z) < r\} \cap Y \\
&= \{x \in Y : d(x, z) < r\} \\
&= S_Y(z, r) \quad \text{since } Y \subseteq X. \qquad \square
\end{aligned}
$$

Let $X = \mathbf{R}^2$ and $Y = \{(x_1, x_2): 0 < x_1 \leq 1, \ 0 \leq x_2 < 1, \ x_1^2 + x_2^2 \geq 1\}$. Here, the open ball in Y with centre $(1, 0)$ and radius $\sqrt{2}$ is the entire space Y. (See Figure 2.6.)

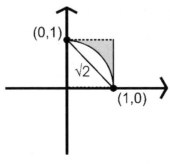

FIGURE 2.6

Theorem 2.2.2. Let (X, d) be a metric space and Y a subspace of X. Let Z be a subset of Y. Then

(i) Z is open in Y if and only if there exists an open set $G \subseteq X$ such that $Z = G \cap Y$;
(ii) Z is closed in Y if and only if there exists a closed set $F \subseteq X$ such that $Z = F \cap Y$.

Proof. (i) Let Z be open in Y. Then if z is any point of Z, there exists an open ball $S_Y(z, r)$ contained in Z. Observe that the radius r of the ball $S_Y(z, r)$ depends on the point $z \in Z$. We then have

$$Z = \bigcup_{z \in Z} S_Y(z, r)$$
$$= \bigcup_{z \in Z} (S_X(z, r) \cap Y) \quad \text{using Lemma 2.2.1}$$
$$= \left(\bigcup_{z \in Z} S_X(z, r) \right) \cap Y$$
$$= G \cap Y,$$

where $G = \bigcup_{z \in Z} S_X(z, r)$ is open in X.

On the other hand, suppose that $Z = G \cap Y$, where G is open in X. If $z \in Z$, then z is a point of G and so there exists an open ball $S_X(z, r)$ such that $S_X(z, r) \subseteq G$. Hence,

$$S_Y(z, r) = S_X(z, r) \cap Y \quad \text{by Lemma 2.2.1}$$
$$\subseteq G \cap Y = Z,$$

so that z is an interior point of the subset Z of Y. As z is an arbitrary point of Z, it follows that Z is open in Y.

(ii) Z is closed in Y if and only if $(X \backslash Z) \cap Y$ is open in Y. Hence, Z is closed in Y if and only if there exists an open set G in X such that

$$(X \backslash Z) \cap Y = G \cap Y \quad \text{using (i) above.}$$

On taking complements in X on both sides, we have

$$Z \cup (X \backslash Y) = (X \backslash G) \cup (X \backslash Y).$$

Hence

$$Z = Z \cap Y = (Z \cup (X \backslash Y)) \cap Y$$
$$= ((X \backslash G) \cup (X \backslash Y)) \cap Y.$$
$$= (X \backslash G) \cap Y$$

So, Z is the intersection of the closed set $X \backslash G$ and Y.

Conversely, let $Z = F \cap Y$, where F is closed in X. Then $X \backslash Z = (X \backslash F) \cup (X \backslash Y)$ and so

$$(X \backslash Z) \cap Y = ((X \backslash F) \cup (X \backslash Y)) \cap Y = (X \backslash F) \cap Y,$$

where $X \backslash F$ is open in X. Hence $(X \backslash Z) \cap Y$ is open in Y, i.e., Z is closed in Y. $\square$

Proposition 2.2.3. Let Y be a subspace of a metric space (X, d).

(i) Every subset of Y that is open in Y is also open in X if and only if Y is open in X.

(ii) Every subset of Y that is closed in Y is also closed in X if and only if Y is closed in X.

Proof. (i) Suppose every subset of Y open in Y is also open in X. We want to show that Y is open in X. Since Y is an open subset of Y, it must be open in X. Conversely, suppose Y is open in X. Let Z be an open subset of Y. By Theorem 2.2.2(i), there exists an open subset G of X such that $Z = G \cap Y$. Since G and Y are both open subsets of X, their intersection must be open in X, i.e., Z must be open in X.

(ii) The proof is equally easy and is, therefore, not included. $\square$

Proposition 2.2.4. Let (X, d) be a metric space and $Z \subseteq Y \subseteq X$. If $\mathrm{cl}_X Z$ and $\mathrm{cl}_Y Z$ denote, respectively, the closures of Z in the metric spaces X and Y, then

$$\mathrm{cl}_Y Z = Y \cap \mathrm{cl}_X Z.$$

Proof. Obviously, $Z \subseteq Y \cap \mathrm{cl}_X Z$. Since $Y \cap \mathrm{cl}_X Z$ is closed in Y (see Theorem 2.2.2(ii)), it follows that $\mathrm{cl}_Y Z \subseteq Y \cap \mathrm{cl}_X Z$. On the other hand, by Theorem 2.2.2(ii), $\mathrm{cl}_Y Z = Y \cap F$, where F is a closed subset of X. But then

$$Z \subseteq \mathrm{cl}_Y Z \subseteq F,$$

and hence, by Corollary 2.1.27(ii),

$$\mathrm{cl}_X Z \subseteq F.$$

Therefore,

$$\mathrm{cl}_Y Z = Y \cap F \supseteq Y \cap \mathrm{cl}_X Z.$$

This completes the proof. $\square$

In contrast to the relative properties discussed above, there are some properties that are intrinsic. In fact, the property of x being a limit point of F holds in any subspace containing x and F as soon as it holds in the whole space, and conversely. Another such property is that of being complete. The following propositions describe relations between closed sets and complete sets.

Proposition 2.2.5. If Y is a nonempty subset of a metric space (X, d), and (Y, d_Y) is complete, then Y is closed in X.

Proof. Let x be any limit point of Y. Then x is the limit of a sequence $\{y_n\}_{n\geq 1}$ in Y. In view of Proposition 1.4.3, the sequence $\{y_n\}_{n\geq 1}$ is Cauchy, and hence, by assumption, converges to a point y of Y. But by Remark 3 following Definition 1.3.2, $y = x$. Therefore, $x \in Y$. This shows that Y is closed in X. $\square$

Proposition 2.2.6. Let (X, d) be a complete metric space and Y a closed subset of X. Then (Y, d_Y) is a complete space.

Proof. Let $\{y_n\}_{n\geq 1}$ be a Cauchy sequence in (Y, d_Y). Then $\{y_n\}_{n\geq 1}$ is also a Cauchy sequence in (X, d); so there exists an $x \in X$ such that $\lim_{n\to\infty} y_n = x$. If follows (see Proposition 2.1.28) that $x \in \bar{Y}$, which is the same set as Y by Corollary 2.1.27(i). $\square$

2.3. Countability Axioms and Separability

Definition 2.3.1. Let (X, d) be a metric space and $x \in X$. Let $\{G_\lambda\}_{\lambda\in\Lambda}$ be a family of open sets, each containing x. The family $\{G_\lambda\}_{\lambda\in\Lambda}$ is said to be a **local base at x** if, for every nonempty open set G containing x, there exists a set G_μ in the family $\{G_\lambda\}_{\lambda\in\Lambda}$ such that $x \in G_\mu \subseteq G$.

Examples 2.3.2. (i) In the metric space $\mathbf{R}^2$ with the Euclidean metric, let $G_\lambda = S(x, \lambda)$, where $x=(x_1,x_2)\in\mathbf{R}^2$ and $0<\lambda\in\mathbf{R}$. The family $\{G_\lambda:0<\lambda\in\mathbf{R}\}=\{S(x,\lambda):0<\lambda\in\mathbf{R}\}$ is a family of balls and is a local base at x. Note that $S(x,\lambda)$, where $x=(x_1,x_2)$, can also be described as $\{(y_1,y_2)\in\mathbf{R}^2 : (y_1-x_1)^2+ (y_2-x_2)^2 <\lambda^2\}$.

(ii) Let $x = (x_1, x_2) \in \mathbf{R}^2$ and $G_\lambda = \{(y_1, y_2) \in \mathbf{R}^2 : (y_1 - x_1)^2 + 2(y_2 - x_2)^2 < \lambda\}$, where $0 < \lambda \in \mathbf{R}$. Then the family $\{G_\lambda : 0 < \lambda \in \mathbf{R}\}$ is a local base at x. To see why, consider any open set $G \subseteq \mathbf{R}^2$ such that $x \in G$. Since G is open, there exists $r > 0$ such that $S(x, r) \subseteq G$. Now $S(x, r) = \{(y_1, y_2) \in \mathbf{R}^2 : (y_1 - x_1)^2 + (y_2 - x_2)^2 < r^2\}$. Let $\lambda = r^2$. Then $y \in G_\lambda \Rightarrow (y_1 - x_1)^2 + 2(y_2 - x_2)^2 < \lambda \Rightarrow (y_1 - x_1)^2 + (y_2 - x_2)^2 < \lambda \Rightarrow (y_1 - x_1)^2 + (y_2 - x_2)^2 < r^2 \Rightarrow y \in S(x, r)$, so that $G_\lambda \subseteq S(x, r) \subseteq G$. In this example, the sets G_λ are ellipses.

(iii) Let $x \in \mathbf{R}$. Consider the family of all open intervals (r,s) containing x and having rational endpoints r and s. This family is a local base at x. It consists of open balls, not necessarily centred at x. Moreover, the family is countable and thus constitutes what is called a **countable base at x**.

Proposition 2.3.3. In any metric space, there is a countable base at each point.

Proof. Let (X, d) be a metric space and $x \in X$. The family of open balls centred at x and having rational radii, i.e., $\{S(x, \rho): \rho$ rational and positive$\}$ is a countable base at x. In fact, if G is an open set and $x \in G$, then by the definition of an open set, there exists an $\varepsilon > 0$ (ε depending on x) such that $x \in S(x, \varepsilon) \subseteq G$. Let ρ be a positive rational number less than ε. Then

$$x \in S(x, \rho) \subseteq S(x, \varepsilon) \subseteq G. \qquad \square$$

Definition 2.3.4. A family $\{G_\lambda\}_{\lambda \in \Lambda}$ of nonempty open sets is called a **base for the open sets** of (X, d) if every open subset of X is a union of a subfamily of the family $\{G_\lambda\}_{\lambda \in \Lambda}$.

The condition of the above definition can be expressed in the following equivalent form: If G is an arbitrary nonempty open set and $x \in G$, then there exists a set G_μ in the family such that $x \in G_\mu \subseteq G$.

Proposition 2.3.5. The collection $\{S(x, \varepsilon) : x \in X, \ \varepsilon > 0\}$ of all open balls in X is a base for the open sets of X.

Proof. Let G be a nonempty open subset of X and let $x \in G$. By the definition of an open subset, there exists a positive $\varepsilon(x)$ (depending upon x) such that

$$x \in S(x, \varepsilon(x)) \subseteq G.$$

This completes the proof. $\square$

Generally speaking, an open base is useful if its sets are simple in form. A space that has a countable base for the open sets has pleasant properties and goes by the name of "second countable".

Definition 2.3.6. A metric space is said to be **second countable** (or satisfy **the second axiom of countability**) if it has a countable base for its open sets.

The reason for the name *second* countable is that the property of having a countable base at each point, as in Proposition 2.3.3, is usually called *first* countability.

Examples 2.3.7. (i) Let $(\mathbf{R}, d)$ be the real line with the usual metric. The collection $\{(x, y) : x, y \text{ rational}\}$ of all open intervals with rational endpoints form a countable base for the open sets of $\mathbf{R}$.

(ii) The collection

$\{S(x, r) : x = (x_1, x_2, \ldots, x_n), x_i \text{ rationals}, \ 1 \le i \le n, \text{ and } r \text{ positive rational}\}$

of all r-balls with rational centres and rational radii is a countable base for the open sets of the metric space $(\mathbf{R}^n, d)$, where d may be any of the metrics on $\mathbf{R}^n$ described in Example 1.2.2(iii).

(iii) Let X have the discrete metric. Then any set $\{x\}$ containing a single point x is also the open ball $S(x, 1/2)$ and therefore must be a union of nonempty sets of any base. So any base has to contain each set $\{x\}$ as one of the sets in it. If X is nondenumerable, then the sets $\{x\}$ are also nondenumerable, forcing every base to be nondenumerable as well. Consequently, X does not satisfy the second axiom of countability when it is nondenumerable.

It is easy to see that any subspace of a second countable space is also a second countable space. In fact, the class of all intersections with the subspace of the sets of a base form a base for the open sets of the subspace.

Definition 2.3.8. Let (X, d) be a metric space and $\mathcal{G}$ be a collection of open sets in X. If for each $x \in X$ there is a member $G \in \mathcal{G}$ such that $x \in G$, then $\mathcal{G}$ is called an **open cover** (or **open covering**) of X. A subcollection of $\mathcal{G}$ which is itself an open cover of X is called a **subcover** (or **subcovering**).

Examples 2.3.9. (i) The union of the family $\{\ldots, (-3, -1), (-2, 0), (-1, 1), (0, 2), \ldots\}$ of open intervals is $\mathbf{R}$. The family is therefore an open covering of $\mathbf{R}$. However, the family of open intervals $\{\ldots, (-2, -1), (-1, 0), (0, 1), (1, 2), \ldots\}$ is not an open covering, because the intervals' union does not contain the integers. The aforementioned cover contains no subcovering besides itself, because, if we delete any interval from the family, the midpoint of the deleted interval will not belong to the union of the remaining intervals.

(ii) Let X be the discrete metric space consisting of the five elements a, b, c, d, e. The union of the family of subsets $\{\{a\}, \{b, c\}, \{c, d\}, \{a, d, e\}\}$ is X and all subsets are open. Therefore the family is an open cover. The family $\{\{b, c\}, \{c, d\}, \{a, d, e\}\}$ is a proper subcover.

(iii) Consider the set $\mathbf{Z}$ of all integers with the discrete metric. As in any discrete metric space, all subsets are open. Consider the family consisting of the three subsets

$$\{3n : n \in \mathbf{Z}\}, \{3n + 1 : n \in \mathbf{Z}\} \text{ and } \{3n + 2 : n \in \mathbf{Z}\}.$$

Since every integer must be of the form $3n$, $3n + 1$ or $3n + 2$, the above three subsets form an open cover of $\mathbf{Z}$. There is no proper subcover.

(iv) The family of intervals $\{(-n, n) : n \in \mathbf{N}\}$ is an open cover of $\mathbf{R}$ and the family consisting of the open balls $\{z \in \mathbf{C} : |z + 17| < n^{3/2}, n \in \mathbf{N}\}$ is an open cover of $\mathbf{C}$. If we extract a subfamily by restricting n to be greater than some integer n_0, the subfamily is also an open cover. Indeed, if we delete a finite number of sets in the family, the remaining subfamily is an open cover. Thus, there are infinitely many open subcovers.

Definition 2.3.10. A metric space is said to be **Lindelöf** if each open covering of X contains a countable subcovering.

Proposition 2.3.11. Let (X, d) be a metric space. If X satisfies the second axiom of countability, then every open covering $\{U_\alpha\}_{\alpha \in \Lambda}$ of X contains a countable subcovering. In other words, a second countable metric space is Lindelöf.

Proof. Let $\{G_i : i = 1, 2, \ldots\}$ be a countable base of open sets for X. Since each U_α is a union of sets G_i, it follows that a subfamily $\{G_{i_j} : j = 1, 2, \ldots\}$ of the base $\{G_i : i = 1, 2, \ldots\}$ is a covering of X. Choose $U_{i_j} \supseteq G_{i_j}$ for each j. Then $\{U_{i_j} : j = 1, 2, \ldots\}$ is the required countable subcovering. $\qquad\square$

Definition 2.3.12. A subset X_0 of a metric space (X, d) is said to be **everywhere dense** or simply **dense** if $\overline{X_0} = X$, i.e., if every point of X is either a point or a limit point of X_0. This means that, given any point x of X, there exists a sequence of points of X_0 that converges to x.

It follows easily from this definition and the definition of interior (see Definition 2.1.12) that a subset of X_0 is dense if and only if X_0^c has empty interior.

It may be noted that X is always a dense subset of itself; interest centres around what *proper* subsets of a metric space are dense.

Examples 2.3.13. (i) The set of rationals is a dense subset of $\mathbf{R}$ (usual metric) and so is the set of irrationals. Note that the former is countable whereas the latter is not.

(ii) Consider the metric space $(\mathbf{R}^n, d)$ with any of the metrics described in Example 1.2.2(iii). Within any neighbourhood of any point in $\mathbf{R}^n$, there is a point with rational coordinates. Thus,

$$\mathbf{Q}^n = \mathbf{Q} \times \mathbf{Q} \times \ldots \times \mathbf{Q}$$

is dense in $\mathbf{R}^n$.

(iii) In the space $C[0, 1]$ of Example 1.2.2(ix), we consider the set C_0 consisting of all polynomials with rational coefficients. We shall check that C_0 is dense in $C[0,1]$. Let $x(t) \in C[0,1]$. By Weierstrass' theorem (Theorem 0.8.4), there exists a polynomial $P(t)$ such that

$$\sup\{|x(t) - P(t)| : 0 \le t \le 1\} < \frac{\varepsilon}{2},$$

where $\varepsilon > 0$ is given. Corresponding to $P(t)$ there is a polynomial $P_0(t)$ with rational coefficients such that

$$\sup\{|P(t) - P_0(t)| : 0 \le t \le 1\} < \frac{\varepsilon}{2}.$$

So,

$$\sup\{|x(t) - P_0(t)| : 0 \le t \le 1\} < \varepsilon.$$

It is easy to see that C_0 is countable. In fact, if $\mathcal{P}_n$ denotes the set of all polynomials of degree n and having rational coefficients, then the cardinality of $\mathcal{P}_n$ is the same as that of $\mathbf{Q}^{n+1} = \mathbf{Q} \times \mathbf{Q} \times \ldots \times \mathbf{Q}$, which is countable. The assertion now follows from the fact that a countable union of countable sets is countable.

(iv) Let (X, d) be a discrete metric space. Since every subset is closed, the only dense subset is X itself.

(v) Let $X = \ell_p$ of Example 1.2.2(vii). Recall that the metric is given by

$$d(x,y) = \left(\sum_{i=1}^{\infty} |x_i - y_i|^p\right)^{1/p}.$$

Let E denote the set of all elements of the form $(r_1, r_2, \ldots, r_n, 0, 0 \ldots)$, where r_i are rational numbers and n is an arbitrary natural number. We shall show that E is dense in ℓ_p. Let $x = (x_1, x_2, \dots)$ be an element in ℓ_p and let $\varepsilon > 0$ be given. There exists a natural number n_0 such that

$$\sum_{j=n_0+1}^{\infty} |x_j|^p < \frac{\varepsilon^p}{2}.$$

Choose an element $x_0 = (r_1, r_2, \dots, r_{n_0}, 0, 0, \dots)$ in E such that

$$\sum_{j=1}^{n_0} |x_j - r_j|^p < \frac{\varepsilon^p}{2}.$$

We then obtain

$$(d(x, x_0))^p = \sum_{j=1}^{n_0} |x_j - r_j|^p + \sum_{j=n_0+1}^{\infty} |x_j|^p < \varepsilon^p,$$

and this implies

$$d(x, x_0) < \varepsilon.$$

Thus, E is dense in (ℓ_p, d). Also, E is countable (in fact, if E_n denotes the subset of all those elements $x = \{r_i\}_{i \geq 1}$ such that $r_j = 0$ for $j \geq n+1$, then E_n is countable and $E = \bigcup_{n=1}^{\infty} E_n$).

(vi) By Definition 1.5.1, any metric space is dense in its completion.

Definition 2.3.14. The metric space X is said to be **separable** if there exists a countable, everywhere dense set in X. In other words, X is said to be separable if there exists in X a sequence

$$\{x_1, x_2, \dots\} \tag{2.1}$$

such that for every $x \in X$, some sequence in the range of (2.1) converges to x.

Examples 2.3.15. In Examples 2.3.13(i)–(iii) and (v), we saw dense sets that are countable. Therefore, the spaces concerned are separable. In (iv) however, the space is separable if and only if the set X is countable.

There are metric spaces other than the discrete metric space mentioned above which fail to satisfy the separability criterion. The next example is one such case. Let X denote the set of all bounded sequences of real numbers with metric

$$d(x, y) = \sup\{|x_i - y_i| : i = 1, 2, 3, \dots\},$$

as in Example 1.2.2(vi). We shall show that X is **inseparable**.

First we consider the set A of elements $x = (x_1, x_2, \dots)$ of X for which each x_i is either 0 or 1 and show that it is uncountable. If E is any countable subset of A, then the elements of E can be arranged in a sequence $s_1, s_2, \dots$. We construct a sequence s as follows. If the mth element of s_m is 1, then the mth element of s is 0, and vice versa. Then the element s of X differs from each s_m in the mth place and is therefore equal to none of them. So, $s \notin E$ although $s \in A$. This shows that any countable subset of A must be a proper subset of A. It follows that A is uncountable, for if it were to be countable, then it would have to be a proper subset of itself, which is absurd. We proceed to use the uncountability of the subset A to argue that X must be inseparable.

The distance between two distinct elements $x = (x_1, x_2, \ldots)$ and $y = (y_1, y_2, \ldots)$ of A is $d(x, y) = \sup\{|x_i - y_i| : i = 1, 2, 3, \ldots\} = 1$. Suppose, if possible, that E_0 is a countable, everywhere dense subset of X. Consider the balls of radii $1/3$ whose centres are the points of E_0. Their union is the entire space X, because E_0 is everywhere dense, and in particular contains A. Since the balls are countable in number while A is not, in at least one ball there must be two distinct elements x and y of A. Let x_0 denote the centre of such a ball. Then

$$1 = d(x, y) \leq d(x, x_0) + d(x_0, y) < \frac{1}{3} + \frac{1}{3} < 1,$$

which is, however, impossible. Consequently, (X, d) cannot be separable.

Proposition 2.3.16. Let (X, d) be a metric space and $Y \subseteq X$. If X is separable, then Y with the induced metric is separable, too.

Proof. Let $E = \{x_i : i = 1, 2, \ldots\}$ be a countable dense subset of X. If E is contained in Y, then there is nothing to prove. Otherwise, we construct a countable dense subset of Y whose points are arbitrarily close to those of E. For positive integers n and m, let $S_{n, m} = S(x_n, 1/m)$ and choose $y_{n, m} \in S_{n, m} \cap Y$ whenever this set is nonempty. We show that the countable set $\{y_{n, m} : n \text{ and } m \text{ positive integers}\}$ of Y is dense in Y.

For this purpose, let $y \in Y$ and $\varepsilon > 0$. Let m be so large that $1/m < \varepsilon/2$ and find $x_n \in S(y, 1/m)$. Then $y \in S_{n, m} \cap Y$ and

$$d(y, y_{n, m}) \leq d(y, x_n) + d(x_n, y_{n, m}) < \frac{1}{m} + \frac{1}{m} < \frac{\varepsilon}{2} + \frac{\varepsilon}{2} = \varepsilon.$$

Thus, $y_{n, m} \in S(y, \varepsilon)$. Since $y \in Y$ and $\varepsilon > 0$ are arbitrary, the assertion is proved. ☐

The main result of this section is the following.

Theorem 2.3.17. Let (X, d) be a metric space. The following statements are equivalent:

(i) (X, d) is separable;
(ii) (X, d) satisfies the second axiom of countability;
(iii) (X, d) is Lindelöf.

Proof. (i)$\Rightarrow$(ii). Let $E = \{x_i : i = 1, 2, \ldots\}$ be a countable, dense subset of X and let $\{r_j : j = 1, 2, \ldots\}$ be an enumeration of positive rationals. Consider the countable collection of balls with centres at $x_i, i = 1, 2, \ldots$ and radii $r_j, j = 1, 2, \ldots$; i.e.,

$$\{S(x_i, r_j) : x_i \in E \text{ for } i = 1, 2, \ldots \text{ and } r_j, \text{ is rational } j = 1, 2, \ldots\}.$$

If G is any open set and $x \in G$, we want to show that for some i and some j, $x \in S(x_i, r_j) \subseteq G$. Since G is open, there is a ball $S(x, \delta)$ such that $S(x, \delta) \subseteq G$. Let $r_k > 0$ be a rational number such that $0 < r_k < \delta$. Since x is a point of closure of E, there is a point $x_i \in E$ such that $d(x, x_i) < 1/2 r_k$. Hence,

$$x \in S(x_i, \frac{1}{2} r_k) \subseteq S(x, r_k) \subseteq G.$$

In fact, if $y \in S(x_i, 1/2 r_k)$ then $d(y, x) \leq d(y, x_i) + d(x_i, x) < 1/2 r_k + 1/2 r_k = r_k$.

(ii)$\Rightarrow$(iii). See Proposition 2.3.11.

(iii)$\Rightarrow$(i). From each open covering $\{S(x, \varepsilon): x \in X\}$, we extract a countable subcovering $\{S(x_i, \varepsilon): x_i \in X, i = 1, 2, \ldots\}$ and let $A(\varepsilon) = \{x_1, x_2, \ldots\}$. Define $E = \bigcup_{n=1}^{\infty} A(\frac{1}{n})$. Then E is a countable, dense subset of X. $\square$

2.4. Baire's Category Theorem

Definition 2.4.1. Let (X, d) be a metric space. A subset $Y \subseteq X$ is said to be **nowhere dense** if $(\bar{Y})^{\circ}$ is empty, i.e., $(\bar{Y})^{\circ}$ contains no interior point. A subset $F \subseteq X$ is said to be of **category I** if it is a countable union of nowhere dense subsets. Subsets that are not of category I are said to be of **category II**.

Remarks 2.4.2. (i) A subset Y of X is nowhere dense if and only if the complement $(\bar{Y})^c$ is dense in X, or $\overline{(X - \bar{Y})} = X$. This follows easily from the remark immediately after Definition 2.3.12.

(ii) If d denotes the discrete metric, the only nowhere dense set is the null set.

(iii) The notion of being nowhere dense is not the opposite of being everywhere dense, i.e., not being nowhere dense does not imply that the set is everywhere dense. For an example of a set which is neither, let $\mathbf{R}$ denote the real line with the usual metric and consider the set $Y = \{x \in \mathbf{R} : 1 < x < 2\}$. Then

$$(\bar{Y})^{\circ} = Y \neq \varnothing \text{ and } (\bar{Y})^c = \{x \in \mathbf{R} : x < 1 \text{ or } x > 2\} = (-\infty, 1) \cup (2, \infty),$$

which is not dense in $\mathbf{R}$.

(iv) Every subset must be either of category I or of category II.

(v) It is clear that the null set is of category I. Also, the subset $\mathbf{Q}$ of rationals in $\mathbf{R}$ is a set of category I. Indeed, if $x_1, x_2, \ldots$ is an enumeration of the rationals, each $\{x_i\}$ is closed and $\{x_i\}^{\circ} = \varnothing$; it follows that $\cup \{x_i\}$, the set of all rationals in $\mathbf{R}$, is of category I.

(vi) Since a denumerable union of denumerable sets is again a denumerable set, it follows that, if $Y_1, Y_2, \ldots$ are each of category I, then so must be $\bigcup_i Y_i$.

(vii) If $X = Y_1 \cup Y_2$ and it is known that Y_1 is of category I while X is of category II, then Y_2 must be category II. For, if Y_2 is of category I, then it follows from (vi) above that X, too, is of category I, which is a contradiction.

(viii) A subset of a nowhere dense set is nowhere dense and, therefore, a subset of a set of category I is again of category I.

Theorem 2.4.3. (Baire Category Theorem) Any complete metric space is of category II.

Proof. We assume the contrary, i.e., we suppose that (X, d) is a complete metric space and

$$X = \bigcup_{n=1}^{\infty} E_n,$$

where each of the E_n is nowhere dense. Since each E_n is nowhere dense, each $(\overline{E_n})^c$ is everywhere dense. So we can assert the existence of points in each of these sets $(\overline{E_n})^c$ (i.e., none of them can be empty). In the case of $(\overline{E_1})^c$, let $x_1 \in (\overline{E_1})^c$. Since $(\overline{E_1})^c$ is open, there exists $r > 0$ such that $S(x_1, r) \subseteq (\overline{E_1})^c$. For $\varepsilon_1 < r$, we have

$$\bar{S}(x_1, \varepsilon_1) \subseteq S(x_1, r) \subseteq (\overline{E_1})^c \subseteq E_1^c.$$

This, in turn, implies

$$\bar{S}(x_1, \varepsilon_1) \cap E_1 = \varnothing.$$

We make the following induction hypothesis: There exist balls $S(x_k, \varepsilon_k)$ for $k = 1, 2, \ldots, n-1$ such that

$$\bar{S}(x_k, \varepsilon_k) \cap E_k = \varnothing, \qquad \text{where } x_k \in (\overline{E_k})^c$$

and

$$\varepsilon_k \leq \frac{1}{2}\varepsilon_{k-1} \qquad \text{for } k = 2, \ldots, n-1.$$

Using this information, we can construct the nth ball with the above properties. To this end, choose

$$x_n \in S(x_{n-1}, \varepsilon_{n-1}) \cap (\overline{E_n})^c.$$

Such an element must exist, because otherwise

$$S(x_{n-1}, \varepsilon_{n-1}) \subseteq \overline{E_n}$$

and this implies $x_{n-1} \in (\overline{E_n})^\circ$, contradicting the fact that $(\overline{E_n})^\circ$ is empty. Since the intersection $S(x_{n-1}, \varepsilon_{n-1}) \cap (\overline{E_n})^c$ is open, there exists $\varepsilon > 0$ such that

$$S(x_n, \varepsilon) \subseteq S(x_{n-1}, \varepsilon_{n-1}) \cap (\overline{E_n})^c.$$

Now we choose a positive $\varepsilon_n < \min\{\varepsilon, (1/2)\varepsilon_{n-1}\}$. Then

$$\bar{S}(x_n, \varepsilon_n) \subseteq S(x_n, \varepsilon) \subseteq S(x_{n-1}, \varepsilon_{n-1}) \cap (\overline{E_n})^c,$$

which says

$$\bar{S}(x_n, \varepsilon_n) \cap E_n = \varnothing.$$

As we also have

$$\varepsilon_n \le \frac{1}{2}\varepsilon_{n-1},$$

the nth ball with the requisite properties has been constructed.

As $\bar{S}(x_n, \varepsilon_n) \subseteq \bar{S}(x_{n-1}, \varepsilon_{n-1})$, the balls $\{\bar{S}(x_n, \varepsilon_n)\}_{n \ge 1}$ form a nested sequence of nonempty closed balls in a complete metric space with diameters tending to zero. By Theorem 2.1.44, there exists $x_0 \in \bigcap_1^\infty \bar{S}(x_n, \varepsilon_n)$. Since $\bar{S}(x_n, \varepsilon_n) \cap E_n = \varnothing$ for every n, we have $x_0 \notin E_n$ for any n, i.e., $x_0 \in E_n^c$ for all n. However, $\bigcap_1^\infty E_n^c = \varnothing$. This contradiction shows that X is not of category I. This completes the proof. $\square$

Corollary 2.4.4. The irrationals in **R** are of category II.

Proof. Since **R** is a complete metric space, it follows from Remarks (v) and (vii) prior to the Baire category theorem (Theorem 2.4.3) that the irrationals are of category II in **R**. $\square$

Corollary 2.4.5. A nonempty open interval is of category II.

Proof. If a nonempty open interval is of category I, then so is each of its translates. Since **R** is a countable union of such translates, it follows that **R** is of category I, contradicting the Baire category theorem (Theorem 2.4.3). $\square$

We next take up some applications of the Baire category theorem.

Theorem 2.4.6. (Osgood) Let $\mathcal{F}$ be a collection of continuous real-valued functions on **R** such that for each $x \in \mathbf{R}$, there exists $M_x > 0$ for which $|f(x)| \le M_x$ for all $f \in \mathcal{F}$. Then there exists an $M > 0$ and a nonempty open subset $Y \subseteq \mathbf{R}$ such that

$$|f(x)| \le M \text{ for each } x \in Y \text{ and for each } f \in \mathcal{F}.$$

Proof. For each integer n, let $E_{n, f} = \{x \in X : |f(x)| \le n\} = f^{-1}([-n, n])$. This set $E_{n, f}$ is closed for the following reason: Let x_0 be a limit point of it. Then there exists a sequence $\{x_m\}_{m \ge 1}$ in $E_{n, f}$ such that $\lim_{m \to \infty} x_m = x_0$. For each m, we have $-n \le f(x_m) \le n$, from which it follows that $-n \le f(x_0) \le n$, using the continuity of f. Therefore, $x_0 \in E_{n, f}$, showing that the set is closed. It now follows that the intersection $E_n = \bigcap_{f \in \mathcal{F}} E_{n, f}$ is a closed subset of **R**. Observe that $\mathbf{R} = \bigcup_{n=1}^\infty E_n$. Indeed, if $x \in \mathbf{R}$, by hypothesis there exists $M_x > 0$ such that $|f(x)| \le M_x$ for each $f \in \mathcal{F}$, which shows that $x \in E_{n_0}$ for any integer $n_0 > M_x$. Since **R** is complete, there exists an integer $M > 0$ such that E_M is not nowhere dense (Baire category theorem). Since E_M is closed, it must contain some nonempty open set Y. Then, for each $x \in Y$, we have $|f(x)| \le M$ for all $f \in \mathcal{F}$. $\square$

Another illuminating application of Baire's category theorem is the following. To begin with, we make an observation regarding a continuous real-valued function f defined on $[0,1]$. Let f_1 be an integral of f, that is,

$$f_1'(x) = f(x) \qquad \text{for all } x \in [0,\,1].$$

Let f_2 be an integral of f_1 and so on. If $f_k \equiv 0$ for some k, then obviously the same is true of f. The proof of the following generalisation of this observation uses the Baire category theorem.

Theorem 2.4.7. Let f be a continuous real-valued function on $[0,\,1]$. Let f_1 be an integral of f, that is, $f_1'(x) = f(x)$ for all $x \in [0,\,1]$. Let f_2 be an integral of f_1 and so on. If for each $x \in [0,\,1]$, there is an integer k depending on x such that $f_k(x) = 0$, then f is identically 0 on $[0,1]$.

Proof. Let $Z_n = \{x \in [0,\,1] : f_n(x) = 0\}$. Observe that Z_n is closed. Indeed, if $x \in [0,\,1]$ is the limit of a sequence $\{x_m\}$ in Z_n, then $f_n(x) = f_n(\lim_m x_m) = \lim_m f_n(x_m) = 0$, so that $x \in Z_n$. Also, by hypothesis,

$$\bigcup_{n=1}^{\infty} Z_n = [0,1].$$

Since $[0,1]$ is a complete metric space, there exists a positive integer n such that Z_n is not nowhere dense and so $Z_n^\circ \neq \varnothing$. Let $x_0 \in Z_n^\circ$. Then there exists an $\varepsilon > 0$ such that $[x_0 - \varepsilon,\, x_0 + \varepsilon] \subseteq Z_n^\circ$. Since $f_n(x) = 0$ on $[x_0 - \varepsilon,\, x_0 + \varepsilon]$, it follows that $f(x) = 0$ on $[x_0 - \varepsilon,\, x_0 + \varepsilon]$; in particular, $f(x_0) = 0$, and, hence, $f(x) = 0$ for all $x \in Z_n^\circ$.

Let $Y_n = Z_n \backslash Z_n^\circ = Z_n \cap ([0,\,1] \backslash Z_n^\circ)$. Now Y_n, being the intersection of closed sets, is itself closed. Moreover, $(\overline{Y_n})^\circ = Y_n^\circ = \varnothing$. So, Y_n is nowhere dense. Thus, $f(x) = 0$ for all $x \in [0,\,1]$ except possibly for a set of category I. Since f is continuous, we shall argue that $f(x) = 0$ for all $x \in [0,\,1]$. Let $x_0 \in [0,\,1]$ be such that $f(x_0) \neq 0$. Since f is continuous, there exists a nonempty open interval I_{x_0} containing x_0 such that $f(x) \neq 0$ for $x \in I_{x_0}$. By the argument above, I_{x_0} is contained in a set of category I and hence is itself a set of category I (see Remark (vii) after Definition 2.4.1), which contradicts Corollary 2.4.5 of the Baire category theorem. □

That a continuous function may fail to have a derivative at any point of its domain of definition, though surprising, is nevertheless true. It turns out that "most" continuous functions have this property. More specifically, the set of continuous functions that have a finite derivative even on one side constitute a set of category I in the metric space $C[0,1]$. Thus, the functions that one deals with in calculus form a subset of a set of category I. In what follows, we shall show that functions in $C[0,1]$ that are nowhere differentiable form a set of category II.

Consider the metric space $C[0,1]$ equipped (as usual) with the metric

$$d(f,g) = \sup\{|f(x) - g(x)| : 0 \leq x \leq 1\}, \; f, g \in C[0,1].$$

It is a complete metric space. (See Proposition 1.4.13.)

Let A denote the subset of $C[0,1]$ such that, for some $x \in [0,1]$, f has a finite right hand derivative, i.e., there exists $\ell \in \mathbf{R}$ such that, given any $\varepsilon > 0$, there exists a $\delta > 0$ for which

$$\left| \frac{f(x+h) - f(x)}{h} - \ell \right| < \varepsilon$$

for all h satisfying $x + h \in [0,1]$ and $0 < h < \delta$.

For each positive integer n, let E_n denote the set of all $f \in C[0,1]$ such that for some $x \in [0, 1 - 1/n]$,

$$\left| \frac{f(x+h) - f(x)}{h} \right| \leq n$$

whenever $0 < h < 1/n$. It is clear that $E_n \subseteq E_{n+1}$. Moreover, if f has a finite right hand derivative at x, then $f \in E_n$ for some n. So, $A \subseteq \bigcup_{n=1}^{\infty} E_n$.

We shall show that each E_n is nowhere dense; then the union of the E_n is of category I and, hence, so is A. The space $C[0,1]$ with metric d, being complete, is of category II. Consequently, A^c, which consists of those functions in $C[0,1]$ that do not possess a right hand derivative at any point, is of category II. Since A^c is a subset of those $f \in C[0,1]$ that do not possess a derivative anywhere, it follows that there exist continuous functions that are nowhere differentiable and that the collection of these functions is a subset of category II.

In order to prove that each E_n is nowhere dense, we proceed by showing: (i) $\overline{E_n} = E_n$, and (ii) E_n° is empty.

Let $g \in \overline{E_n}$ and $\{f_j\}_{j \geq 1}$ be a sequence of functions in E_n such that $\lim_{j \to \infty} d(f_j, g) = 0$. Since each of the f_j is in E_n, there exists some point x_j (depending on f_j) such that

$$\left| \frac{f_j(x_j + h) - f_j(x_j)}{h} \right| \leq n \qquad \text{for } 0 < h < 1/n, \ x_j \in [0, 1 - 1/n].$$

The points $\{x_j\}_{j \geq 1}$ constitute a bounded sequence of real numbers and so, by the Bolzano-Weierstrass theorem (Theorem 0.4.2), there exists a subsequence $\{x_{j_k}\}_{k \geq 1}$ such that $x_{j_k} \to x_0$. Since any subsequence of a convergent sequence converges to the same limit, it follows that $\lim_{k \to \infty} d(f_{j_k}, g) = 0$. Now,

$$|g(x_0 + h) - g(x_0)| \leq |g(x_0 + h) - g(x_{j_k} + h)| + |g(x_{j_k} + h) - f_{j_k}(x_{j_k} + h)|$$
$$+ |f_{j_k}(x_{j_k} + h) - f_{j_k}(x_{j_k})| + |f_{j_k}(x_{j_k}) - g(x_{j_k})| + |g(x_{j_k}) - g(x_0)|$$

$$(2.2)$$

By continuity of g, there exists m_1 such that $j_k \geq m_1$ implies

$$|g(x_{j_k}) - g(x_0)| < \frac{1}{4}\varepsilon h \qquad \text{and} \qquad |g(x_0 + h) - g(x_{j_k} + h)| < \frac{1}{4}\varepsilon h,$$

in view of the fact that $x_{j_k} \to x_0$. Since $\lim_{k\to\infty} d(f_{j_k}, g) = 0$, there exists m_2 such that $j_k \geq m_2$ implies

$$\sup\{|f_{j_k}(x) - g(x)| : x \in [0,1]\} < \frac{1}{4}\varepsilon h.$$

Choosing $j_k > \max\{m_1, m_2)$, we obtain from (2.2) that

$$\left|\frac{g(x_0 + h) - g(x_0)}{h}\right| \leq \left|\frac{f_{j_k}(x_{j_k} + h) - f_{j_k}(x_{j_k})}{h}\right| + \varepsilon \leq n + \varepsilon.$$

Since $x_0 \in [0, 1 - 1/n]$ and $\varepsilon > 0$ is arbitrary, it follows that $g \in E_n$. This establishes (i).

We next establish (ii), i.e., that E_n is nowhere dense. Since E_n is closed by (i), it is enough to show that E_n^c is everywhere dense (see Remark (i) after Definition 2.4.1). Let $f \in C[0,1]$. Since f is uniformly continuous on $[0,1]$, there exists $\delta > 0$ such that

$$|f(x) - f(x')| < \frac{1}{2}\varepsilon \qquad \text{whenever } |x - x'| < \delta.$$

Choose a positive integer n such that $(1/n)\varepsilon < \delta$. Let

$$0 = x_0 < x_1 < \ldots < x_n = 1$$

be the partition of $[0, 1]$ that divides the interval into n equal parts. Consider the rectangle with vertices

$$(x_{k-1}, f(x_{k-1}) - \frac{1}{2}\varepsilon), \ (x_k, f(x_k) - \frac{1}{2}\varepsilon), \ (x_k, f(x_k) + \frac{1}{2}\varepsilon), \ (x_{k-1}, f(x_{k-1}) + \frac{1}{2}\varepsilon).$$

Join the points $(x_{k-1}, f(x_{k-1}))$, $(x_k, f(x_k))$ by a sawtooth function that remains inside the rectangle and whose line segments have slopes greater than n in absolute value. Carrying out this process for each subinterval (x_{k-1}, x_k), $k = 1, 2, \ldots, n$, we obtain a function g in $C[0,1]$ such that $|f(x) - g(x)| < \varepsilon$ for all $x \in [0,1]$. Moreover, $g \in E_n^c$. This completes the proof. $\qquad\square$

The above proof of the existence of a continuous function that is nowhere differentiable is nonconstructive in the sense that it does not provide a concrete example of such a function. The first known example, namely, $\sum_{n=0}^{\infty} \frac{\cos 3^n x}{2^n}$, is due to Weierstrass. The following example, due to van der Waerden, of a continuous nowhere differentiable function is the easiest to work with. Although the proof of its continuity uses a result to be proved in the next chapter, we prefer to present it here because of its immediate relevance to the foregoing discussion.

Example 2.4.8. Let $h : \mathbf{R} \to \mathbf{R}$ be defined as

$$h(x) = \begin{cases} x & \text{if } 0 \leq x \leq \frac{1}{2}, \\ 1 - x & \text{if } \frac{1}{2} \leq x \leq 1 \end{cases}$$

extended to all of $\mathbf{R}$ by requiring that

$$h(x + 1) = h(x);$$

in other words, h is periodic of period 1. (See Figure 2.7.) It is easily verified that h is continuous on $\mathbf{R}$. Define

$$f(x) = \sum_{n=0}^{\infty} \frac{h(10^n x)}{10^n}.$$

Since this series is dominated by the convergent series $(1/2) \sum_{n=0}^{\infty} 1/10^n$, it follows by the Weierstrass M-test (see Theorem 3.6.12) that the series converges uniformly. Its sum is therefore a continuous function, as argued in Chapter 1. We shall show that this function is nowhere differentiable. As the function is periodic, we may restrict ourselves to the case when $0 \le x < 1$. Let $a \in [0,1)$ have the decimal representation $a = .a_1 a_2 \ldots a_n \ldots$.

For $n \in \mathbf{N}$, let

$$x_n = .a_1 a_2 \ldots a_{n-1} b_n a_{n+1} \ldots,$$

where $b_n = a_n + 1$ if $a_n \ne 4$ or 9, while $b_n = a_n - 1$ if $a_n = 4$ or 9. Thus $x_n - a = \pm 10^{-n}$ and so $\lim_{n \to \infty} x_n = a$. To complete the proof, it will be suficient to show that

$$\lim_{n \to \infty} \frac{f(x_n) - f(a)}{x_n - a}$$

does not exist.

Now,

$$h(10^m x_n) - h(10^m a) = \begin{cases} 0 & \text{if } m \ge n, \\ \pm 10^{m-n} & \text{if } m < n. \end{cases}$$

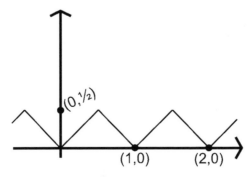

FIGURE 2.7

Thus,

$$\frac{f(x_n) - f(a)}{x_n - a} = \sum_{m=0}^{\infty} \frac{h(10^m x_n) - h(10^m a)}{10^m(x_n - a)}$$

$$= \sum_{m=0}^{n-1} \pm \frac{10^{m-n}}{10^m(\pm 10^{-n})}$$

$$= \sum_{m=0}^{n-1} \pm 1. \tag{2.3}$$

Thus, for each n, the difference quotient on the left of (2.3) is the sum of n terms, each of which is either 1 or -1, so that the sum is an odd integer when n is odd and an even integer when n is even. It follows that

$$\lim_{n \to \infty} \frac{f(x_n) - f(a)}{x_n - a}$$

does not exist.

Finally, we show that the set of points of discontinuity of an arbitrary real-valued function defined on **R** is of a special kind. We begin with the following definition.

Definition 2.4.9. A subset S of **R** is said to be of type F_σ if $S = \bigcup_{n=1}^{\infty} S_n$, where each S_n is a closed subset of **R**.

Examples 2.4.10. (i) If F is a closed subset of **R**, then F is of type F_σ, since $F = \bigcup_{n=1}^{\infty} F_n$, where $F_1 = F$ and $F_2 = F_3 = \ldots = \emptyset$.

(ii) The set **Q** of rationals in **R** is of type F_σ. Indeed, if $x_1, x_2, \ldots$ is an enumeration of **Q**, then each set $\{x_i\}$ is closed and we have **Q** $= \bigcup_{i=1}^{\infty} \{x_i\}$.

(iii) Each open interval (a,b) is of type F_σ. This is because, if m is a positive integer such that $2/m < b - a$, then

$$(a, b) = \bigcup_{n=m}^{\infty} \left[a + \frac{1}{n}, b - \frac{1}{n}\right].$$

The statement now follows, as $[a + 1/n, \ b - 1/n]$ is closed for each n.

Let f be a real-valued function defined on **R**. We shall show that the set of points of **R** at which f is discontinuous is always of type F_σ.

Definition 2.4.11. Let $f: \mathbf{R} \to \mathbf{R}$. If I is any bounded open interval of **R**, we define $\omega(f, I)$, called the **oscillation over** I of the function f, as

$$\omega(f, I) = \sup_{x \in I} f(x) - \inf_{x \in I} f(x).$$

If $a \in \mathbf{R}$ is arbitrary, the **oscillation at** a of the function f, $\omega(f, a)$, is defined as

$$\omega(f, a) = \inf \omega(f, I),$$

where the inf is taken over all bounded open intervals containing a.

Clearly, $\omega(f,I)$ and $\omega(f,a)$ are both nonnegative.

The following criterion of continuity is well known from real analysis:

Proposition 2.4.12. Let f be a real-valued function defined on **R**. Then $\omega(f,a)=0$ if and only if f is continuous at a.

Proof. Suppose f is continuous at a. Let $\varepsilon > 0$ be arbitrary. There exists a $\delta > 0$ such that $|x-a|<\delta \Rightarrow |f(x)-f(a)|<\varepsilon/2$. If $I=(a-\delta,a+\delta)$, then for $x\in I$, $f(a)-\varepsilon/2<f(x)<f(a)+\varepsilon/2$. So, $\omega(f,I)=\sup_{x\in I}f(x)-\inf_{x\in I}f(x)<\varepsilon$ and consequently,

$$\omega(f,a)=\inf\omega(f,I)<\varepsilon.$$

Since $\varepsilon > 0$ is arbitrary, and $\omega(f,a)\geq 0$, it follows that $\omega(f,a)=0$.

On the other hand, suppose that $\omega(f,a)=0$. If f is not continuous at a, there exists $\varepsilon > 0$ such that in every bounded open interval containing a, there exists an x for which $|f(x)-f(a)|\geq\varepsilon$, that is,

$$f(x)\geq f(a)+\varepsilon \text{ or } f(x)\leq f(a)-\varepsilon.$$

So, for every bounded open interval I containing a,

$$\omega(f,I)=\sup_{x\in I}f(x)-\inf_{x\in I}f(x)\geq 2\varepsilon,$$

which, in turn, implies

$$\omega(f,a)\geq 2\varepsilon,$$

and this contradicts the supposition that $\omega(f,a)=0$. $\qquad\square$

Theorem 2.4.13. Let $f:\mathbf{R}\to\mathbf{R}$ and $S_n=\{x\in\mathbf{R}: \omega(f,x)\geq 1/n\}$. Denote by S the set of points of **R** at which f is not continuous. Then, for each n the set S_n is closed. Moreover,

$$S=\bigcup_{n=1}^{\infty}S_n.$$

Thus, the points of **R** at which f is not continuous form a set of type F_σ.

Proof. Let x be a limit point of S_n. We need to show that $x\in S_n$. Let I be a bounded open interval containing x. Then I contains a point $y\in S_n$. But then $\omega(f,I)\geq\omega(f,y)\geq 1/n$. As I is any bounded open interval containing x, we have $\omega(f,x)\geq 1/n$, that is, $x\in S_n$.

It remains to show that $S=\bigcup_{n=1}^{\infty}S_n$. Let $x\in S$. Then by Proposition 2.4.12, $\omega(f,x)>0$. So, there exists a positive integer n such that $\omega(f,x)\geq 1/n$. Hence $x\in S_n$. On the other hand, if $x\in S_n$, then clearly, $x\in S$. $\qquad\square$

The irrational numbers in **R** form a set A of category II (see Corollary 2.4.4). We shall show that A is not of type F_σ. Suppose that, on the contrary,

$$A = \bigcup_{i=1}^\infty F_i,$$

where each F_i is closed. Since each closed set F_i contains only irrational numbers, it cannot contain an interval. Thus, F_i is nowhere dense and so A is of category I. This contradicts the fact that A is of category II. We have thus proved the following theorem:

Theorem 2.4.14. There is no real-valued function defined on **R** that is continuous at each rational point and is discontinuous at each irrational point.

We give an example of a function that is continuous at every irrational number and discontinuous at every rational number.

Example 2.4.15. The function f defined as

$$f(x) = \begin{cases} 1/n & \text{where } n \text{ is least in } \mathbf{N} \text{ such that } x = m/n, \\ 0 & \text{if } x \text{ is irrational} \end{cases}$$

has the required property, as we shall argue.

Let $c \in \mathbf{R}$ be rational, so that $f(c) = 1/n$, where n is the least integer in **N** such that $c = m/n$ and $m \in \mathbf{Z}$. Choose $\varepsilon = 1/2n$. For any $\delta > 0$, the interval $(c - \delta, c + \delta)$ contains an irrational number x, so that $|f(x) - f(c)| = |0 - 1/n| = 1/n > \varepsilon$. Therefore, when $\varepsilon = 1/2n$, no positive number δ can have the property that $|x - c| < \delta \Rightarrow |f(x) - f(c)| < \varepsilon$.

On the other hand, if $c \in \mathbf{R}$ is an irrational number and ε any positive number whatsoever, there exists (by the Archimedean property of **R**) some $n_0 \in \mathbf{N}$ such that $1/n_0 < \varepsilon$. Now consider the interval $(c - 1/2n_0^2, c + 1/2n_0^2)$. For any p and q in this interval, $|p - q| < 1/n_0^2$. It follows that this interval can contain at most one rational number of the form $r = m/n$ with $n \leq n_0$ because, when $m_1 n_2 - m_2 n_1 \neq 0$, we have

$$n_1 \leq n_0, \ n_2 \leq n_0 \Rightarrow \left| \frac{m_1}{n_1} - \frac{m_2}{n_2} \right| = \frac{|m_1 n_2 - m_2 n_1|}{n_1 n_2} \geq \frac{1}{n_1 n_2} \geq \frac{1}{n_0^2}.$$

If there is any such r in the interval $(c - 1/2n_0^2, c + 1/2n_0^2)$, let $\delta = |c - r|$. If there is no such r, let $\delta = 1/2n_0^2$. In both cases, no number x in the interval $(c - \delta, c + \delta)$ can be of the form $x = m/n$ with $n \leq n_0$. Thus, every number in this interval is either irrational or is a rational number of the form m/n with $n > n_0$. Therefore,

$$|x - c| < \delta \Rightarrow \text{ either } f(x) = 0 \text{ or } f(x) = \frac{1}{n} \text{ with } n > n_0$$

$$\Rightarrow \text{ either } |f(x) - f(c)| = 0 \text{ or } |f(x) - f(c)| = \frac{1}{n} \text{ with } n > n_0$$

$$\Rightarrow |f(x) - f(c)| < \frac{1}{n_0} < \varepsilon.$$

2.5. Exercises

1. Let $S(x, \delta)$ be a ball with centre x and radius δ in a metric space (X, d). Prove that if $0 < \varepsilon < \delta - d(x, z)$, then $S(z, \varepsilon) \subseteq S(x, \delta)$.
   Hint: If $y \in S(z, \varepsilon)$, then $d(x, y) \leq d(x, z) + d(z, y) < d(x, z) + \varepsilon < \delta$.

2. Prove that $\overline{S(x, \varepsilon)} \subseteq \{y : d(x, y) \leq \varepsilon\}$ and give an example of a metric space containing a ball for which the inclusion is proper.
   Hint: $\{y : d(x, y) \leq \varepsilon\}$ is closed and contains $S(x, \varepsilon)$; use Corollary 2.1.27(iii). Let (X, d) be discrete, X contain more than one point, and let $\varepsilon = 1$. Then $\overline{S(x, 1)} = \overline{\{x\}} = \{x\}$, whereas $\{y : d(x, y) \leq 1\} = X$.

3. Show that for any two points x and y of a metric space there exist disjoint open balls such that one is centred at x and the other at y.
   Hint: Let $r = d(x, y)$. Then $r > 0$, and $S(x, r/2)$ and $S(y, r/2)$ are the desired balls.

4. Let (X, d) be a metric space and let $S(x, r_1)$ and $S(y, r_2)$ be two intersecting balls containing a common point z. Show that there exists an $r_3 > 0$ such that $S(z, r_3) \subseteq S(x, r_1) \cap S(y, r_2)$.
   Hint: Since $z \in S(x, r_1)$ and $S(x, r_1)$ is open, there exists an open ball $S(z, r_1')$ centred at z and with radius r_1' such that $S(z, r_1') \subseteq S(x, r_1)$. Similarly, there exists an open ball $S(z, r_2')$ centred at z and with radius r_2' such that $S(z, r_2') \subseteq S(y, r_2)$. Let $r_3 = \min\{r_1', r_2'\}$. Then $S(z, r_3) \subseteq S(x, r_1) \cap S(y, r_2)$ since $S(z, r_3) \subseteq S(z, r_1')$ as well as $S(z, r_3) \subseteq S(z, r_2')$.

5. Let $S(x, r)$ be an open ball in a metric space (X, d) and A be closed subset of X such that $d(A) \leq r$ and $A \cap S(x, r) \neq \varnothing$. Show that $A \subseteq S(x, 2r)$.
   Hint: Let $y \in A \cap S(x, r)$. For $z \in A$,

$$d(z, x) \leq d(z, y) + d(y, x) < r + r = 2r.$$

6. Let $A \subseteq [0, 1]$ and $F = \{f \in C[0, 1] : f(t) = 0 \text{ for every } t \in A\}$. Show that F is a closed subset of $C[0, 1]$ equipped with the uniform metric.
   Hint: Let $t \in A$ be fixed. The set $\{f \in C[0, 1] : f(t) = 0\}$ can be shown to be a closed subset of $C[0, 1]$ as follows: If f is a limit point of the set, then there exists a sequence $\{f_n\}_{n \geq 1}$ in the set such that $\lim_{n \to \infty} f_n = f$ uniformly. Since uniform convergence implies pointwise convergence, $\lim_{n \to \infty} f_n(t) = f(t)$. Since F is an intersection of such sets, Theorem 2.1.34(i) applies.

7. Let $C[0, 1]$ be equipped with the metric defined by

$$d(f, g) = \int_0^1 |f(t) - g(t)| dt, \quad f, g \in C[0, 1].$$

With notations as in Exercise 6, show that F is not necessarily closed.
Hint: Let $A = \{0\}$. Consider the sequence

$$f_n(t) = \begin{cases} nt & 0 \le t \le \dfrac{1}{n} \\ 1 & t > \dfrac{1}{n}. \end{cases}$$

If $f \equiv 1$, then

$$d(f_n, f) = \int_0^{1/n} (1 - nt) dt = \frac{1}{2n} \to 0 \text{ as } n \to \infty.$$

The functions of the sequence $\{f_n\}_{n \ge 1}$ are in F, but $f \notin F$.

8. Let X denote the space of all bounded sequences with

$$d(x, y) = \sup_i |x_i - y_i|,$$

where $x = \{x_i\}_{i \ge 1}$ and $y = \{y_i\}_{i \ge 1}$ are in X. Show that the subset Y of convergent sequences is closed in X.
Hint: Let $z \in X$ be a limit point of Y. Then there exists a sequence $\{y^{(n)}\}_{n \ge 1}$ in Y satisfying the following condition: For every $\varepsilon > 0$ there exists $n_0(\varepsilon)$ such that $n \ge n_0(\varepsilon)$ implies

$$\sup_k |y_k^{(n)} - z_k| < \frac{\varepsilon}{3}.$$

The sequence $\{y_j^{(n_0)}\}_{j \ge 1}$, being convergent, is Cauchy. So there exists l such that $i, j \ge l$ implies

$$|y_i^{(n_0)} - y_j^{(n_0)}| < \frac{\varepsilon}{3}.$$

Now,

$$|z_i - z_j| \le |z_i - y_i^{(n_0)}| + |y_i^{(n_0)} - y_j^{(n_0)}| + |y_j^{(n_0)} - z_j| < \varepsilon \qquad \text{for } i, j \ge l.$$

The bounded sequence $\{z_i\}_{i \ge 1}$ is Cauchy and is, therefore, convergent and, hence, belongs to Y.

9. Let A be a subset of a metric space (X, d). Show that

$$\bar{A} = \bigcap \{F \subseteq X : F \text{ is closed and } F \supseteq A\}$$

Hint: $\bar{A}$ is a closed set and $\bar{A} \supseteq A$. Therefore, on the one hand, $\bar{A}$ is one of the sets in the intersection, while on the other hand, by Corollary 2.1.27(iii), it is a subset of every set in the intersection.

10. Let $X = \mathbf{C}$ with the usual metric and $A = \{(x, y): y = \sin(1/x),\ 0 < x \le 1\}$. Show that

$$\bar{A} = A \cup \{(0, y) : -1 \le y \le 1\}$$

Hint: Each open ball centred at $(0, y)$, $-1 \le y \le 1$, has nonempty intersection with A. It may be seen that every point outside $A \cup \{(0, y) : -1 \le y \le 1\}$ is the centre of a ball having an empty intersection with A.

11. Let $X = \{(x_1, x_2) \in \mathbf{R}^2 : |x_1| < 2$ and $|x_2| < 1\}$ be equipped with the metric induced from $\mathbf{R}^2$. For any $x = (x_1, x_2) \in X$ and $r \ge 2\sqrt{5}$, show that

$$S_X(x, r) = X.$$

Hint: $S_X(x, r) = S(x, r) \cap X$, where

$$S(x, r) = \{(y_1, y_2) \in \mathbf{R}^2 : \sqrt{[(y_1 - x_1)^2 + (y_2 - x_2)^2]} < r\}.$$

It is enough to show that $X \subseteq S(x, r)$. For $x \in X$ as well as $y \in X$, we have $d(y, x) \le d(X) = 2\sqrt{5} \le r$.

12. Let $A = \{z \in \mathbf{C}: |z + 1|^2 \le 1\}$ and $B = \{z \in \mathbf{C}: |z - 1|^2 < 1\}$, and let $A \cup B$ be equipped with the metric induced from $\mathbf{C}$. Identify $\mathrm{cl}_{A \cup B}(B)$.
    Hint: $\mathrm{cl}_{A \cup B}(B) = (A \cup B) \cap \mathrm{cl}_{\mathbf{C}}(B) = (A \cup B) \cap \{z \in \mathbf{C}: |z - 1|^2 \le 1\} = B \cup \{0\}$.

13. Let (X, d) be a metric space and A be a subset of X. Show that (i) $X \backslash \bar{A} = (X \backslash A)^\circ$;
    (ii) $X \backslash A^\circ = \overline{(X \backslash A)}$.
    Hint: (i) $x \in (X \backslash A)^\circ$ iff there exists a ball $S(x, \varepsilon))$ centred at x with suitable radius ε such that $S(x, \varepsilon) \subseteq X \backslash A$ iff $S(x, \varepsilon) \cap A = \varnothing$ iff $x \notin \bar{A}$.
    (ii) Replace A by $X \backslash A$ in (i) and take complements.

14. Give an example of a subset Y of a metric space (X, d) for which $(\bar{Y})^\circ \ne \overline{(Y^\circ)}$.
    Hint: Let $(\mathbf{R}, d)$ be the usual real line and Y denote the set of rationals in $\mathbf{R}$. Then

$$(\bar{Y})^\circ = (\mathbf{R})^\circ = \mathbf{R} \text{ whereas } \overline{(Y^\circ)} = \bar{\varnothing} = \varnothing.$$

15. For a subset Y of a metric space (X, d), $(X \backslash Y)^\circ$ is called the **exterior** of Y and is denoted by **ext(Y)**. The **boundary** of Y is defined to be $\bar{Y} \cap \overline{(X \backslash Y)}$ and is denoted by $\partial(Y)$. Show that
    (i) $\partial(Y) = \partial(X \backslash Y)$;
    (ii) $\bar{Y} = Y^\circ \cup \partial(Y)$;
    (iii) $Y^\circ \cap \partial(Y) = \varnothing$;
    (iv) $(X \backslash Y)^\circ \cap \partial(Y) = \varnothing$;
    (v) $X = Y^\circ \cup \partial(Y) \cup (X \backslash Y)^\circ$;
    (vi) $Y \backslash \partial(Y) = Y^\circ$.

Hint: (i) $\partial(X\backslash Y) = \overline{(X\backslash Y)} \cap \overline{(X\backslash(X\backslash Y))} = \overline{(X\backslash Y)} \cap \bar{Y} = \partial(Y)$.

(ii) $\partial(Y) \subseteq \bar{Y}$, $Y^\circ \subseteq Y \subseteq \bar{Y}$. So $Y^\circ \cup \partial(Y) \subseteq \bar{Y}$. Let $y \in \bar{Y}$. If $y \in Y^\circ$, then $y \in Y^\circ \cup \partial(Y)$. If $y \notin Y^\circ$, then for all $\varepsilon > 0$, $S(y, \varepsilon) \not\subseteq Y$, i.e., $S(y, \varepsilon) \cap (X\backslash Y) \neq \emptyset$. Hence, $y \in \overline{(X\backslash Y)}$. So $y \in \bar{Y} \cap \overline{(X\backslash Y)} = \partial(Y)$. Consequently, $y \in Y^\circ \cup \partial(Y)$.

(iii) $Y^\circ \cap \partial(Y) = Y^\circ \cap (\bar{Y} \cap \overline{(X\backslash Y)}) = (Y^\circ \cap \bar{Y}) \cap (Y^\circ \cap \overline{(X\backslash Y)}) = Y^\circ \cap (Y^\circ \cap \overline{(X\backslash Y^\circ)}) = Y^\circ \cap \emptyset = \emptyset$, using Exercise 13(ii).

(iv) Replace Y by $X\backslash Y$ in (iii) and use (i).

(v) $Y^\circ \cup \partial(Y) \cup (X\backslash Y)^\circ = \bar{Y} \cup (X\backslash Y)^\circ = \bar{Y} \cup (X\backslash\bar{Y}) = X$, using (ii) and Exercise 13(i).

(vi) $Y\backslash\partial(Y) = Y \cap (X\backslash\partial(Y)) = Y \cap (Y^\circ \cup (X\backslash Y)^\circ) = (Y \cap Y^\circ) \cup (Y \cap (X\backslash Y)^\circ) = Y^\circ$, using (iii), (iv) and (v) above.

16. Show that the Cantor set P is nowhere dense.
Hint: No segment of the form

$$\left(\frac{3k+1}{3^m}, \frac{3k+2}{3^m}\right), \tag{2.3}$$

where k and m are positive integers, has a point in common with P. Since every interval (α, β) contains an interval of the form (2.3) whenever

$$3^{-m} < \frac{\beta - \alpha}{6},$$

it follows that P contains no interval.

17. Consider the rationals $\mathbf{Q}$ as a subset of the complete metric space $\mathbf{R}$. Prove that $\mathbf{Q}$ cannot be expressed as the intersection of a countable collection of open sets.
Hint: Suppose $\mathbf{Q} = G_1 \cap G_2 \cap \ldots$, where each G_i is open in $\mathbf{R}$. Then the set of irrationals is $\bigcup_{i=1}^\infty G_i^c$, where each G_i^c is closed. Since each G_i^c contains only irrationals, no G_i^c contains a nonempty interval. Thus, G_i^c is closed and nowhere dense for each $i = 1, 2 \ldots$.

18. Consider a real valued function f on $[0,1]$. If f has an nth derivative that is identically zero, it easily follows by using the mean value theorem that f coincides on $[0,1]$ with a polynomial of degree at most $n - 1$. The following generalisation is valid: If f has derivatives of all orders on $[0,1]$, and if at each x there is an integer $n(x)$ such that $f^{(n(x))}(x) = 0$, then f coincides on $[0,1]$ with some polynomial.
Hint: See [3; p. 58].

19. Let A be either an open subset or a closed subset of (X, d). Then $(\partial(A))^\circ = \emptyset$, so that $\partial(A)$ is nowhere dense. Is this true if we drop the requirement that either A is open or A is closed?
Hint: One need prove only the case of a closed set, because A is open iff A^c is open and $\partial(A) = \partial(A^c)$ by Exercise 15(i) above. If A is closed, then $\partial(A) = A \cap \overline{(X\backslash A)}$. Let G be an open subset of (X, d) such that $G \subseteq \partial(A)$.

Then $G \subseteq A \cap \overline{(X \backslash A)}$. This implies not only that $G \subseteq A$, so that $G \subseteq A^\circ$ (because A° is the largest open set contained in A), but also that (see Exercise 13(ii)) $G \subseteq \overline{(X \backslash A)} = X \backslash A^\circ$. Hence, $G = \emptyset$. When A is neither open nor closed, $\partial(A)$ need not be nowhere dense: Consider the set A of rational numbers in the metric space $(\mathbf{R}, d)$. For this set, $\partial(A) = \bar{A} \cap \overline{(X \backslash A)} = \mathbf{R} \cap \mathbf{R} = \mathbf{R}$, and hence $(\partial(A))^\circ = \mathbf{R} \neq \emptyset$.

20. Let G_1, $G_2, \ldots$ be a sequence of open subsets of $\mathbf{R}$, each of which is dense. Prove that $\bigcap_{n=1}^\infty G_n$ is dense.
Hint: Suppose not. Then there exists $x \in \mathbf{R}$ and an open interval I_x containing x such that $I_x \cap \bigcap_{n=1}^\infty G_n = \emptyset$. Thus, $x \in I_x \subseteq \bigcup_{n=1}^\infty G_n^c$. But each G_n^c is nowhere dense, and, hence, $\bigcup_{n=1}^\infty G_n^c$ is of category I (see Definition 2.4.1). But I_x is of category II by Corollary 2.4.4. Since a subset of a set of category I must be of category I (see Remark (viii) just before Theorem 2.4.3), we arrive at a contradiction. (The reader may note that the argument is valid in any complete metric space.)

21. Let E be a closed subset of a metric space (X, d). Prove that E is nowhere dense if and only if for every open subset G there is a ball contained in $G \backslash E$.
Hint: Suppose E is nowhere dense. Then $G \backslash E \neq \emptyset$ because, otherwise, $G \subseteq E$, and this contradicts the supposition that E is nowhere dense. Let $x \in G \backslash E = G \cap E^c$. Since G and E^c are both open, there exists an $r > 0$ such that $S(x, r) \subseteq G \backslash E$. For the converse, the hypothesis implies that every open set has nonempty intersection with E^c. It follows that $E^c = (\bar{E})^c$ is dense in X, so that E is nowhere dense.

22. Let $(\mathbf{R}, d_1)$ be the metric space where

$$d_1(x, y) = \begin{cases} |x| + |x - y| + |y| & \text{if } x \neq y, \\ 0 & \text{if } x = y. \end{cases}$$

Show that the ε-ball about 0 with the metric d_1 is the same as the $(\varepsilon/2)$-ball about 0 with the usual metric. Also, if $0 < \varepsilon < |y|$, then the ε-ball about any nonzero element y with the metric d_1 consists of y alone. Describe a base for the open sets of $(\mathbf{R}, d_1)$.

3 **Continuity**

One of the main aims in considering metric spaces is the study of continuous functions in a context more general than that of classical analysis. This approach does not distinguish between functions of one variable or several variables.

Early mathematicians considered defining a real-valued continuous function with an interval domain as one that maps every subinterval in its domain onto an interval or a point (intermediate value property). However, it was soon discovered that this definition was flawed. In fact, the function $f(x) = \sin(1/x), x \neq 0$, and $f(0) = 0$, possesses the intermediate value property but fails to be continuous at zero.

We give below the definition of a continuous function, which was eventually adopted in the form suitable to the present context, and provide several other characterisations. The problem of extension of a continuous or uniformly continuous function defined on a subspace to the whole space is discussed. Also discussed in the chapter is the uniform convergence of sequences and series of continuous functions. The contraction mapping principle and its applications to a system of algebraic equations and Picard's theorem on first order differential equations are dealt with in the final section of this chapter.

3.1. Continuous Mappings

For a real-valued function f with domain $A \subseteq \mathbf{R}$, a rough and rather inaccurate description of continuity at a point $a \in A$ is the statement "$f(x)$ is close to $f(a)$ when x is close to a". The measure of "closeness" of two numbers, or distance between them, is the absolute value of the difference of the numbers. In terms of the standard metric d on $\mathbf{R}$, continuity involves a relationship between $d(x, a)$ and $d(f(x), f(a))$. This observation makes it possible to extend the concept of continuity to functions with domain and range in metric spaces.

Definition 3.1.1. Let (X, d_X) and (Y, d_Y) be metric spaces and $A \subseteq X$. A function $f : A \to Y$ is said to be **continuous at** $a \in A$, if for every $\varepsilon > 0$, there exists some $\delta > 0$ such that

$$d_Y(f(x), f(a)) < \varepsilon \text{ whenever} x \in A \text{ and } d_X(x, a) < \delta.$$

If f is continuous at every point of A, then it is said to be **continuous on** A.

Remarks 3.1.2. (i) If one positive number δ satisfies this condition, then every positive number $\delta_1 < \delta$ also satisfies it. This is obvious because whenever $x \in A$ and $d_X(x, a) < \delta_1$, it is also true that $x \in A$ and $d_X(x, a) < \delta$. Therefore, such a number δ is far from being unique.

(ii) In the definition of continuity, we have placed no restriction whatever on the nature of the domain A of the function. It may happen that a is an isolated point of A, i.e., there is a neighbourhood of a that contains no point of A other than a. In this case, the function f is continuous at a irrespective of how it is defined at other points of the set A. However, if a is a limit point of A and $\{x_n\}$ is a sequence of points of A such that $x_n \to a$, it follows from the continuity of f at a that $f(x_n) \to f(a)$. In fact, we have the following theorem:

Theorem 3.1.3. Let (X, d_X) and (Y, d_Y) be metric spaces and $A \subseteq X$. A function $f \colon A \to Y$ is continuous at $a \in A$ if and only if whenever a sequence $\{x_n\}$ in A converges to a, the sequence $\{f(x_n)\}$ converges to $f(a)$.

Proof. First suppose the function $f \colon A \to Y$ is continuous at $a \in A$ and let $\{x_n\}$ be a sequence in A converging to a. We shall show that $\{f(x_n)\}$ converges to $f(a)$. Let ε be any positive real number. By continuity of f at a, there exists some $\delta > 0$ such that $x \in A$ and $d_X(x, a) < \delta \Rightarrow d_Y(f(x), f(a)) < \varepsilon$. Since $\lim_{n \to \infty} x_n = a$, there exists some n_0 such that $n \geq n_0 \Rightarrow d_X(x_n, a) < \delta$. Therefore $n \geq n_0 \Rightarrow d_Y(f(x_n), f(a)) < \varepsilon$. Thus, $\lim_{n \to \infty} f(x_n) = f(a)$.

Now suppose that every sequence $\{x_n\}$ in A converging to a has the property that $\lim_{n \to \infty} f(x_n) = f(a)$. We shall show that f is continuous at a. Suppose, if possible, that f is not continuous at a. There must exist $\varepsilon > 0$ for which no positive δ can satisfy the requirement that $x \in A$ and $d_X(x, a) < \delta \Rightarrow d_Y(f(x), f(a)) < \varepsilon$. This means that for every $\delta > 0$, there exists $x \in A$ such that $d_X(x, a) < \delta$ but $d_Y(f(x), f(a)) \geq \varepsilon$. For every $n \in \mathbf{N}$, the number $1/n$ is positive and therefore there exists $x_n \in A$ such that $d_X(x_n, a) < 1/n$ but $d_Y(f(x_n), f(a)) \geq \varepsilon$. The sequence $\{x_n\}$ then converges to a but the sequence $\{f(x_n)\}$ does not converge to $f(a)$. This contradicts the assumption that every sequence $\{x_n\}$ in A converging to a has the property that $\lim_{n \to \infty} f(x_n) = f(a)$. Therefore, the supposition that f is not continuous at a must be false. $\square$

Definition 3.1.4. Let (X, d_X) and (Y, d_Y) be metric spaces and $A \subseteq X$. Let $f \colon A \to Y$ and a be a limit point of A. We write $\boldsymbol{\lim_{x \to a} f(x) = b}$, where $b \in Y$, if for every $\varepsilon > 0$ there exists $\delta > 0$ such that

$$d_Y(f(x), b) < \varepsilon \text{ whenever } x \in A \text{ and } 0 < d_X(x, a) < \delta.$$

Remark. In the definition of limit, the point a in X need only be a limit point of A and does not have to belong to A. In addition, if $a \in A$, we may have $\lim_{x \to a} f(x) \neq f(a)$.

Proposition 3.1.5. Let (X, d_X), (Y, d_Y), A, f and a be as in the definition above. Then

$$\lim_{x \to a} f(x) = b$$

if and only if

$$\lim_{n \to \infty} f(x_n) = b$$

for every sequence $\{x_n\}$ in A such that $x_n \neq a$ and $\lim_{n \to \infty} x_n = a$.

Proof. The argument is similar to that of Theorem 3.1.3 and is therefore not included. $\qquad\qquad\square$

Lemma 3.1.6. Let $f \colon X \to Y$ be an arbitrary function and let $A \subseteq X$ and $B \subseteq Y$. Then $f(A) \subseteq B$ if and only if $A \subseteq f^{-1}(B)$.

The next characterisation of continuity follows immediately from Definitions 3.1.1 and 3.1.4.

Proposition 3.1.7. Let (X, d_X) and (Y, d_Y) be metric spaces and $A \subseteq X$. Let $f \colon A \to Y$ and a be a limit point of A. Then f is continuous at a if and only if $\lim_{x \to a} f(x) = f(a)$. If a is an isolated point of A, the function f is continuous at a irrespective of how it is defined at other points of A.

The following reformulation of the definition of continuity at a point a in terms of neighbourhoods is useful.

Proposition 3.1.8. A mapping f of a metric space (X, d_X) into a metric space (Y, d_Y) is continuous at a point $a \in X$ if and only if for every $\varepsilon > 0$, there exists $\delta > 0$ such that

$$S(a, \delta) \subseteq f^{-1}(S(f(a), \varepsilon)),$$

where $S(x, r)$ denotes the open ball of radius r with centre x.

Proof. The mapping $f \colon X \to Y$ is continuous at $a \in X$ if and only if for every $\varepsilon > 0$ there exists $\delta > 0$ such that

$$d_Y(f(x), f(a)) < \varepsilon \qquad \text{for all } x \text{ satisfying } d_X(x, a) < \delta,$$

i.e.,

$$x \in S(a, \delta) \Rightarrow f(x) \in S(f(a), \varepsilon)$$

or

$$f(S(a,\delta)) \subseteq S(f(a),\varepsilon).$$

This is equivalent to the condition

$$S(a,\delta) \subseteq f^{-1}(S(f(a),\varepsilon)).$$

(See Lemma 3.1.6.)   □

Theorem 3.1.9. A mapping $f: X \to Y$ is continuous on X if and only if $f^{-1}(G)$ is open in X for all open subsets G of Y.

Proof. Suppose f is continuous on X and let G be an open subset of Y. We have to show $f^{-1}(G)$ is open in X. Since $\varnothing$ and X are open, we may suppose that $f^{-1}(G) \neq \varnothing$ and $f^{-1}(G) \neq X$. Let $x \in f^{-1}(G)$. Then $f(x) \in G$. Since G is open, there exists $\varepsilon > 0$ such that $S(f(x),\varepsilon) \subseteq G$. Since f is continuous at x, by Proposition 3.1.8, for this ε there exists $\delta > 0$ such that

$$S(x,\delta) \subseteq f^{-1}(S(f(x),\varepsilon)) \subseteq f^{-1}(G).$$

Thus, every point x of $f^{-1}(G)$ is an interior point, and so $f^{-1}(G)$ is open in X.

Suppose, conversely, that $f^{-1}(G)$ is open in X for all open subsets G of Y. Let $x \in X$. For each $\varepsilon > 0$, the set $S(f(x),\varepsilon)$ is open (see Theorem 2.1.5) and so $f^{-1}(S(f(x),\varepsilon))$ is open in X. Since

$$x \in f^{-1}(S(f(x),\varepsilon)),$$

it follows that there exists $\delta > 0$ such that

$$S(x,\delta) \subseteq f^{-1}(S(f(x),\varepsilon)).$$

By Proposition 3.1.8, it follows that f is continuous at x.   □

Theorem 3.1.10. A mapping $f: X \to Y$ is continuous on X if and only if $f^{-1}(F)$ is closed in X for all closed subsets F of Y.

Proof. Let F be a closed subset of Y. Then $Y\backslash F$ is open in Y so that $f^{-1}(Y\backslash F)$ is open in X by Theorem 3.1.9. But

$$f^{-1}(Y\backslash F) = X\backslash f^{-1}(F).$$

So $f^{-1}(F)$ is closed in X.

Suppose, conversely, that $f^{-1}(F)$ is closed in X for all closed subsets F of Y. Then, by Theorem 2.1.31, $X\backslash f^{-1}(F)$ is open in X and so

$$f^{-1}(Y\backslash F) = X\backslash f^{-1}(F)$$

is open in X. Since every open subset of Y is a set of the type $Y\backslash F$, where F is a suitable closed set, it follows by using Theorem 3.1.9 that f is continuous.   □

The characterisation of continuity in terms of open sets (Theorem 3.1.9) leads to an elegant and brief proof of the fact that a composition of continuous maps is continuous.

Theorem 3.1.11. Let (X, d_X), (Y, d_Y) and (Z, d_Z) be metric spaces and let $f : X \to Y$ and $g : Y \to Z$ be continuous. Then the composition $g \circ f$ is a continuous map of X into Z.

Proof. Let G be an open subset of Z. By Theorem 3.1.9, $g^{-1}(G)$ is an open subset of Y, and another application of the same theorem shows that $f^{-1}(g^{-1}(G))$ is an open subset of X. Since $(g \circ f)^{-1}(G) = f^{-1}(g^{-1}(G))$, it follows from the same theorem again that $g \circ f$ is continuous.   □

Theorem 3.1.12. Let (X, d_X) and (Y, d_Y) be metric spaces and let $f : X \to Y$. Then the following statements are equivalent:

(i) f is continuous on X;
(ii) $\overline{f^{-1}(B)} \subseteq f^{-1}(\overline{B})$ for all subsets B of Y;
(iii) $f(\overline{A}) \subseteq \overline{f(A)}$ for all subsets A of X.

Proof. (i)$\Rightarrow$(ii). Let B be a subset of Y. Since $\overline{B}$ is a closed subset of Y, $f^{-1}(\overline{B})$ is closed in X. Moreover, $f^{-1}(B) \subseteq f^{-1}(\overline{B})$, and so $\overline{f^{-1}(B)} \subseteq f^{-1}(\overline{B})$. (Recall that $\overline{f^{-1}(B)}$ is the smallest closed set containing $f^{-1}(B)$.)

(ii)$\Rightarrow$(iii). Let A be a subset of X. Then, if $B = f(A)$, we have $A \subseteq f^{-1}(B)$ and $\overline{A} \subseteq \overline{f^{-1}(B)} \subseteq f^{-1}(\overline{B})$. Thus $f(\overline{A}) \subseteq f(f^{-1}(\overline{B})) = \overline{B} = \overline{f(A)}$.

(iii)$\Rightarrow$(i) Let F be a closed set in Y and set $f^{-1}(F) = F_1$. By Theorem 3.1.10, it is sufficient to show that F_1 is closed in X, that is, $F_1 = \overline{F_1}$. Now,

$$f(\overline{F_1}) \subseteq \overline{f(f^{-1}(F))} \subseteq \overline{F} = F,$$

so that

$$\overline{F_1} \subseteq f^{-1}(f(\overline{F_1})) \subseteq f^{-1}(F) = F_1.$$ □

Examples 3.1.13. (i) Let (X, d) be a metric space and suppose $f_1 : X \to \mathbf{R}$ and $f_2 : X \to \mathbf{R}$ are continuous functions. Then the function $f : X \to \mathbf{R}^2$ defined by $f(x) = (f_1(x), f_2(x))$ is continuous. In fact, given $\varepsilon > 0$, there exist $\delta_1 > 0$ and $\delta_2 > 0$ such that $d(x, a) < \delta_1$ implies $|f_1(x) - f_1(a)| < \varepsilon/\sqrt{2}$ and $d(x, a) < \delta_2$ implies $|f_2(x) - f_2(a)| < \varepsilon/\sqrt{2}$. Take $\delta = \min\{\delta_1, \delta_2\}$. Then $d(x, a) < \delta$ implies

$$|f(x) - f(a)| = \{(f_1(x) - f_1(a))^2 + (f_2(x) - f_2(a))^2\}^{1/2} < \varepsilon.$$

More generally, if $f_k : X \to \mathbf{R}$ are continuous functions, $k = 1, 2, \ldots, n$, then the function $f : X \to \mathbf{R}^n$ defined by $f(x) = (f_1(x), f_2(x), \ldots, f_n(x))$ is continuous.

Any map $f : X \to \mathbf{R}^n$ can be written as $f(x) = (f_1(x), f_2(x), \ldots, f_n(x))$, $x \in X$. If f is continuous, so is each f_k. In fact,

$$|f_k(x) - f_k(y)| \leq |f(x) - f(y)|, k = 1, 2, \ldots, n,$$

where $|f(x) - f(y)| = (\sum_{k=1}^{n} |f_k(x) - f_k(y)|^2)^{1/2}$.

(ii) Since the functions $x \rightarrow \cos x$ and $x \rightarrow \sin x$ are continuous from $\mathbf{R}$ to $\mathbf{R}$, it follows from (i) above that the function $x \rightarrow (\cos x, \sin x)$ from $\mathbf{R}$ to $\mathbf{R}^2$ is continuous.

(iii) Let $X = \mathbf{C}$ with the usual metric. The mappings

$$z \rightarrow \frac{1}{2} \cos (\text{Im } z) \quad \text{and} \quad z \rightarrow \frac{1}{2} \sin (\text{Re } z) + 1$$

from $\mathbf{C}$ to $\mathbf{R}$ are continuous, and it follows from (i) above that the mapping

$$z \rightarrow \left(\frac{1}{2} \cos (\text{Im } z), \frac{1}{2} \sin (\text{Re } z) + 1 \right)$$

from $\mathbf{C}$ to $\mathbf{C}$ is continuous.

(iv) Let $X = \mathbf{C}$ with the usual metric. The mappings

$$z \rightarrow \frac{2\text{Re } z}{1 + |z|^2}, z \rightarrow \frac{2\text{Im } z}{1 + |z|^2} \text{ and } z \rightarrow \frac{|z|^2 - 1}{|z|^2 + 1}$$

from $\mathbf{C}$ to $\mathbf{R}$ are continuous. It follows from (i) above that the mapping

$$z \rightarrow \left(\frac{2\text{Re } z}{1 + |z|^2}, \frac{2\text{Im } z}{1 + |z|^2}, \frac{|z|^2 - 1}{|z|^2 + 1} \right)$$

from $\mathbf{C}$ to $\mathbf{R}^3$ is continuous.

The mapping is the familiar representation of the complex plane by points of the unit sphere in $\mathbf{R}^3$.

(v) Let $X = Y = C[0, 1]$ with the uniform metric $d(x, y) = \sup\{|x(t) - y(t)|: t \in [0, 1]\}$. The mapping $\varphi \colon X \rightarrow Y$ defined by

$$(\varphi(x))(t) = \int_0^t x(s)\,ds \, , x \in C[0, 1]$$

is continuous. In fact,

$$|(\varphi(x))(t) - (\varphi(y))(t)| = \left| \int_0^t x(s)\,ds - \int_0^t y(s)\,ds \right|$$
$$\leq \int_0^t |x(s) - y(s)|\,ds$$
$$\leq d(x, y).$$

Hence,

$$d(\varphi(x), \varphi(y)) \leq d(x, y)$$

for $x, y \in C[0, 1]$. On choosing $\delta = \varepsilon$, the assertion follows.

(vi) If (X, d) is a discrete metric space, then every function $f: X \to Y$, where Y is any metric space, is continuous. Let $a \in X$ and $S(f(a), \varepsilon)$ be an open ball centred at $f(a)$ with radius ε. Choose $\delta < 1$. Then $S(a, \delta) = \{a\}$ and so $f(S(a, \delta)) = \{f(a)\} \subseteq S(f(a), \varepsilon)$.

3.2. Extension Theorems

Consider the function $f: (0, \infty) \to \mathbf{R}$ defined by $f(x) = 1/x$. There is no continuous function g defined on $[0, \infty)$ that agrees with f. In other words, f has no continuous "extension" to $[0, \infty)$. The term "extension" is formally defined below.

Definition 3.2.1. Let X and Y be abstract sets and let A be a proper subset of X. If f is a mapping of A into Y, then a mapping $g: X \to Y$ is called an **extension** of f if $g(x) = f(x)$ for each $x \in A$; the function f is then called the **restriction** of g to A.

   If X and Y are metric spaces, $A \subseteq X$ and $f: A \to Y$ is continous, then we might ask whether there exists a continuous extension g of f. Extension problems abound in analysis and have attracted the attention of many celebrated mathematicians. Below, we deal with some simple extension techniques.

Theorem 3.2.2. Let (X, d_X) and (Y, d_Y) be metric spaces and let $f: X,\ g: X \to Y$ be continuous maps. Then the set $\{x \in X : f(x) = g(x)\}$ is a closed subset of X.

Proof. Let $F = \{x \in X : f(x) = g(x)\}$. Then $X \backslash F = \{x \in X : f(x) \neq g(x)\}$. We shall show that $X \backslash F$ is open. If $X \backslash F = \emptyset$, then there is nothing to prove. So let $X \backslash F \neq \emptyset$ and let $a \in X \backslash F$. Then $f(a) \neq g(a)$. Let $r > 0$ be the distance $d_Y(f(a), g(a))$. For $\varepsilon = r/3$, there exists a $\delta > 0$ such that

$$d_X(x, a) < \delta \text{ implies } d_Y(f(x), f(a)) < r/3 \text{ and } d_Y(g(x), g(a)) < r/3.$$

By the triangle inequality, we have

$$d_Y(f(a), g(a)) \leq d_Y(f(a), f(x)) + d_Y(f(x), g(x)) + d_Y(g(x), g(a)),$$

which implies

$$d_Y(f(x), g(x)) \geq d_Y(f(a), g(a)) - d_Y(f(a), f(x)) - d_Y(g(x), g(a)) > r/3$$

for all x satisfying $d_X(x, a) < \delta$. Thus, for each $x \in S(a, \delta)$, $d_Y(f(x), g(x)) > 0$, i.e., $f(x) \neq g(x)$. So,

$$S(a, \delta) \subseteq X \backslash F.$$

Hence, $X \backslash F$ is open and thus F is closed. $\qquad\square$

Corollary 3.2.3. Let (X, d_X) and (Y, d_Y) be metric spaces and let $f: X \to Y$, $g: X \to Y$ be continuous maps. If $F = \{x \in X : f(x) = g(x)\}$ is dense in X, then $f = g$.

Proof. By Theorem 3.2.2, F is closed. Since F is assumed dense in X, we have $X = \overline{F} = F$, i.e., $f(x) = g(x)$ for all $x \in X$. $\qquad\square$

Theorem 3.2.4. Let (X, d_X) and (Y, d_Y) be metric spaces, A a dense subset of X and f a map from A to Y. Then f has a continuous extension $g\colon X \to Y$ if and only if for every $x \in X$ that is a limit point of A, the limit $\lim_{y\to x} f(y)$ not only exists in Y but also equals $f(x)$ in case $x \in A$. When the extension exists, it is unique. (Note that the stipulation $\lim_{y\to x} f(y) = f(x)$ when $x \in A$ says that f is continuous on A.)

Proof. Suppose that f has a continuous extension g, and consider any $x \in X$ that is a limit point of X. Since A is dense, x must be a limit point of A as well, as we now argue. Any ball $S(x, \varepsilon)$ contains a point $y \in X, y \neq x$. There exists $S(y, \varepsilon') \subseteq S(x, \varepsilon)$ such that $x \notin S(y, \varepsilon')$. Since A is dense, $S(y, \varepsilon')$ contains a point $a \in A$. Thus, $S(x, \varepsilon)$ contains the point $a \in A$ and $a \neq x$.

Now

$$g(x) = \lim_{y\to x} g(y) \qquad\qquad (g \text{ is continuous})$$
$$= \lim_{y\to x} g(y) \text{ with } y \in A \quad (x \text{ is a limit point of } A)$$
$$= \lim_{y\to x} f(y) \qquad\qquad (g \text{ is an extension of } f).$$

Thus, $\lim_{y\to x} f(y)$ exists and equals $g(x)$.

Conversely, suppose that for every limit point $x \in X$, $\lim_{y\to x} f(y)$ exists and that it equals $f(x)$ when $x \in A$. Define $g(x)$ by

$$g(x) = \begin{cases} f(x) & \text{if } x \in A, \\ \lim_{y\to x} f(y) & \text{if } x \notin A \text{ but } x \in A'. \end{cases}$$

Since A is dense in X, the function g is defined on the whole of X. We need to show that g is continuous. By the definition of a limit, for every positive number ε, there exists a positive number $\delta > 0$ such that

$$f(y) \in S\left(g(x), \frac{1}{2}\varepsilon\right) \text{ whenever } y \neq x \text{ and } y \in S(x, \delta) \cap A.$$

Consider any $z \in S(x, \delta)$. In case z is an isolated point of X, then $g(z) \in S(g(x), \varepsilon/2)$, in view of the observation above. If z is not an isolated point of X, then $g(z)$ is the limit of $f(y)$ as $y \to z$ in $S(x, \delta) \cap A$. Therefore,

$$g(z) \in \overline{f(A \cap S(x, \delta))} \subseteq \overline{S(g(x), \varepsilon/2)} \subseteq S(g(x), \varepsilon),$$

so that g is continuous at x. Hence, g is continuous on X. By Corollary 3.2.3, it follows that g is the unique continuous extension of f. $\qquad\square$

Examples 3.2.5. (i) Let $f(x) = \sin(1/x), x \in \mathbf{R}\backslash\{0\}$. We shall show that $\lim_{x\to 0} \sin(1/x)$ does not exist. Hence, the function f cannot be extended to a continuous function on $\mathbf{R}$.

Assume that there exists $L \in \mathbf{R}$ such that $\lim_{x \to 0} \sin(1/x) = L$. Then, for $\varepsilon = 1$, there would exist $\delta > 0$ such that $0 < |x| < \delta$ implies

$$\left| \sin\left(\frac{1}{x}\right) - L \right| < 1. \tag{3.1}$$

Observe that $\sin(1/x) = 1$ for $x = \frac{2}{\pi(4n+1)}$, $n \in \mathbf{N}$; since $\frac{2}{\pi(4n+1)} \to 0$ as $n \to \infty$, we see that $\sin(1/x) = 1$ for some $x \in (0, \delta)$. For this x, (3.1) implies

$$|1 - L| < 1, \qquad \text{i.e., } L \in (0, 2). \tag{3.2}$$

Similarly, $\sin(1/x) = -1$ for $x = \frac{2}{\pi(4n+3)}$, $n \in \mathbf{N}$; since $\frac{2}{\pi(4n+3)} \to 0$ as $n \to \infty$, we see that $\sin(1/x) = -1$ for some $x \in (0, \delta)$. For this x, (3.1) implies

$$|-1 - L| < 1, \qquad \text{i.e., } L \in (-2, 0). \tag{3.3}$$

As the intervals $(0, 2)$ and $(-2, 0)$ are disjoint, they cannot have a point L in common.

(ii) From elementary calculus, we know that $\lim_{x \to 0} \frac{\sin x}{x}$ exists and equals 1. Hence, the function $f(x) = \left(\frac{\sin x}{x}\right)$, $x \neq 0$, has a continuous extension g defined by

$$g(x) = \begin{cases} \dfrac{\sin x}{x} & x \neq 0, \\ 1 & x = 0. \end{cases}$$

(iii) Let $\{r_n\}_{n \geq 1}$ be an enumeration of the rationals in $[0,1]$. Define a function $f: [0, 1] \to \mathbf{R}$ by putting $f(x) = \sum_{r_n < x} 1/2^n$, the infinite sum being extended to only those n such that $r_n < x$. If there are no points r_n to the left of x, the sum is, of course, empty and we define it to be zero. Since the series in the defining equation is absolutely convergent, the order in which the terms are arranged is immaterial.

It is clear that f is a monotonically increasing function on $[0, 1]$. For any $c \in (0, 1]$,

$$f(c-) = \lim_{h \to 0+} f(c - h) = \lim_{h \to 0+} \sum_{r_n < c - h} \frac{1}{2^n} = \sum_{r_n < c} \frac{1}{2^n} = f(c);$$

i.e., the function f is left continuous at each point $c \in (0, 1]$. For any $c \in [0, 1)$, we have

$$f(c+) = \lim_{h \to 0+} f(c + h) = \lim_{h \to 0+} \sum_{r_n < c + h} \frac{1}{2^n} = \sum_{r_n \leq c} \frac{1}{2^n}$$

$$= \begin{cases} f(c) & \text{if } c \text{ is irrational,} \\ > f(c) & \text{if } c \text{ is rational.} \end{cases}$$

Thus, f is continuous at every irrational point of $[0,1]$ and $\lim_{y \to x} f(y)$ does not exist at rational points x of $[0,1]$. Let $\tilde{f}$ denote the restriction of f to all irrationals $x \in [0, 1]$. The function $\tilde{f}$ is continuous on its domain of definition, which is dense in $[0, 1]$. However, it cannot be extended to a function continuous on $[0, 1]$ since $\lim_{y \to x} \tilde{f}(y)$ does not exist at rational points $x \in [0, 1]$.

3.3. Real and Complex-valued Continuous Functions

Definition 3.3.1. Let X be a nonempty set. Given mappings f and g of X into $\mathbf{C}$ and $\alpha \in \mathbf{C}$, we define the mappings $f + g, \alpha f, fg$ and $|f|$ into $\mathbf{C}$ as follows:

$$(f + g)(t) = f(t) + g(t)$$
$$(\alpha f)(t) = \alpha f(t)$$
$$(fg)(t) = f(t)g(t)$$
$$|f|(t) = |f(t)|$$

for all $t \in X$. Further, if $f(t) \neq 0$ for all $t \in X$, we define the mapping $1/f$ of X into $\mathbf{C}$ by

$$(1/f)(t) = \frac{1}{f(t)} \qquad \text{for all } t \in X.$$

The proofs of the assertions in the following theorem are direct generalisations of the familiar proofs in the case where X is the real line.

Theorem 3.3.2. Let f and g be continuous mappings of a metric space (X, d_X) into $\mathbf{C}$ and let $\alpha \in \mathbf{C}$. Then the mappings $f + g, \alpha f, fg$ and $|f|$ are continuous on X, and so is the mapping $1/f$, if it is defined.

Examples 3.3.3. (i) Let $f \colon \mathbf{R} \to \mathbf{C}$ be defined by

$$f(x) = x + ix^2.$$

We shall argue that f is continuous at $2 \in \mathbf{R}$. Consider any $\varepsilon > 0$. Upon using the fact that the functions $g \colon \mathbf{R} \to \mathbf{R}$ and $h \colon \mathbf{R} \to \mathbf{R}$ defined by $g(x) = x$ and $h(x) = x^2$ are continuous at 2, it follows that there exist $\delta_1 > 0$ and $\delta_2 > 0$ such that

$$|x - 2| < \delta_1 \Rightarrow |g(x) - g(2)| < \frac{\varepsilon}{\sqrt{2}} \text{ and } |x - 2| < \delta_2 \Rightarrow |h(x) - h(2)| < \frac{\varepsilon}{\sqrt{2}}.$$

The positive number $\delta = \min\{\delta_1, \delta_2\}$ then has the property that $|x - 2| < \delta$ implies

$$|g(x) - g(2)| < \frac{\varepsilon}{\sqrt{2}} \text{ as well as } |h(x) - h(2)| < \frac{\varepsilon}{\sqrt{2}}.$$

These two inequalities together imply

$$(g(x) - g(2))^2 + (h(x) - h(2))^2 < \frac{\varepsilon^2}{2} + \frac{\varepsilon^2}{2} = \varepsilon^2.$$

Consequently,

$$|(x + ix^2) - (2 + 4i)|^2 = |x - 2|^2 + |x^2 - 4|^2$$
$$= |g(x) - g(2)|^2 + |h(x) - h(2)|^2 < \varepsilon^2$$

whenever $|x - 2| < \delta$.

(ii) Let $(X, d_X) = (Y, d_Y) = (\mathbf{R}, d)$, where d denotes the usual metric on $\mathbf{R}$. Define f on $\mathbf{R}$ by

$$f(x) = \begin{cases} 1 & x \text{ is rational,} \\ 0 & x \text{ is irrational.} \end{cases}$$

Then f is continuous at no point of $\mathbf{R}$. It is enough to show that $\lim_{x \to a} f(x)$ does not exist for any a. Assume the contrary, that $\lim_{x \to a} f(x) = L$ for some $L \in \mathbf{R}$. Given $\varepsilon = 1/2$, there must exist $\delta > 0$ such that $|f(x) - L| < 1/2$ if $0 < |x - a| < \delta$. But the interval $(a - \delta, a)$ contains both a rational number and an irrational number. If $x \in (a - \delta, a)$ is rational we get $|1 - L| < 1/2$, while if $x \in (a - \delta, a)$ is irrational we get $|0 - L| < 1/2$. This leads to the contradiction that $|0 - 1| < 1$.

(iii) Let $X = C[0,1]$ with the uniform metric. Define $f: X \to \mathbf{C}$ by $f(x) = x(0)$ whenever $x \in X$. We shall show that f is continuous on X. Let $\{x_n\}_{n \geq 1}$ be a sequence in X, i.e., in $C[0,1]$ such that $\lim_n x_n = x$. Since uniform convergence implies pointwise convergence, we have

$$\lim_n f(x_n) = \lim_n x_n(0) = x(0) = f(x).$$

Thus, f is continuous on $X = C[0,1]$.

(iv) Let $X = C[0,1]$ with the uniform metric. Define $f: X \to \mathbf{C}$ by

$$f(x) = \int_0^1 x(t)\,dt.$$

Let $\{x_n\}_{n \geq 1}$ be a sequence in X, i.e., in $C[0,1]$ such that $\lim_n x_n = x$. So, for $\varepsilon > 0$, there exists n_0 such that $n \geq n_0$ implies

$$\sup\{|x_n(t) - x(t)| : 0 \leq t \leq 1\} < \varepsilon.$$

Now,

$$
\begin{aligned}
|f(x_n) - f(x)| &= \left| \int_0^1 x_n(t)\,dt - \int_0^1 x(t)\,dt \right| \\
&\leq \int_0^1 |x_n(t) - x(t)|\,dt \\
&\leq \sup\{|x_n(t) - x(t)| : 0 \leq t \leq 1\} \cdot \int_0^1 dt \\
&< \varepsilon
\end{aligned}
$$

for $n \geq n_0$. Thus, f is continuous on $X = C[0,1]$, in view of Theorem 3.1.3.

(v) Let (X, d) be a metric space. Define $f: X \to \mathbf{R}$ by

$$f(x) = d(x, x_0),\, x \in X$$

where x_0 is fixed. We shall show that f is continuous on X. In fact,

$$
\begin{aligned}
|f(x) - f(y)| &= |d(x, x_0) - d(y, x_0)| \\
&\leq d(x, y)
\end{aligned}
$$

for $x, y \in X$. The continuity of f now follows on choosing $\delta = \varepsilon$.

3.4. Uniform Continuity

Let (X, d_X) and (Y, d_Y) be two metric spaces and let f be a function continuous at each point x_0 of X. In the definition of continuity, when x_0 and ε are specified, we make a definite choice of δ so that

$$d_Y(f(x), f(x_0)) < \varepsilon \qquad \text{whenever} \quad d_X(x, x_0) < \delta.$$

This describes δ as dependent upon x_0 and ε, say $\delta = \delta(x_0, \varepsilon)$. If $\delta(x_0, \varepsilon)$ can be chosen in such a way that its values have a lower positive bound when ε is kept fixed and x_0 is allowed to vary over X, and if this happens for each positive ε, then we have the notion of "uniform continuity". More precisely, we have the following definition:

Definition 3.4.1. Let (X, d_X) and (Y, d_Y) be two metric spaces. A function $f \colon X \to Y$ is said to be **uniformly continuous on** X if, for every $\varepsilon > 0$, there exists a $\delta > 0$ (depending on ε alone) such that

$$d_Y(f(x_1), f(x_2)) < \varepsilon \quad \text{whenever} \quad d_X(x_1, x_2) < \delta$$

for all $x_1, x_2 \in X$.

Every function $f \colon X \to Y$ which is uniformly continuous on X is necessarily continuous on X. However, the converse may not be true. We shall see later (see Theorem 5.4.10) that these two concepts agree on certain kinds of metric spaces called "compact".

Examples 3.4.2. (i) Let $(X, d_X) = (Y, d_Y) = (\mathbf{R}, d)$, where d denotes the usual metric on $\mathbf{R}$. Define $f \colon \mathbf{R} \to \mathbf{R}$ by

$$f(x) = x^2, x \in \mathbf{R}.$$

It is well known from elementary analysis that f is continuous. We shall prove that it is not uniformly continuous by exhibiting an ε for which no δ works. Take $\varepsilon = 1$ and let $\delta > 0$ be arbitrary. Choose $x = \delta/2 + 1/\delta$ and $y = 1/\delta$. Then

$$|x - y| = \left| \frac{\delta}{2} + \frac{1}{\delta} - \frac{1}{\delta} \right| = \frac{\delta}{2} < \delta$$

but

$$|f(x) - f(y)| = \left| \left(\frac{\delta}{2} + \frac{1}{\delta} \right)^2 - \left(\frac{1}{\delta} \right)^2 \right|.$$

$$= 1 + \frac{\delta^2}{4} > 1$$

Thus, whatever $\delta > 0$ may be, there exist points x and y such that $|x - y| < \delta$ but $|f(x) - f(y)| > 1$. (The reader may attempt to prove the result starting with $\varepsilon = 1/2$.)

(ii) Let A be a subset of the metric space (X, d). Define

$$f(x) = d(x, A) = \inf\{d(x, y) : y \in A\}, \quad x \in X.$$

We shall prove that f is uniformly continuous over X. For $y \in A$ and $x, z \in X$, the triangle inequality gives

$$d(x, y) \leq d(x, z) + d(z, y).$$

On taking the infimum as y varies over A, we get

$$\inf_{y \in A} d(x, y) \leq \inf_{y \in A} [d(x, z) + d(z, y)] = d(x, z) + \inf_{y \in A} d(z, y).$$

Thus,

$$d(x, A) - d(z, A) \leq d(x, z), \quad x, z \in X.$$

Interchanging x and z and observing that $d(x, z) = d(z, x)$, we get

$$d(z, A) - d(x, A) \leq d(x, z), \quad x, z \in X.$$

Hence,

$$|f(x) - f(z)| = |d(x, A) - d(z, A)| \leq d(x, z), \quad x, z \in X.$$

The uniform continuity of f results on choosing $\delta = \varepsilon$.

(iii) The function $f : (0, 1) \to \mathbf{R}$ defined by $f(x) = 1/x$ is not uniformly continuous. We shall prove the assertion by exhibiting an ε for which no δ works. (See Figure 3.1.)

Take $\varepsilon = 1/2$ and let δ be any positive number. Choose $x = 1/n$ and $y = 1/(n+1)$, where n is a positive integer such that $n > 1/\delta$. Then

$$|x - y| = \left| \frac{1}{n} - \frac{1}{n+1} \right| = \frac{1}{n(n+1)} < \frac{1}{n} < \delta;$$

but

$$|f(x) - f(y)| = |n - (n+1)| = 1 > \varepsilon.$$

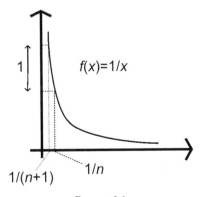

$f(x) = 1/x$

1

$1/(n+1)$

$1/n$

FIGURE 3.1

Thus, whatever $\delta > 0$ may be, there exist x and y such that $|x - y| < \delta$ but $|f(x) - f(y)| > 1/2$.

(iv) The function $f\colon [0,1] \to \mathbf{R}$ defined by $f(x) = x^2$ is uniformly continuous. Let $x, y \in [0,1]$. Then

$$\begin{aligned}
|f(x) - f(y)| &= |x^2 - y^2| \\
&= |x + y||x - y| \\
&\leq (|x| + |y|)|x - y| \\
&\leq 2|x - y|.
\end{aligned}$$

For $\varepsilon > 0$, choose $\delta = \varepsilon/2$. Then $|x - y| < \delta$ implies $|f(x) - f(y)| \leq 2|x - y| < 2\delta = 2(\varepsilon/2) = \varepsilon$. This proves the assertion that $f(x) = x^2, x \in [0,1]$, is uniformly continuous.

Recall from Example 3.4.2(ii) that $d(x, A) = \inf\{d(x,y)\colon y \in A\}$.

Proposition 3.4.3. Let (X, d) be a metric space and let $x \in X$ and $A \subseteq X$ be nonempty. Then $x \in \bar{A}$ if and only if $d(x, A) = 0$.

Proof. Suppose $d(x, A) = 0$. There are two possibilities: $x \in A$ or $x \notin A$. If $x \in A$, then $x \in \bar{A}$. We shall next show that if $x \notin A$, then x is a limit point of A. Let $\varepsilon > 0$ be given. By the definition of $d(x, A)$, there exists a $y \in A$ such that $d(x, y) < \varepsilon$, i.e., $y \in S(x, \varepsilon)$. Thus, every ball with centre x and radius ε contains a point of A distinct from x; so $x \in \bar{A}$. Conversely, suppose $x \in \bar{A}$. If $x \in A$, then obviously $d(x, A) = 0$. We shall next show that if x is a limit point of A, then $d(x, A) = 0$. By the definition of limit point, every ball $S(x, \varepsilon)$ with centre x and radius $\varepsilon > 0$ contains a point $y \in A$ distinct from x. Consequenly, $d(x, A) < \varepsilon$, i.e., $d(x, A) = 0$. $\square$

A more general version of the following theorem is known as Urysohn's lemma.

Theorem 3.4.4. Let A and B be disjoint closed subsets of a metric space (X, d). Then there is a continuous real-valued function f on X such that $f(x) = 0$ for all $x \in A, f(x) = 1$ for all $x \in B$ and $0 \leq f(x) \leq 1$ for all $x \in X$.

Proof. From Example (ii) above, it follows that the mappings $x \to d(x, A)$ and $x \to d(x, B)$ are continuous on X. Since A and B are closed and $A \cap B = \varnothing$, Proposition 3.4.3 shows that $d(x, A) + d(x, B) > 0$ for all $x \in X$. Indeed, if $d(x, A) + d(x, B) = 0$ for some $x \in X$, then $d(x, A) = d(x, B) = 0$; so $x \in \bar{A} = A$ and $x \in \bar{B} = B$, and hence $x \in A \cap B$, a contradiction.

Now define a mapping $f\colon X \to \mathbf{R}$ by

$$f(x) = \frac{d(x, A)}{d(x, A) + d(x, B)}, \quad x \in X.$$

Then f is continuous on X. Moreover,

$$f(x) = \begin{cases} 0 & \text{if } x \in A, \\ 1 & \text{if } x \in B \end{cases}$$

and $0 \leq f(x) \leq 1$. $\square$

Corollary 3.4.5. Let (X, d) be a metric space and A, B be disjoint closed subsets of X. Then there exist open sets G, H such that $A \subseteq G, B \subseteq H$ and $G \cap H = \emptyset$.

Proof. Let $f: X \to [0, 1]$ be any function guaranteed by Theorem 3.4.4, and let

$$G = \left\{ x \in X : f(x) < \frac{1}{2} \right\} \text{ and } H = \left\{ x \in X : f(x) > \frac{1}{2} \right\}.$$

Then $G = f^{-1}([0, 1/2))$ and $H = f^{-1}((1/2, 1])$ are open subsets of X, being inverse images of open subsets of $[0,1]$. Moreover, $A \subseteq G, B \subseteq H$ and $G \cap H = \emptyset$. $\quad\square$

A composition of uniformly continuous mappings is again a uniformly continuous mapping. More precisely, we have the following theorem:

Theorem 3.4.6. If f and g are two uniformly continuous mappings of metric spaces (X, d_X) to (Y, d_Y), and (Y, d_Y) to (Z, d_Z), respectively, then $g \circ f$ is a uniformly continuous mapping of (X, d_X) to (Z, d_Z).

Proof. Since g is uniformly continuous, for each $\varepsilon > 0$, there exists a $\delta > 0$ such that

$$d_Y(f(x), f(y)) < \delta \text{ implies } d_Z((g \circ f)(x), (g \circ f)(y)) < \varepsilon$$

for all $f(x), f(y) \in Y$.

As f is uniformly continuous, corresponding to $\delta > 0$, there exists an $\eta > 0$ such that

$$d_X(x, y) < \eta \text{ implies } d_Y(f(x), f(y)) < \delta$$

for all $x, y \in X$.

Thus, for each $\varepsilon > 0$, there exists an $\eta > 0$ such that

$$d_X(x, y) < \eta \text{ implies } d_Z((g \circ f)(x), (g \circ f)(y)) < \varepsilon$$

for all $x, y \in X$ and so $g \circ f$ is uniformly continuous on X. $\quad\square$

A continuous function may not map a Cauchy sequence into a Cauchy sequence as the following example shows:

Example 3.4.7. Let $X = (0, \infty)$ with the induced usual metric of the reals and Y be the reals with the usual metric. The function $f: X \to Y$ defined by

$$f(x) = \frac{1}{x}, x \in X,$$

is continuous on X. Now $\{1/n\}_{n \geq 1}$ is a Cauchy sequence in X (because it is convergent in $\mathbf{R}$). But $\{f(1/n)\}_{n \geq 1} = \{n\}_{n \geq 1}$ is not a Cauchy sequence in Y. Indeed, the absolute difference of any two distinct terms is at least as large as 1.

However, Cauchy sequences are mapped into Cauchy sequences by uniformly continuous functions.

Theorem 3.4.8. Let (X, d_X) and (Y, d_Y) be two metric spaces and $f: X \to Y$ be uniformly continuous. If $\{x_n\}_{n \geq 1}$ is a Cauchy sequence in X, then so is $\{f(x_n)\}_{n \geq 1}$ in Y.

Proof. Since f is uniformly continuous, for every $\varepsilon > 0$, there exists a $\delta > 0$ such that

$$d_Y(f(x), f(y)) < \varepsilon \quad \text{whenever } d_X(x, y) < \delta \tag{3.4}$$

for all $x, y \in X$.

Because the sequence $\{x_n\}_{n \geq 1}$ is Cauchy, corresponding to $\delta > 0$, there exists n_0 such that

$$n, m \geq n_0 \quad \text{implies } d_X(x_n, x_m) < \delta. \tag{3.5}$$

From (3.4) and (3.5), we conclude that

$$d_Y(f(x_n), f(x_m)) < \varepsilon \qquad \text{for } n, m \geq n_0,$$

and so $\{f(x_n)\}_{n \geq 1}$ is a Cauchy sequence in Y. $\qquad\square$

Theorem 3.4.9. Let f be a uniformly continuous mapping of a set A, dense in the metric space (X, d_X), into a complete metric space (Y, d_Y). Then there exists a unique continuous mapping $g: X \to Y$ such that $g(x) = f(x)$ when $x \in A$; moreover, g is uniformly continuous.

Proof. Since f is uniformly continuous, a fortiori, continuous, therefore, for every $x \in A$ that is a limit point of X, the limit $\lim_{y \to x} f(y)$ not only exists in Y but also equals $f(x)$. Therefore, by Theorem 3.2.4, in order to prove the existence and uniqueness of such a continuous mapping $g: X \to Y$, it is sufficient to show for every $x \in X \backslash A$ that $f(y)$ tends to a limit as $y \to x$. (It is understood that $y \in A$, because the domain of f is A.)

Let $x \in X$ be arbitrary. Since A is dense in X, there exists a sequence $\{x_n\}_{n \geq 1}$ in A such that $\lim_{n \to \infty} d_X(x_n, x) = 0$. Since $\{x_n\}_{n \geq 1}$ is convergent, it is a fortiori Cauchy; so by Theorem 3.4.8, it follows that $\{f(x_n)\}_{n \geq 1}$ is a Cauchy sequence in the complete metric space (Y, d_Y) and hence converges to a limit, which we shall denote by b. Now consider any sequence $\{x'_n\}_{n \geq 1}$ in A with $x'_n \neq x$ for each n and $\lim_{n \to \infty} x'_n = x$. It follows from uniform continuity of f that, for $\varepsilon > 0$, there exists a $\delta > 0$ such that

$$d_Y(f(z), f(y)) < \varepsilon \qquad \text{whenever } d_X(z, y) < \delta. \tag{3.6}$$

Since $\lim_{n \to \infty} x_n = x = \lim_{n \to \infty} x'_n$, there exists an integer n_1 such that $d_X(x_n, x'_n) < \delta$ whenever $n \geq n_1$. Therefore by (3.6)

$$d_Y(f(x_n), f(x'_n)) < \varepsilon \quad \text{whenever } n \geq n_1. \tag{3.7}$$

Letting $n \to \infty$, we get from (3.7) that

$$d_Y(\lim_{n\to\infty} f(x_n),\ \lim_{n\to\infty} f(x'_n)) \le \varepsilon,$$

i.e.,

$$d_Y(b,\ \lim_{n\to\infty} f(x'_n)) \le \varepsilon.$$

Since $\varepsilon > 0$ is arbitrary, it follow that $\lim_{n\to\infty} f(x'_n) = b$ for every sequence $\{x'_n\}_{n \ge 1}$ in A with $x'_n \ne x$ for each n, and $\lim_{n\to\infty} x'_n = x$. It follows by Proposition 3.1.5 that $f(y)$ tends to a limit, namely b, as $y \to x$. As already pointed out earlier, this shows that a unique continuous extension g of f to X exists.

It remains to prove that g is uniformly continuous. Let x and x' be two points of X such that $d_X(x,x') < \delta/3$. Let $\{x_n\}_{n\ge 1}$ and $\{x'_n\}_{n\ge 1}$ be sequences of points in A such that $\lim_{n\to\infty} d_X(x_n, x) = 0$ and that $\lim_{n\to\infty} d_X(x'_n, x') = 0$. We can choose an integer n_2 such that $d_X(x_n, x) < \delta/3$ and $d_X(x'_n, x') < \delta/3$ whenever $n \ge n_2$. Since

$$d_X(x_m, x'_n) \le d_X(x_m, x) + d_X(x, x') + d_X(x', x'_n) < \delta$$

for $m, n \ge n_2$, it follows from (3.6) that

$$d_Y(f(x_m), f(x'_n)) < \varepsilon.$$

Letting $m \to \infty$ and then $n \to \infty$ in the above inequality, we get

$$d_Y(g(x), g(x')) \le \varepsilon$$

whenever $d_X(x, x') < \delta/3$. This proves that g is uniformly continuous. $\square$

Remark 3.4.10. The condition that the metric space (Y, d_Y) is complete in Theorem 3.4.9 cannot be omitted. In fact, let $X = \mathbf{R}$ with the usual metric and $Y = \mathbf{Q}$, the set of rationals with the metric induced from $\mathbf{R}$. Let $A = \mathbf{Q}$. Observe that A is a dense subset of X. The function $f : A \to Y$ defined by $f(x) = x$ for every $x \in A$ is uniformly continuous but it possesses no continuous extension to X, as the only continuous *rational-valued* functions on $X = \mathbf{R}$ are constant functions.

3.5. Homeomorphism, Equivalent Metrics and Isometry

Definition 3.5.1. Let (X, d_X) and (Y, d_Y) be any two metric spaces. A function $f : X \to Y$ which is both one-to-one and onto is said to be a **homeomorphism** if and only if the mappings f and f^{-1} are continuous on X and Y, respectively. Two metric spaces X and Y are said to be **homeomorphic** if and only if there exists a homeomorphism of X onto Y, and in this case, Y is called a **homeomorphic image** of X.

If X and Y are homeomorphic, the homeomorphism puts their points in one-to-one correspondence in such a way that their open sets also correspond to one another.

For metric spaces X and Y, let $X \sim Y$ mean that X and Y are homeomorphic. It is easily verified that the relation is reflexive, symmetric and transitive.

Examples 3.5.2. (i) The metric spaces [0,1] and [0,2] with the usual absolute value metric are homeomorphic. In fact, the mapping $f(x) = 2x$ is a homeomorphism.

(ii) The function $f(x) = \ln x$ is a homeomorphism between $(0, \infty)$ and **R**, where **R** is equipped with the usual metric and $(0, \infty)$ with the usual metric induced from **R**.

(iii) The function $w = \frac{z-a}{1-\bar{a}z}$, $z \in \mathbf{C}$ and $|a| < 1$, maps the closed disc $|z| \leq 1$ onto the closed disc $|w| \leq 1$. It is, in fact, a homeomorphism of the discs involved.

Suppose that whenever a metric space (X, d) has the property "P", every metric space homeomorphic to (X, d) also has the property; then we say that the property is "preserved under homeomorphism". There are a large number of properties that are not preserved under homeomorphism, as the following example shows:

Example 3.5.3. Let $X = \mathbf{N}$ and $Y = \{1/n : n \in \mathbf{N}\}$, each equipped with the usual absolute value metric. The function $f : X \rightarrow Y$ defined by $f(x) = 1/x$ is a homeomorphism of X onto Y. Observe that X is a closed subset of **R** and since **R** is complete, it follows that X is complete. On the other hand, $\{1/n\}_{n \geq 1}$ is a Cauchy sequence in Y that does not converge; so Y is not complete. Besides, the space X is not bounded, whereas Y is bounded.

Recall from Definition 1.5.2 that a mapping f of X into Y is an **isometry** if

$$d_Y(f(x), f(y)) = d_X(x, y)$$

for all $x, y \in X$. It is obvious that an isometry is one-to-one and uniformly continuous. Recall also that X and Y are said to be isometric if there exists an isometry between them that is onto. An isometry is necessarily a homeomorphism, but the converse is not true, as is evident from Examples 3.5.2 (i) and (ii) above.

By definition, it follows that isometric spaces possess the same metric properties. For metric spaces X and Y, let $X \approx Y$ mean that X and Y are isometric. It is easily verified that this relation between metric spaces is reflexive, symmetric and transitive.

Definition 3.5.4. Let d_1 and d_2 be metrics on a nonempty set X such that, for every sequence $\{x_n\}_{n \geq 1}$ in X and $x \in X$,

$$\lim_{n \to \infty} d_1(x_n, x) = 0 \qquad \text{if and only if} \qquad \lim_{n \to \infty} d_2(x_n, x) = 0,$$

i.e., a sequence converges to x in (X, d_1) if and only if it converges to x in (X, d_2). We then say that d_1 and d_2 are **equivalent metrics** on X and that (X, d_1) and (X, d_2) are **equivalent metric spaces**.

Remark 3.5.5. In view of Theorem 3.1.3, two metrics d_1 and d_2 on a nonempty set X are equivalent if and only if the identity maps id: $(X, d_1) \rightarrow (X, d_2)$ and id: $(X, d_2) \rightarrow (X, d_1)$ are both continuous, i.e., if and only if the identity mapping from (X, d_1) to (X, d_2) is a homeomorphism (as Definition in 3.5.1 above). Note that this amounts to saying that the families of open sets are the same in (X, d_1) and (X, d_2).

The following is a sufficient condition for two metrics on a set to be equivalent.

Theorem 3.5.6. Two metrics d_1 and d_2 on a nonempty set X are equivalent if there exists a constant K such that

$$\frac{1}{K} d_2(x, y) \le d_1(x, y) \le K\, d_2(x, y)$$

for all $x, y \in X$.

Proof. Let $x_n \to x$ in (X, d_1). Then we show that $x_n \to x$ in (X, d_2). This follows from the inequality

$$\frac{1}{K} d_2(x, y) \le d_1(x, y).$$

If $x_n \to x$ in (X, d_2), then $x_n \to x$ in (X, d_1) in view of the inequality

$$d_1(x, y) \le K\, d_2(x, y).$$

This completes the proof. □

Examples 3.5.7. (i) The metrics d_1, d_2 and d_∞ defined on $\mathbf{R}^n$ by

$$d_1(x, y) = \sum_{i=1}^{n} |x_i - y_i|$$

$$d_2(x, y) = \left(\sum_{i=1}^{n} (x_i - y_i)^2 \right)^{1/2}$$

$$d_\infty(x, y) = \max\{|x_i - y_i| : 1 \le i \le n\}$$

are equivalent. In fact, for any real numbers $\alpha_1, \alpha_1, \ldots, \alpha_n$, we have (see Corollary 1.4.9)

$$\left(\sum_{i=1}^{n} \alpha_i^2 \right)^{1/2} \le \sum_{i=1}^{n} |\alpha_i| \le n \cdot \max\{|\alpha_i| : 1 \le i \le n\} \le n \left(\sum_{i=1}^{n} \alpha_i^2 \right)^{1/2}.$$

(ii) The metric spaces (X, d) and (X, ρ), where

$$\rho(x, y) = \frac{d(x, y)}{1 + d(x, y)}$$

are equivalent.

Let $d(x_n, x) \to 0$ as $n \to \infty$. It is obvious that $\rho(x_n, x) \to 0$ as $n \to \infty$. On the other hand, if $\rho(x_n, x) \to 0$ as $n \to \infty$, then for $1 > \varepsilon > 0$, we have $2 - \varepsilon > 1$, and also, there exists n_0 such that $\rho(x_n, x) < \varepsilon/2$ for $n \ge n_0$. So

$$\frac{d(x_n, x)}{1 + d(x_n, x)} = \rho(x_n, x) < \frac{1}{2}\varepsilon \qquad \text{for } n \ge n_0.$$

Hence, $d(x_n, x) < \frac{\varepsilon}{2-\varepsilon} < \varepsilon$ for $n \geq n_0$.

(iii) Let $C[0,1]$ be as in Example 1.2.2 (ix). Consider the metrics d and e defined on $C[0,1]$ by

$$d(f, g) = \sup \{|f(x) - g(x)| : x \in [0, 1]\}$$

$$e(f, g) = \int_0^1 |f(x) - g(x)| dx.$$

These metrics are not equivalent.

The sequence $\{f_n\}_{n \geq 1}$ with $f_n(x) = x^n, 0 \leq x \leq 1$, is such that $e(f_n, f) \to 0$ as $n \to \infty$, where $f(x) = 0$ for all $x \in [0, 1]$. In fact,

$$e(f_n, f) = \int_0^1 x^n dx = \frac{1}{n+1} \to 0 \text{ as } n \to \infty.$$

However, $d(f_n, f) = \sup\{|f_n(x) - f(x)| : x \in [0, 1]\} = 1$ for all n. So, $\lim_{n \to \infty} d(f_n, f) \neq 0$. Thus, the sequence $\{f_n\}_{n \geq 1}$ converges to f in the metric e but not in the metric d. Thus, the metrics d and e are not equivalent.

We note however that for any sequence $\{f_n\}_{n \geq 1}$ in $C[0,1]$ and any $f \in C[0,1]$, when $d(f_n, f) \to 0$ as $n \to \infty$, it does follow that $e(f_n, f) \to 0$ as $n \to \infty$. To see why, observe that

$$e(f_n, f) = \int_0^1 |f_n(x) - f(x)| dx \leq \sup_x |f_n(x) - f(x)| = d(f_n, f).$$

(iv) Let m be the set of all bounded sequences with the following two metrics:

$$d(x, y) = \sup\{|x_n - y_n| : 1 \leq n < \infty\}$$

$$e(x, y) = \sum_{n=1}^{\infty} \frac{1}{2^n} \frac{|x_n - y_n|}{1 + |x_n - y_n|},$$

where the sequences $x = (x_1, x_2, \dots)$ and $y = (y_1, y_2, \dots)$ are elements of m. Consider the sequence $\{e_k\}_{k \geq 1}$ in m whose respective terms are the sequences

$$e_1 = (1, 0, 0, \dots), e_2 = (0, 1, 0, 0, \dots), \dots, e_k = (0, 0, \dots, 0, 1, 0, \dots), \dots,$$

where 1 is in the kth place. Let $e_0 = (0, 0, 0, \dots)$. Then

$$d(e_k, e_0) = 1 \text{ while } e(e_k, e_0) = \frac{1}{2^k} \frac{1}{1+1} = \frac{1}{2^{k+1}} \to 0 \text{ as } k \to \infty.$$

Thus, the sequence $\{e_k\}_{k \geq 1}$ converges in e but not in d. Hence, the metrics d and e on the space of all bounded sequences are not equivalent.

(v) Let $X = \mathbf{C}$ be equipped with the metrics

$$d(z_1, z_2) = |z_1 - z_2|$$

$$e(z_1, z_2) = \frac{2|z_1 - z_2|}{\sqrt{1 + |z_1|^2}\sqrt{1 + |z_2|^2}}.$$

Then d and e are equivalent metrics on $\mathbf{C}$. Let $\{z_n\}_{n\geq 1}$ be a sequence of complex numbers converging to z in the metric d, i.e., $\lim_{n\to\infty} d(z_n, z) = 0$. Then $\lim_{n\to\infty} e(z_n, z) = 0$, since $e(z_n, z) \leq 2d(z_n, z)$.

Let (ξ_n, η_n, ζ_n) and (ξ, η, ζ) be the points on the Riemann sphere corresponding to z_n and z, respectively, and let $\lim_{n\to\infty} \xi_n = \xi$, $\lim_{n\to\infty} \eta_n = \eta$ and $\lim_{n\to\infty} \zeta_n = \zeta$. It follows upon using Example 1.2.2(xiii) that

$$\lim_{n\to\infty} z_n = \lim_{n\to\infty} \frac{\xi_n + i\eta_n}{1 - \zeta_n} = \frac{\xi + i\eta}{1 - \zeta} = z,$$

i.e.,

$$\lim_{n\to\infty} d(z_n, z) = 0.$$

3.6. Uniform Convergence of Sequences of Functions

In this section, we discuss the convergence of sequences of functions defined on a metric space (X, d_X) taking values in another metric space (Y, d_Y). (In Definitions 3.6.1 and 3.6.3, the metric of X plays no role; the concepts make sense when X is any nonempty set as long as Y has a metric. Similarly, in Definitions 3.6.7 and 3.6.9 to come.)

Definition 3.6.1. Let $f: X \to Y$ and $f_n: X \to Y, n = 1, 2, \ldots$ be given. We say that $\{f_n\}_{n\geq 1}$ **converges pointwise** to the function f if and only if

$$\lim_{n\to\infty} d_Y(f_n(x), f(x)) = 0, x \in X. \tag{3.8}$$

The word "pointwise" is sometimes omitted when it is clear from the context.

According to the above definition, a sequence $\{f_n\}_{n\geq 1}$ converges pointwise to f on X if, for a given $\varepsilon > 0$ and given $x \in X$, there exists an integer n_0 such that

$$d_Y(f_n(x), f(x)) < \varepsilon \qquad \text{for } n \geq n_0. \tag{3.9}$$

In general, the integer n_0 depends upon ε as well as x. It is not always possible to find an n_0 such that (3.9) holds for all $x \in X$ simultaneously.

Examples 3.6.2. (i) Let $X = Y = \mathbf{R}$ with the usual metric and let f_n be defined by

$$f_n(x) = \begin{cases} 0 & \text{if } x \leq 0, \\ nx & \text{if } 0 \leq x \leq 1/n, \\ -nx + 2 & \text{if } 1/n \leq x \leq 2/n, \\ 0 & \text{if } x \geq 2/n, \end{cases}$$

and $f(x) = 0$ for all $x \in \mathbf{R}$ (see Figure 3.2). Then $f_n \to f$ pointwise on $\mathbf{R}$. In fact, if x is close to 0 and positive, there exists a positive integer n_0 such that $2/n_0 < x$; so $f_n(x) = 0$ for $n \geq n_0$. However, there exists no n_0 such that

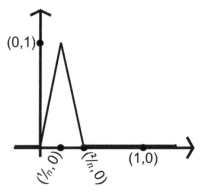

FIGURE 3.2

$$|f_n(x) - f(x)| < \frac{1}{2}$$

for all $n \geq n_0$ and all $x \in \mathbf{R}$. For, if such an n_0 were to exist, we would have

$$f_n(x) < \frac{1}{2} \text{ for } n \geq n_0$$

and for all $x \in \mathbf{R}$. We would then have $n_0 x < 1/2$ for $0 \leq x \leq 1/n_0$; but for $x = 2/3 n_0$, we obtain the contradiction $2/3 < 1/2$.

(ii) Let $X = Y = [0,1]$ with the metric induced from $\mathbf{R}$ and

$$f_n(x) = x^n, \quad 0 \leq x \leq 1.$$

Then $\{f_n\}_{n \geq 1}$ converges pointwise to f on X, where

$$f(x) = \begin{cases} 0 & 0 \leq x < 1, \\ 1 & x = 1, \end{cases}$$

as illustrated in Figure 3.3. The statement is obviously true if $x = 0$. For $0 < x < 1, x^n < \varepsilon$ provided $n \geq n_0$, where n_0 is the smallest integer greater than $(\ln \varepsilon)/(\ln x)$. However, there exists no n_0 such that

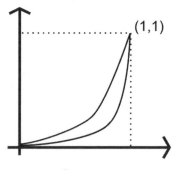

FIGURE 3.3

$$|f_n(x) - f(x)| < \frac{1}{2}$$

for all $n \geq n_0$ and all $x \in [0,1]$. For, if such an n_0 were to exist, we would have

$$x^n < \frac{1}{2} \qquad \text{whenever } n \geq n_0$$

for all $x \in [0,1]$. We would then have $x^{n_0} < 1/2$ for $0 \leq x < 1$ and upon letting $x \to 1$, would get the contradiction that $1 \leq 1/2$.

(iii) Let $X = [0, \infty)$, $Y = [0,1)$ with the metric induced from $\mathbf{R}$. Let

$$g_n(x) = \frac{x}{1 + nx}, \qquad 0 \leq x < \infty.$$

If $x > 0$ then $0 < g_n(x) \leq x/nx = 1/n \to 0$ as $n \to \infty$. Also, since $g_n(0) = 0$ for each n, it is clear that $\{g_n\}_{n \geq 1}$ converges to the function identically zero on $[0, \infty)$. Also $0 \leq g_n(x) \leq 1/n$ for $0 \leq x < \infty$. Hence, for $\varepsilon > 0$, the statement

$$|g_n(x) - 0| < \varepsilon$$

is true for all $x \in [0, \infty)$ simultaneously, provided $n_0 > 1/\varepsilon$, for, in this case,

$$|g_n(x) - 0| \leq \frac{1}{n} \leq \frac{1}{n_0} < \varepsilon \qquad \text{whenever } n \geq n_0$$

for all $x \in [0, \infty)$. In this case, n_0 depends only on ε and not on x.

Definition 3.6.3. Let $\{f_n\}_{n \geq 1}$ be a sequence of mappings of (X, d_X) into (Y, d_Y). We say that the sequence $\{f_n\}_{n \geq 1}$ **converges uniformly on X to** a mapping $f: X \to Y$ if, for every $\varepsilon > 0$, there exists an n_0 (depending on ε only) such that

$$d_Y(f_n(x), f(x)) < \varepsilon$$

for all $n \geq n_0$ and all $x \in X$; i.e.,

$$\lim_{n \to \infty} \left(\sup_{x \in X} d_Y(f_n(x), f(x)) \right) = 0.$$

It is clear that uniform convergence implies pointwise convergence; the converse is not true (see Examples 3.6.2(i) and (ii) above).

Example 3.6.4. The sequence $\{f_n\}_{n \geq 1}$ defined by

$$f_n(x) = \tan^{-1}(nx), x \geq 0,$$

is uniformly convergent on $[\alpha, \infty)$ when $\alpha > 0$, but is not uniformly convergent on $[0, \infty)$. The pointwise limit function is

$$f(x) = \lim_{n \to \infty} f_n(x) = \begin{cases} \dfrac{\pi}{2} & \text{if } x > 0, \\ 0 & \text{if } x = 0. \end{cases}$$

We shall show that $f_n \to f$ uniformly on $[\alpha, \infty)$ when $\alpha > 0$. For $x > 0$,

$$|f_n(x) - f(x)| = \left|\tan^{-1}(nx) - \frac{\pi}{2}\right| = \cot^{-1}(nx),$$

as we shall now prove. Since $0 < \tan^{-1}\theta < \pi/2$ for any $\theta > 0$, therefore when $x > 0$, we have $0 < \tan^{-1}(nx) < \pi/2$ and hence

$$0 < \frac{\pi}{2} - \tan^{-1}(nx) < \frac{\pi}{2}. \tag{3.10}$$

Also,

$$\cot\left(\frac{\pi}{2} - \tan^{-1}(nx)\right) = nx. \tag{3.11}$$

Now, it follows from (3.10) and (3.11) that $\pi/2 - \tan^{-1}(nx) = \cot^{-1}(nx)$. It also follows from the first inequality in (3.10) that $|\tan^{-1}(nx) - \pi/2| = \pi/2 - \tan^{-1}(nx)$ for $x > 0$. Thus, $|\tan^{-1}(nx) - \pi/2| = \cot^{-1}(nx)$.

Let $\varepsilon > 0$ be arbitrary. When $x \geq \alpha$, the inequality $n > (\cot \varepsilon)/\alpha$ implies that $n > (\cot \varepsilon)/x$, so that $nx > \cot\varepsilon$ and hence $\cot^{-1} nx < \varepsilon$ in view of the fact that $\cot^{-1}$ is a decreasing function. It follows that if n_0 is an integer greater than or equal to $(\cot\varepsilon)/\alpha$, then $|f_n(x) - f(x)| = |\tan^{-1}(nx) - \pi/2| = \cot^{-1} nx < \varepsilon$ whenever $n \geq n_0$ and $x \geq \alpha$. However, $(\cot\varepsilon)/x \to \infty$ as $x \to 0$, so that no integer n_0 exists for which $|f_n(x) - f(x)| < \varepsilon$ for all $n \geq n_0$ and all $x \in [0, \infty)$. Actually this proves that the convergence fails to be uniform even on the smaller set $(0, \infty)$.

The following basic result about transmission of the property of being continuous will be needed in the sequel.

Theorem 3.6.5. Let (X, d_X) and (Y, d_Y) be metric spaces, $\{f_n\}_{n \geq 1}$ a sequence of functions, each defined on X with values in Y, and let $f: X \to Y$. Suppose that $f_n \to f$ uniformly over X and that each f_n is continuous over X. Then f is continuous over X. Briefly put, a uniform limit of continuous functions is continuous.

Proof. Let $x_0 \in X$ be arbitrary and let $\varepsilon > 0$ be given. Since $f_n \to f$ uniformly over X, there exists n_0 (depending on ε only) such that for each $x \in X$,

$$d_Y(f_n(x), f(x)) < \frac{\varepsilon}{3} \qquad \text{for } n \geq n_0. \tag{3.12}$$

Since f_{n_0} is continuous at x_0, we can choose $\delta > 0$ such that $x \in S(x_0, \delta) = \{x \in X : d_X(x, x_0) < \delta\}$ implies

$$d_Y(f_{n_0}(x), f_{n_0}(x_0)) < \frac{\varepsilon}{3}. \tag{13.13}$$

Using (3.12) and (3.13), $x \in S(x_0, \delta)$ implies

$$d_Y(f(x), f(x_0)) \leq d_Y(f(x), f_{n_0}(x)) + d_Y(f_{n_0}(x), f_{n_0}(x_0)) + d_Y(f_{n_0}(x_0), f(x_0))$$
$$< \frac{\varepsilon}{3} + \frac{\varepsilon}{3} + \frac{\varepsilon}{3} = \varepsilon,$$

and therefore, f is continuous at x_0. $\qquad\qquad\qquad\qquad\qquad\qquad\qquad\square$

Proposition 3.6.6. (Cauchy Criterion) Let $\{f_n\}_{n \geq 1}$ a sequence of functions defined on a metric space (X, d_X) with values in a complete metric space (Y, d_Y). Then there exists a function $f : X \to Y$ such that

$$f_n \to f \text{ uniformly on } X$$

if and only if the following condition is satisfied: For every $\varepsilon > 0$, there exists an integer n_0 such that

$$m, n \geq n_0 \quad \text{implies } d_Y(f_m(x), f_n(x)) < \varepsilon$$

for every $x \in X$.

Proof. Suppose that $f_n \to f$ uniformly on X. Then, given $\varepsilon > 0$, there exists n_0 such that $n \geq n_0$ implies $d_Y(f_n(x), f(x)) < \varepsilon/2$ for all $x \in X$. Therefore, $m, n \geq n_0$ implies $d_Y(f_m(x), f(x)) < \varepsilon/2$ as well as $d_Y(f_n(x), f(x)) < \varepsilon/2$ for all $x \in X$, and hence

$$d_Y(f_m(x), f_n(x)) \leq d_Y(f_m(x), f(x)) + d_Y(f_n(x), f(x))$$

$$< \frac{\varepsilon}{2} + \frac{\varepsilon}{2} = \varepsilon \quad \text{for all } x \in X.$$

Conversely, suppose that $m, n \geq n_0$ implies $d_Y(f_m(x), f_n(x)) < \varepsilon$ for all $x \in X$. Then for each $x \in X$, the sequence $\{f_n(x)\}_{n \geq 1}$ is Cauchy in the complete space Y and therefore converges. Let $f(x) = \lim_{n \to \infty} f_n(x)$ for each $x \in X$. We must show that $f_n \to f$ uniformly over X. If $\varepsilon > 0$ is given, we can choose n_0 such that $n \geq n_0$ implies

$$d_Y(f_n(x), f_{n+k}(x)) < \frac{\varepsilon}{2}$$

for every $k = 1, 2, \ldots$ and every $x \in X$. Letting $k \to \infty$, we get

$$d_Y(f_n(x), f(x)) = \lim_{k \to \infty} d_Y(f_n(x), f_{n+k}(x)) \leq \frac{\varepsilon}{2}.$$

Hence, $n \geq n_0$ implies $d_Y(f_n(x), f(x)) < \varepsilon$ for every $x \in X$, and so $f_n \to f$ uniformly over X. $\qquad \square$

We next consider convergence and uniform convergence of series of functions.

Definition 3.6.7. Let $f_1, f_2, \ldots$ be a sequence of real-valued functions defined on a set X. We say that $\sum_{n=1}^{\infty} f_n$ **converges on X to a function** $f : X \to \mathbf{R}$ if the sequence of functions $\{S_n\}_{n \geq 1}$ converges to f on X, where $S_n = f_1 + f_2 + \ldots + f_n$. In this case, we write

$$\sum_{n=1}^{\infty} f_n = f$$

or

$$\sum_{n=1}^{\infty} f_n(x) = f(x), x \in X.$$

Example 3.6.8. If $f_n(x) = x^n$, $-1 < x < 1$, then $\sum_{n=1}^{\infty} f_n$ converges to f on $(-1, 1)$, where $f(x) = x/(1-x)$. This is because

$$\sum_{n=1}^{\infty} f_n(x) = \sum_{n=1}^{\infty} x^n = \frac{x}{1-x} = f(x), x \in (-1, 1).$$

Note that the series $\sum_{n=1}^{\infty} x^n$ does not converge at $x = \pm 1$.

For *uniform convergence* of a series, the sequence $\{S_n\}_{n \geq 1}$ of partial sums is required to converge uniformly. More precisely, we have the following.

Definition 3.6.9. Let $f_1, f_2, \ldots$ be a sequence of real-valued functions defined on a set X. We say that $\sum_{n=1}^{\infty} f_n$ **converges uniformly on** X **to a function** $f: X \to \mathbf{R}$ if the sequence of functions $\{S_n\}_{n \geq 1}$ converges uniformly to f on X, where $S_n = f_1 + f_2 + \ldots + f_n$. In this case, we write

$$\sum_{n=1}^{\infty} f_n = f \text{ uniformly}$$

or

$$\sum_{n=1}^{\infty} f_n(x) = f(x) \text{ uniformly, } x \in X.$$

Corollary 3.6.10. Let $f_1, f_2, \ldots$ be a sequence of real-valued functions defined on a metric space (X, d_X). If $\sum_{n=1}^{\infty} f_n$ converges uniformly to f on X, and if each f_n is continuous on X, then so is f.

Proof. The sequence of functions $\{S_n\}_{n \geq 1}$ converges uniformly to f on X, where $S_n = f_1 + f_2 + \ldots + f_n$. Since each S_n, being a *finite* sum of continuous functions on X, is continuous by Theorem 3.3.2, it follows, using Theorem 3.6.5, that f is continuous on X. $\square$

Example 3.6.11. The series

$$\sum_{n=0}^{\infty} x(1-x)^n$$

converges to the function $f(x) = \begin{cases} 0 & \text{if } x = 0, \\ 1 & \text{if } 0 < x \leq 1. \end{cases}$

In fact, if $0 < x < 1$,

$$\sum_{n=0}^{\infty} x(1-x)^n = x \frac{1}{1-(1-x)} = 1.$$

Since the function f is not continuous, the convergence is not uniform by Corollary 3.6.10.

The following theorem provides sufficient conditions for uniform convergence of series of functions.

Theorem 3.6.12. (Weierstrass M-test) Let $f_1, f_2, \ldots$ be a sequence of real-valued functions defined on a set X and suppose that

$$|f_n(x)| \leq M_n$$

for all $x \in X$ and all $n = 1, 2, \ldots$. If $\sum_{n=1}^{\infty} M_n$ converges, then $\sum_{n=1}^{\infty} f_n$ converges uniformly.

Proof. If $\sum_{n=1}^{\infty} M_n$ converges, then for arbitrary $\varepsilon > 0$,

$$\left| \sum_{k=n}^{m} f_k(x) \right| \leq \sum_{k=n}^{m} |f_k(x)| \leq \sum_{k=n}^{m} M_k < \varepsilon$$

for all x provided that m and n are sufficiently large. Uniform convergence now follows from Proposition 3.6.6. $\qquad\square$

Examples 3.6.13. (i) The series $\sum_{n=1}^{\infty} \frac{\sin nx}{n^p}$, $-\infty < x < \infty$, $p > 1$, is such that

$$\left| \frac{\sin nx}{n^p} \right| \leq \frac{1}{n^p}$$

for all $x \in (-\infty, \infty)$ and $n = 1, 2, \ldots$. Since $\sum_{n=1}^{\infty} \frac{1}{n^p} < \infty$ when $p > 1$, it follows upon using the Weierstrass M-test (Theorem 3.6.12) that $\sum_{n=1}^{\infty} \frac{\sin nx}{n^p}$ is uniformly convergent. Hence, by Corollary 3.6.10, $\sum_{n=1}^{\infty} \frac{\sin nx}{n^p}$ represents a continuous function on $(-\infty, \infty)$.

(ii) For any integer n, let g_n be the continuous function defined on $\mathbf{R}$ by

$$g_n(t) = \begin{cases} 0 & \text{if } t \leq 0, \\ nt & \text{if } 0 \leq t \leq \frac{1}{n}, \\ -nt + 2 & \text{if } \frac{1}{n} \leq t \leq 2/n, \\ 0 & \text{if } t \geq 2/n. \end{cases}$$

We have already seen in Example 3.6.2 (i) that $g_n \to 0$ pointwise on $\mathbf{R}$ and that the convergence is not uniform. Let $\{r_m\}_{m \geq 1}$ be an enumeration of the rationals and let

$$f_n(t) = \sum_{m=1}^{\infty} 2^{-m} g_n(t - r_m), \qquad t \in \mathbf{R}.$$

Since $|2^{-m} g_n(t - r_m)| \leq 2^{-m}$ and since $\sum_{m=1}^{\infty} 2^{-m}$ is convergent, it follows, using the Weierstrass M-test, that the series converges uniformly. As each of the terms of the series is a continuous function, so is $f_n(t)$, by Corollary 3.6.10. Observe that the pointwise limit of the sequence $\{f_n\}_{n \geq 1}$ is the function identically zero. In fact, for fixed $t \in \mathbf{R}$,

$$\lim_{n\to\infty} f_n(t) = \lim_{n\to\infty} \sum_{m=1}^{\infty} 2^{-m} g_n(t - r_m)$$

$$= \sum_{m=1}^{\infty} 2^{-m} \lim_{n\to\infty} g_n(t - r_m).$$

However, the sequence $\{f_n\}_{n\geq 1}$ does not converge uniformly in any interval of **R**. Observe that $0 \leq g_n(t) \leq 1$ and for $t = r_m + 1/n, g_n(t - r_m) = g_n(1/n) = 1$ for $m = 1, 2, \ldots$. So,

$$f_n(t) = \sum_{m=1}^{\infty} 2^{-m} g_n(t - r_m)$$

$$\geq 2^{-1} g_n(t - r_1) = \frac{1}{2}$$

for $t = r_1 + 1/n, n = 1, 2, \ldots$. Hence, the convergence of f_n to the identically zero function is not uniform.

We encountered extension theorems in Section 3.2. Those were extension theorems from subsets to their closures. The Tietze-Urysohn extension theorem extends a continuous real-valued function defined on a proper closed subset of a metric space to the whole space. First we establish the following proposition, which is needed in the proof of the main theorem.

Proposition 3.6.14. Let (X, d_X) be a metric space and Y a nonempty closed subset of (X, d_X); let f be a bounded, continuous real-valued function defined on Y, for which there exists $M > 0$ such that

$$\inf_{x\in Y} f(x) = -M, \quad \sup_{x\in Y} f(x) = M.$$

Then there exists a continuous real-valued function g defined on X satisfying the following properties:

(i) $|g(x)| \leq (1/3)M, x \in X,$
(ii) $|g(x)| < (1/3)M, x \in X \backslash Y,$
(iii) $|f(x) - g(x)| \leq (2/3)M, x \in Y.$

Proof. Let

$$A = \{x \in Y : f(x) \leq -\frac{1}{3}M\} \text{ and } B = \{x \in Y : f(x) \geq \frac{1}{3}M\}.$$

Observe that A and B are nonempty, disjoint and closed in Y. (If $x \in Y$ is a limit point of A and $\{x_n\}_{n\geq 1}$ is a sequence in A that converges to x, it follows, using the continuity of f, that

$$f(x) = f(\lim_{n\to\infty} x_n) = \lim_{n\to\infty} f(x_n) \leq -\frac{1}{3}M,$$

i.e., $x \in A$.) Since Y is itself a closed subset of X, it follows that A and B are closed in X, by Proposition 2.2.3. Define g on X by

$$g(x) = \frac{1}{3} M \frac{d(x, A) - d(x, B)}{d(x, A) + d(x, B)}.$$

By Example 3.4.2 (ii) and Theorem 3.3.2, g is continuous.

Moreover, $|g(x)| \leq M/3$, $x \in X$. If $x \in X \backslash Y$, then $x \notin A$ and $x \notin B$; so $|g(x)| < M/3$. Suppose $x \in Y$. If $x \in A$ then $g(x) = -M/3, -M \leq f(x) \leq -M/3$ and so $|f(x) - g(x)| \leq 2M/3$. Similarly, for $x \in B$. If $x \in Y \backslash (A \cup B)$, then $|g(x)| < M/3$ and $|f(x)| < M/3$ and so $|f(x) - g(x)| < 2M/3$. $\square$

Theorem 3.6.15. (Tietze's Extension Theorem) If f is a bounded, continuous real-valued function defined on a closed subset Y of a metric space (X, d_X), there exists a continuous real-valued function g defined on X which extends f (as in Definition 3.2.1). Moreover, if

$$m = \inf \{f(y) : y \in Y\} < \sup \{f(y) : y \in Y\} = M,$$

then g satisfies $m < g(x) < M$ for all $x \in X \backslash Y$.

Proof. We assume without loss of generality that $m = -M$, for otherwise we may replace f by $f - \frac{m+M}{2}$. Then $|f(x)| \leq M$ on Y. By Proposition 3.6.14, we can find a continuous real-valued function g_1 on X such that

$$|g_1(x)| \leq \frac{1}{3} M, \quad x \in X$$
$$|g_1(x)| < \frac{1}{3} M, \quad x \in X \backslash Y$$
$$|f(x) - g_1(x)| \leq \frac{2}{3} M, \quad x \in Y.$$

Let f_2 be defined in Y by $f_2(x) = f(x) - g_1(x)$. By Proposition 3.6.14, we next find a continuous real-valued function g_2 defined on X such that

$$|g_2(x)| \leq \frac{1}{3}\left(\frac{2}{3}M\right), \quad x \in X$$
$$|g_2(x)| < \frac{1}{3}\left(\frac{2}{3}M\right), \quad x \in X \backslash Y$$
$$|f_2(x) - g_2(x)| \leq \left(\frac{2}{3}\right)^2 M, \quad x \in Y.$$

This can be continued indefinitely. An easy induction argument shows that this leads to a sequence of functions $\{g_n\}_{n \geq 1}$, each continuous on X, such that

$$|g_n(x)| \leq \frac{1}{3}\left(\frac{2}{3}\right)^{n-1} M, \quad x \in X$$
$$|g_n(x)| < \frac{1}{3}\left(\frac{2}{3}\right)^{n-1} M, \quad x \in X | Y$$
$$|f_n(x) - g_n(x)| \leq \left(\frac{2}{3}\right)^n M, \quad x \in Y,$$

where $f_n = f_{n-1} - g_{n-1} = f_{n-2} - (g_{n-2} + g_{n-1}) = \ldots = f - (g_1 + g_2 + \ldots + g_{n-1})$;
it follows that

$$|f(x) - \sum_{k=1}^{n} g_k(x)| \leq \left(\frac{2}{3}\right)^n M, \quad x \in Y. \tag{3.15}$$

Define g by

$$g(x) = \sum_{k=1}^{\infty} g_k(x), \quad x \in X.$$

Since $|g_n(x)| \leq (1/3)(2/3)^{n-1} M$ and since $\sum_{n=1}^{\infty} (1/3)(2/3)^{n-1} M$ is convergent, it follows from the Weierstrass M-test (see Theorem 3.6.12) that the series $\sum_{k=1}^{\infty} g_k(x)$ is uniformly convergent. As each term of the series is continuous, g represents a continuous function on X.

Letting $n \to \infty$ in (3.15), it follows that $f(x) = g(x), x \in Y$. For $x \in X \backslash Y$, we have

$$|g(x)| \leq \sum_{k=1}^{\infty} |g_k(x)| < \sum_{n=1}^{\infty} \frac{1}{3}\left(\frac{2}{3}\right)^{n-1} M = M.$$

This completes the proof. $\qquad \square$

Remark 3.6.16. The restriction that f is bounded on Y can be dispensed with by considering the continuous bounded function $\tan^{-1} \circ f$ in place of f. An application of the Theorem 3.6.15 with $M = \pi/2$ yields a continuous extension H of $\tan^{-1} \circ f$, where H is defined on all of X. Observe that, in view of the last part of the above theorem, H does not assume the values $\pm \pi/2$. Define g on X by

$$g = \tan \circ H;$$

then g is continuous on X and $g(x) = f(x), x \in Y$.

3.7. Contraction Mappings and Applications

The concept of completeness of metric spaces has interesting and important applications in classical analysis. In this section, we show how various existence and uniqueness theorems in the theory of differential and integral equations follow from very simple facts about mappings in a complete metric space. The simple fact alluded to above is called the **contraction mapping principle**, which we now consider.

Definition 3.7.1. Let (X, d) be a metric space. A mapping T of X into itself is said to be a contraction (or contraction mapping) if there exists a real number $\alpha, 0 < \alpha < 1$, such that

$$d(Tx, Ty) \leq \alpha d(x, y)$$

for all $x, y \in X$.

It is obvious that a contraction mapping is uniformly continuous (see Definition 3.4.1).

Examples 3.7.2. (i) Let $f: [a, b] \to [a, b]$ be differentiable. Assume that there exists a number $K < 1$ such that $|f'(x)| \leq K$ for all $x \in (a, b)$. Then f is a contraction mapping. Indeed, if $x, y \in [a, b], x \neq y$, then

$$\frac{f(x) - f(y)}{x - y} = f'(c),$$

where c lies between x and y. Since $|f'(c)| \leq K$, we have

$$|f(x) - f(y)| \leq K|x - y|.$$

(ii) If $x = \{x_n\} \in \ell_2$ then $Tx = \{\frac{x_n}{2}\}$ is a contraction mapping of ℓ_2 into itself. For, if $y = \{y_n\} \in \ell_2$ and $x \neq y$, then

$$d(Tx, Ty) = \left(\sum_{n=1}^{\infty} \left(\frac{x_n}{2} - \frac{y_n}{2} \right)^2 \right)^{1/2} = \frac{1}{2} \left(\sum_{n=1}^{\infty} (x_n - y_n)^2 \right)^{1/2} = \frac{1}{2} d(x, y).$$

(iii) If $Tx = x^n, 0 \leq x \leq 1/(n+1)$, where $n > 1$, then T is a contraction mapping of $[0, 1/(n+1)]$ with the usual metric d. This is because

$$d(Tx, Ty) = |x^n - y^n| = |x - y||x^{n-1} + x^{n-2}y + \ldots + y^{n-1}| \leq \frac{n}{(n+1)^{n-1}} |x - y|$$

for all $x, y \in [0, 1/(n+1)]$. This is actually a contraction mapping of $[0,a]$, where $0 < a < n^{-(1/(n-1))}$.

Definition 3.7.3. A point $x \in X$ is called a **fixed point** of the mapping $T: X \to X$ if $Tx = x$.

The key result of this section is the following theorem, due to S. Banach.

Theorem 3.7.4. (Contraction Mapping Principle) Let $T: X \to X$ be a contraction of the complete metric space (X, d). Then T has a unique fixed point.

Proof. Let $x_0 \in X$ and let $\{x_n\}_{n \geq 1}$ be the sequence defined iteratively by $x_{n+1} = Tx_n$ for $n = 0, 1, 2, \ldots$. We shall prove that $\{x_n\}_{n \geq 1}$ is a Cauchy sequence. For $p = 1, 2, \ldots$, we have

$$d(x_{p+1}, x_p) = d(Tx_p, Tx_{p-1}) \leq \alpha d(x_p, x_{p-1}), \tag{3.16}$$

where $0 < \alpha < 1$ is such that

$$d(Tx, Ty) \leq \alpha d(x, y)$$

for all $x, y \in X$.

Repeated application of the inequality (3.16) gives

$$d(x_{p+1}, x_p) \le \alpha d(x_p, x_{p-1})$$
$$\le \alpha^2 d(x_{p-1}, x_{p-2}) \le \ldots \le \alpha^p d(x_1, x_0).$$

Now, let m, n be positive integers with $m > n$. By the triangle inequality,

$$d(x_m, x_n) \le d(x_m, x_{m-1}) + d(x_{m-1}, x_{m-2}) + \ldots + d(x_{n+1}, x_n)$$
$$\le (\alpha^{m-1} + \alpha^{m-2} + \ldots + \alpha^n) d(x_1, x_0)$$
$$= \alpha^n (\alpha^{m-n-1} + \alpha^{m-n-2} + \ldots + 1) d(x_1, x_0)$$
$$\le \frac{\alpha^n}{1 - \alpha} d(x_1, x_0).$$

But $\lim_{n \to \infty} \alpha^n = 0$. It follows that $\{x_n\}$ is a Cauchy sequence in (X, d), which is complete. Let $y = \lim_{n \to \infty} x_n$. Since T is a contraction, it is continuous. It follows that $Ty = T(\lim_{n \to \infty} x_n) = \lim_{n \to \infty} Tx_n = \lim_{n \to \infty} x_{n+1} = y$. Thus, y is a fixed point of T. Moreover, it can be shown to be unique: If $y \ne z$ are such that $Ty = y$ and $Tz = z$, then $d(y, z) = d(Ty, Tz) \le \alpha d(y, z) < d(y, z)$. This implies $d(y, z) = 0$, i.e., $y = z$. $\square$

Remark 3.7.5. Let $T: X \to X$, where (X, d) is a complete metric space, satisfy the inequality

$$d(Tx, Ty) < d(x, y)$$

for all $x, y \in X$. Then T need not have a fixed point. The map $T: [1, \infty) \to [1, \infty)$ such that $Tx = x + 1/x$ has no fixed point, although

$$|Tx - Ty| = |x - y| \left(1 - \frac{1}{xy} \right) < |x - y|$$

and $[1, \infty)$ is complete.

Proposition 3.7.6. Suppose $\{x_n\}_{n \ge 1}$ is a sequence in any metric space (X, d) and k a positive integer such that each of the k subsequences

$$\{x_{mk+j}\}_{m \ge 1}, j = 0, 1, 2, \ldots, k - 1$$

converges to the same limit x. Then $\{x_n\}_{n \ge 1}$ converges to x.

Proof. Given $\varepsilon > 0$, there exist positive integers $m_0, m_1, \ldots, m_{k-1}$ such that

$$m \ge m_j \quad \text{implies} \quad d(x_{mk+j}, x) < \varepsilon, \quad j = 0, 1, 2, \ldots, k - 1.$$

Take $N = \max\{m_0, m_1, \ldots, m_{k-1}\}$. Consider any $n \ge Nk$. By the division algorithm, $n = mk + j$, where $0 \le j \le k - 1$. Since $m \le N - 1$ would imply that $n \le (N-1)k + j \le (N-1)k + (k-1) = Nk - 1$, we must have $m > N - 1$, so that $m \ge N \ge m_j$ for $j = 0, 1, 2, \ldots, k - 1$. It follows that $d(x_{mk+j}, x) < \varepsilon$, i.e., $d(x_n, x) < \varepsilon$. This has been proved for any $n \ge Nk$. Therefore, $\{x_n\}_{n \ge 1}$ converges to x. $\square$

The following extension of the contraction mapping principle is often useful.

Corollary 3.7.7. Let T be a continuous mapping of a complete metric space into itself. Define T^n inductively by $T^1 = T$ and $T^{n+1} = T \circ T^n$. If T^k is a contraction for some positive integer k, then the equation

$$Tx = x$$

has one and only one solution. Moreover, for any $x \in X$, the sequence $\{T^n x\}_{n \geq 1}$ converges to that solution.

Proof. Let $x \in X$ be arbitrary and consider the sequence $T^{nk}x, n = 0, 1, 2, \ldots$. A repetition of the argument in the contraction mapping principle yields the convergence of the sequence $\{T^{nk}x\}_{n \geq 1}$. Let $x_0 = \lim_{n \to \infty} T^{nk}x$. We shall show that $Tx_0 = x_0$.

Since the mapping T^k is a contraction,

$$d(T^{nk} Tx, T^{nk}x) \leq \alpha d(T^{(n-1)k} Tx, T^{(n-1)k}x) \leq \ldots \leq \alpha^n d(Tx, x),$$

where $0 < \alpha < 1$, is such that $d(T^k x, T^k y) \leq \alpha d(x, y)$ for all $x, y \in X$. Using the continuity of T, we have,

$$d(Tx_0, x_0) = \lim_{n \to \infty} d(TT^{nk}x, T^{nk}x) = \lim_{n \to \infty} d(T^{nk} Tx, T^{nk}x) \leq \limsup_n \alpha^n d(Tx, x) = 0$$

since $0 < \alpha < 1$. This completes the proof of the existence of a solution of the equation $Tx = x$. The proof of uniqueness is no different from the one given in Theorem 3.7.4 and is, therefore, not repeated here.

Since T^k is a contraction, the sequence $\{T^{mk}\xi\}_{m \geq 1}$ converges to x_0 for any choice of ξ. By successively choosing ξ to be $x, Tx, \ldots, T^{k-1}x$, we get the k sequences

$$\{T^{mk+j}x\}_{m \geq 1}, \quad j = 0, 1, 2, \ldots, k - 1.$$

Since each one of these converges to the same limit x_0, it follows that the sequence $\{T^n x\}_{n \geq 1}$ converges to x_0. $\qquad \square$

I. Linear Equations

Consider the mapping $y = Tx$ of the space $\mathbf{C}^n$ into itself given by the system of equations

$$y_i = \sum_{j=1}^{n} a_{ij}x_j + b_i, \quad i = 1, 2, \ldots, n.$$

(a) If $d_\infty(x, y) = \max\{|x_i - y_i| : 1 \leq i \leq n\}$ is the metric on $\mathbf{C}^n$, then $\sum_{j=1}^{n} |a_{ij}| \leq \alpha < 1, i = 1, 2, \ldots, n$, implies that the system of equations

$$x_i = \sum_{j=1}^{n} a_{ij}x_j + b_i, \quad i = 1, 2, \ldots, n,$$

has exactly one solution.

To prove (a) we only need to show that T is a contraction mapping. To this end, consider

$$x' = (x'_1, x'_2, \ldots, x'_n) \text{ and } x'' = (x''_1, x''_2, \ldots, x''_n).$$

We have

$$d_\infty(Tx', Tx'') = \max_i |\sum_{j=1}^{n} a_{ij}(x'_j - x''_j)|$$

$$\le \max_i \sum_{j=1}^{n} |a_{ij}||x'_j - x''_j|$$

$$\le \max_i (\max_j |x'_j - x''_j|) \sum_{j=1}^{n} |a_{ij}|$$

$$\le \alpha d_\infty(x', x'').$$

This completes the proof of (a). $\square$

(b) If $d_1(x, y) = \sum_{i=1}^{n} |x_i - y_i|$ is the metric on $\mathbf{C}^n$, then $\sum_{i=1}^{n} |a_{ij}| \le \alpha < 1, j = 1, 2, \ldots, n$, implies that the system of equations

$$x_i = \sum_{j=1}^{n} a_{ij}x_j + b_i, i = 1, 2, \ldots, n,$$

has exactly one solution.

Here again, we need only show that the mapping T described above is a contraction in order to prove (b). As before, consider

$$x' = (x'_1, x'_2, \ldots, x'_n) \quad \text{and} \quad x'' = (x''_1, x''_2, \ldots, x''_n).$$

We have

$$d_1(Tx', Tx'') = \sum_{i=1}^{n} |\sum_{j=1}^{n} a_{ij}(x'_j - x''_j)|$$

$$\le \sum_{i=1}^{n} \sum_{j=1}^{n} |a_{ij}||x'_j - x''_j|$$

$$= \sum_{j=1}^{n} \sum_{i=1}^{n} |a_{ij}||x'_j - x''_j|$$

$$= \sum_{j=1}^{n} \left\{ |x'_j - x''_j| \left(\sum_{i=1}^{n} |a_{ij}| \right) \right\}$$

$$\le \max_j \sum_{i=1}^{n} |a_{ij}| d_1(x', x'')$$

$$\le \alpha d_1(x', x'').$$

This completes the proof of (b). $\square$

(c) If $d_2(x, y) = [\sum_{i=1}^{n} |x_i - y_i|^2]^{1/2}$ is the metric on $\mathbf{C}^n$, then $\sum_{i=1}^{n} \sum_{j=1}^{n} |a_{ij}|^2 \le \alpha^2 < 1$ implies that the system of equations

$$x_i = \sum_{j=1}^{n} a_{ij} x_j + b_i, i = 1, 2, \ldots, n,$$

has exactly one solution.

Once again, we need only show that T is a contraction. For $x' = (x_1', x_2', \ldots, x_n')$ and $x'' = (x_1'', x_2'', \ldots, x_n'')$, we have

$$[d_2(Tx', Tx'')]^2 = \sum_{i=1}^{n} \left| \sum_{j=1}^{n} a_{ij}(x_j' - x_j'') \right|^2$$

$$\le \sum_{i=1}^{n} \left(\sum_{j=1}^{n} |a_{ij}| |x_j' - x_j''| \right)^2$$

$$\le \sum_{i=1}^{n} \left(\sum_{j=1}^{n} |a_{ij}|^2 \sum_{j=1}^{n} |x_j' - x_j''|^2 \right)$$

using the Schwarz inequality. So,

$$[d_2(Tx', Tx'')]^2 \le \sum_{i=1}^{n} \sum_{j=1}^{n} |a_{ij}|^2 d_2(x', x'')^2 \le \alpha^2 d_2(x', x'')^2.$$

The proof of (c) is complete. $\qquad\qquad\qquad\qquad\qquad\qquad\qquad\qquad\qquad\qquad$ $\square$

II. Differential Equations (Picard's Theorem)

The contraction mapping principle will now be used to obtain a general result about the existence of a unique solution to a differential equation of the form

$$\frac{dy}{dx} = f(x, y)$$

satisfying certain conditions. First we show that, under some milder conditions, the equation is equivalent to an integral equation.

Definition 3.7.8. Let f be a continuous real-valued function on some rectangle

$$R = \{(x, y): |x - x_0| \le a, |y - y_0| \le b\},$$

where $a > 0$ and $b > 0$ in the (x, y)-plane. A real-valued function φ defined on an interval I is said to be a **solution of the initial value problem**

$$y' = f(x, y), y(x_0) = y_0 \qquad\qquad\qquad\qquad (3.17)$$

if and only if, firstly, $(x, \varphi(x))$ is in R with $\varphi'(x) = f(x, \varphi(x))$ whenever $x \in I$, and secondly, $\varphi(x_0) = y_0$.

Our preliminary step will be to show that the initial value problem is equivalent to the integral equation

$$y(x) = y_0 + \int_{x_0}^{x} f(t, y(t)) dt \tag{3.18}$$

on I.

Definition 3.7.9. By a **solution of (3.18) on** I is meant a real-valued function φ such that $(x, \varphi(x))$ is in R for all $x \in I$ and

$$\varphi(x) = y_0 + \int_{x_0}^{x} f(t, \varphi(t)) dt$$

for all $x \in I$. (We note that this equality implies that φ must be continuous.)

Proposition 3.7.10. A function φ is a solution of the initial value problem (3.17) on an interval I if and only if it is a solution of the integral equation (3.18) on I.

Proof. Suppose φ is a solution of the initial value problem (3.17) on I. Then

$$\varphi'(t) = f(t, \varphi(t)) \tag{3.19}$$

on I. Since φ is continuous on I and f is continuous on R, the function F defined on I by $F(t) = f(t, \varphi(t))$ is continuous on I. Integrating (3.19) from x_0 to x, we get

$$\varphi(x) = \varphi(x_0) + \int_{x_0}^{x} f(t, \varphi(t)) dt.$$

Since $\varphi(x_0) = y_0$, it follows that φ is a solution of (3.18).

Conversely, suppose that φ is a solution of (3.18) on I. Differentiating with respect to x and using the fundamental theorem of integral calculus, we see that

$$\varphi'(x) = f(x, \varphi(x))$$

for all $x \in I$. Moreover, it is clear from (3.18) that

$$\varphi(x_0) = y_0.$$

So φ is a solution of the initial value problem (3.17). $\square$

We shall also need the following proposition for the proof of Picard's theorem.

Proposition 3.7.11. Let X consist of continuous mappings from $[a, b]$ into $[l, m]$, so that X is a subspace of the space $C[a, b]$ of all continuous real-valued functions defined on $[a, b]$. Then X is a complete metric space.

Proof. Since $X \subset C[a, b]$ and the latter is complete, it is enough to show that X is a closed subset of it. Accordingly, we assume that $f \in C[a, b]$ is a limit point of X and proceed to show that $f \in X$. There exists a sequence $\{f_n\}_{n \geq 1}$ in X that converges to f with respect to the metric of $C[a, b]$. For each $x \in [a, b]$, we have

$$|f_n(x) - f(x)| \leq d(f_n, f)$$

and hence $\lim_{n \to \infty} f_n(x) = f(x)$. But for each $x \in [a, b]$, we have

$$l \leq f_n(x) \leq m$$

and so,

$$l \leq f(x) \leq m.$$

This shows that $f \in X$. This completes the proof.   □

Theorem 3.7.12. (Picard's Theorem) Let f be continuous on

$$R = \{(x, y) : |x - x_0| \leq a, |y - y_0| \leq b\}$$

where $a > 0$ and $b > 0$. Let

$$K = \sup\{|f(x, y)| : (x, y) \in R\}. \tag{3.19}$$

If there exists $M > 0$ such that

$$|f(x, y_1) - f(x, y_2)| \leq M|y_1 - y_2| \tag{3.20}$$

for $(x, y_1), (x, y_2) \in R$, then the positive number $\delta = \min\{a, b/K\}$ has the property that there exists a unique function φ on $[x_0 - \delta, x_0 + \delta]$ such that

$$\varphi'(x) = f(x, \varphi(x)) \quad \text{whenever } x \in [x_0 - \delta, x_0 + \delta] \quad \text{and } \varphi(x_0) = y_0. \tag{3.21}$$

The function φ is the uniform limit of any sequence $\{\varphi_n\}$ of continuous functions on $[x_0 - \delta, x_0 + \delta]$ such that

$$\varphi_n(x) = y_0 + \int_{x_0}^{x} f(t, \varphi_{n-1}(t))dt, \quad n = 1, 2, \ldots$$

and φ_0 is any continuous function on $[x_0 - \delta, x_0 + \delta]$ with values in $[y_0 - b, y_0 + b]$.

Proof. Let $X = \{\varphi \in C[x_0 - \delta, x_0 + \delta] : |\varphi(x) - y_0| \leq b$ for $x \in [x_0 - \delta, x_0 + \delta]\}$. Then X is a complete metric space by Proposition 3.7.11. In view of the choice of δ, it is clear that $(x, \varphi(x)) \in R$ if $|x - x_0| \leq \delta$ and $\varphi \in X$. Therefore, we can define T on X as follows:

$$(T\varphi)(x) = y_0 + \int_{x_0}^{x} f(t, \varphi(t))dt, \quad |x - x_0| \leq \delta.$$

Observe that, in view of Proposition 3.7.10, φ satisfies (3.21) if and only if $T\varphi = \varphi$, i.e., if and only if φ is a fixed point of T. Thus, in order to prove the existence of the requisite function φ, it is sufficient to show that T maps X into itself and that it has a unique fixed point.

If $|x - x_0| \leq \delta$, then by (3.19) and the definition of δ,

$$|(T\varphi)(x) - y_0| = \left| \int_{x_0}^{x} f(t, \varphi(t)) dt \right| \leq \int_{x_0}^{x} |f(t, \varphi(t))| dt \leq K|x - x_0| \leq K\delta \leq b$$

and so $T\varphi \in X$. Hence, T is a mapping of X into itself.

We next show that T^k is a contraction for some positive integer k. For this purpose, consider any φ_1 and φ_2 in X. Then

$$(T\varphi_1)(x) - (T\varphi_2)(x) = \int_{x_0}^{x} [f(t, \varphi_1(t)) - f(t, \varphi_2(t))] dt$$

and so, when $x_0 \leq x$, we have

$$|(T\varphi_1)(x) - (T\varphi_2)(x)| \leq \int_{x_0}^{x} |f(t, \varphi_1(t)) - f(t, \varphi_2(t))| dt$$

$$\leq M \int_{x_0}^{x} |\varphi_1(t) - \varphi_2(t)| dt \qquad \text{by (3.20)}$$

$$\leq M(x - x_0) d(\varphi_1, \varphi_2).$$

Since this argument is valid for any φ_1 and φ_2 in X, we can repeat it with $T\varphi_1$ and $T\varphi_2$ in place of φ_1 and φ_2. This leads to

$$|(T^2\varphi_1)(x) - (T^2\varphi_2)(x)| \leq M \int_{x_0}^{x} |T\varphi_1(t) - T\varphi_2(t)| dt$$

$$\leq M \int_{x_0}^{x} M(t - x_0) d(\varphi_1, \varphi_2) dt.$$

$$\leq \frac{M^2(x - x_0)^2}{2} d(\varphi_1, \varphi_2)$$

After $n - 1$ repetitions, we get

$$|(T^n\varphi_1)(x) - (T^n\varphi_2)(x)| \leq \frac{M^n(x - x_0)^n}{n!} d(\varphi_1, \varphi_2)$$

$$\leq \frac{M^n|x - x_0|^n}{n!} d(\varphi_1, \varphi_2)$$

$$\leq \frac{(Ma)^n}{n!} d(\varphi_1, \varphi_2).$$

It is easy to verify that a similar argument when $x \leq x_0$ leads to the same inequality. Since $\lim_{n \to \infty} \frac{(Ma)^n}{n!} = 0$, there exists k such that $\frac{(Ma)^k}{k!} < 1$, which makes T^k a contraction. The result now follows from Corollary 3.7.7. $\qquad \square$

In the above theorem, it is essential that R be bounded, so that the finite supremum K exists. For the important case of linear equations, it is vitally important to replace R by an infinite vertical strip and to show that a unique solution exists on the entire interval covered by the strip.

Theorem 3.7.13. (Picard's Theorem in a Vertical Strip) Let f be continuous on

$$R = \{(x, y) : x \in I\},$$

where I is a closed bounded interval. If there exists $M > 0$ such that

$$|f(x, y_1) - f(x, y_2)| \le M|y_1 - y_2| \tag{3.22}$$

for $(x, y_1), (x, y_2) \in R$, then for any $(x_0, y_0) \in R$, there exists a unique function φ on I such that

$$\varphi'(x) = f(x, \varphi(x)) \quad \text{whenever } x \in I \text{ and } \varphi(x_0) = y_0 \tag{3.23}$$

The function φ is the uniform limit of any sequence $\{\varphi_n\}$ of continuous functions on I such that

$$\varphi_n(x) = y_0 + \int_{x_0}^{x} f(t, \varphi_{n-1}(t))dt, \quad n = 1, 2, \ldots$$

and φ_0 is any continuous function on I.

Proof. Let $X = C(I)$. Then X is a complete metric space and we can define T on X as follows:

$$(T\varphi)(x) = y_0 + \int_{x_0}^{x} f(t, \varphi(t))dt, \quad x \in I$$

As before, φ satisfies (3.23) if and only if $T\varphi = \varphi$, i.e., if and only if φ is a fixed point of T. Thus, in order to prove the existence of the requisite function φ, it is sufficient to show that T maps X into itself and that it has a unique fixed point.

It is clear that $T\varphi$ is a continuous function on I, so that T maps $X = C(I)$ into itself. It follows that T^k is a contraction for some k by exactly the same argument as in the preceding theorem. $\square$

Example 3.7.14. Consider the initial value problem

$$\frac{dy}{dx} = x + y, \quad y(0) = 0.$$

By Proposition 3.7.10, φ is a solution of this initial value problem if and only if

$$\varphi(x) = \int_0^{x} [t + \varphi(t)]dt.$$

We construct the solution using the iteration process. Set $\varphi_0(x) \equiv 0$. If $\varphi_1 = T\varphi_0$, then

$$\varphi_1(x) = \int_0^x t\, dt = \frac{x^2}{2!}.$$

If $\varphi_2 = T\varphi_1$, then

$$\varphi_2(x) = \int_0^x \left(t + \frac{t^2}{2!}\right) dt = \frac{x^2}{2!} + \frac{x^3}{3!}.$$

If $\varphi_n = T\varphi_{n-1}$, then

$$\varphi_n(x) = \frac{x^2}{2!} + \frac{x^3}{3!} + \ldots + \frac{x^{n+1}}{(n+1)!}.$$

Therefore,

$$\lim_{n\to\infty} \varphi_n(x) = \lim_{n\to\infty} \left[\frac{x^2}{2!} + \frac{x^3}{3!} + \ldots + \frac{x^{n+1}}{(n+1)!}\right]$$

$$= \lim_{n\to\infty} \left[1 + x + \frac{x^2}{2!} + \frac{x^3}{3!} + \ldots + \frac{x^{n+1}}{(n+1)!}\right] - x - 1$$

$$= e^x - x - 1.$$

It may be easily verified by substitution that $e^x - x - 1$ is a solution of the initial value problem on the whole of **R** and not merely on a bounded interval as permitted by Theorem 3.7.13. (The reader may note that the hypotheses of Theorem 3.7.13 are satisfied with $M = 1$ on *any* bounded interval.) For an independent verification that it is the only one, the equation may be reinterpreted as $\frac{du}{dx} = u$, where $u = x + y + 1$ and $u(0) = 1$; it is well known that the only such function u is $u(x) = e^x$.

We give below an example of a mapping T that is not a contraction, but T^2 is.

Example 3.7.15. Consider the metric space $C[0, \pi/2]$ with the uniform metric. Define $T: C[0, \pi/2] \to C[0, \pi/2]$ by

$$(Tx)(t) = \int_0^t x(u) \sin u\ du.$$

If $x(t) = -t$ and $y(t) = 1 - t$, then

$$d(x, y) = \sup_{0 \le t \le \frac{\pi}{2}} |-t - (1 - t)| = 1$$

and

$$d(Tx, Ty) = \sup_{0 \le t \le \frac{\pi}{2}} \left| \int_0^t (x(u) - y(u)) \sin u\ du \right| = \sup_{0 \le t \le \frac{\pi}{2}} (1 - \cos t) = 1.$$

So there exists no K such that $0 \le K < 1$ and $d(Tx, Ty) \le Kd(x, y)$ for all $x, y \in C[0, \pi/2]$ and hence T is not a contraction. We shall next show that T^2 is a contraction. Consider any $x, y \in C[0, \pi/2]$. We have

$$|(Tx)(t) - (Ty)(t)| = \left| \int_0^t (x(u) - y(u)) \sin u \, du \right| \leq d(x, y) \int_0^t \sin u \, du$$

$$= d(x, y)(1 - \cos t)$$

and, therefore,

$$|(T^2 x)(t) - (T^2 y)(t)| = \left| \int_0^t [(Tx)(u) - (Ty)(u)] \sin u \, du \right|$$

$$\leq d(x, y) \int_0^t (1 - \cos u) \sin u \, du \leq \frac{1}{2} d(x, y),$$

so that

$$d(T^2 x, T^2 y) \leq \frac{1}{2} d(x, y).$$

3.8. Exercises

1. The real function $f : \mathbf{R}^2 \to \mathbf{R}$ defined by

$$f(x, y) = \begin{cases} \dfrac{xy}{x^2 + y^2} & (x, y) \neq (0, 0), \\ 0 & x = y = 0 \end{cases}$$

is not continuous at $(0, 0)$.
Hint: The sequence $x_n = (1/n, 1/n) \to (0, 0)$ as $n \to \infty$ but $f(x_n) = 1/2$ for all $n \geq 1$, so that $f(x_n) \to 1/2 \neq f(0, 0)$.

2. The real function $f : \mathbf{R}^2 \to \mathbf{R}$ defined by

$$f(x, y) = \begin{cases} \dfrac{x^3 - y^3}{x^2 + y^2} & (x, y) \neq (0, 0), \\ 0 & x = y = 0 \end{cases}$$

is continuous at $(0,0)$.
Hint: In fact,

$$\left| \frac{x^3 - y^3}{x^2 + y^2} \right| = \left| \frac{(x - y)(x^2 + xy + y^2)}{x^2 + y^2} \right| = \left| (x - y) \left(1 + \frac{xy}{x^2 + y^2} \right) \right|$$

$$\leq \frac{3}{2} |x - y|,$$

since $\left| \dfrac{xy}{x^2 + y^2} \right| \leq \dfrac{1}{2}$. Hence, $\left| \dfrac{x^3 - y^3}{x^2 + y^2} \right| \leq \dfrac{3}{2} (|x| + |y|).$

3. Let f be a mapping of (X_1, d_1) into (X_2, d_2). Prove that f is continuous on X_1 if and only if for every subset $Y_2 \subseteq X_2$, $(f^{-1}(Y_2))^\circ \supseteq f^{-1}(Y_2^\circ)$.
   Hint: Suppose $(f^{-1}(Y_2))^\circ \supseteq f^{-1}(Y_2^\circ)$ whenever $Y_2 \subseteq X_2$ and consider an open subset Y_2 of X_2. Then $(f^{-1}(Y_2))^\circ \supseteq f^{-1}(Y_2^\circ) = f^{-1}(Y_2)$. Since $(f^{-1}(Y_2))^\circ \subseteq f^{-1}(Y_2)$ by the definition of interior, it follows that $f^{-1}(Y_2)^\circ = f^{-1}(Y_2)$. Therefore, $f^{-1}(Y_2)$ is open. Conversely, suppose f is continuous and $Y_2 \subseteq X_2$. Then $f^{-1}(Y_2^\circ) \subseteq f^{-1}(Y_2)$ and is open in X_1. By Theorem 2.1.14 (i), $(f^{-1}(Y_2))^\circ \supseteq f^{-1}(Y_2^\circ)$.

4. Let X and Z be metric spaces and let Y_1 and Y_2 be subsets of X such that $Y_1 \cup Y_2 = X$. Let the functions $f_1: Y_1 \to Z$ and $f_2: Y_2 \to Z$ be continuous and $f_1(x) = f_2(x)$ for all $x \in Y_1 \cap Y_2$. Define $h: X \to Z$ by $h(x) = f_1(x)$ if $x \in Y_1$ and $h(x) = f_2(x)$ if $x \in Y_2$. Give an example to show that h is not necessarily continuous. If, however, Y_1 and Y_2 are assumed to be both open or both closed, then h is continuous.
   Hint: Let Y_1 denote the set of rationals and $f_1(x) = 1$ for $x \in Y_1$; let Y_2 denote the set of all irrationals and $f_2(x) = 0$ for $x \in Y_2$. The function

$$h(x) = \begin{cases} 1 & x \text{ rational,} \\ 0 & x \text{ irrational} \end{cases}$$

   is not continuous (see Example 3.3.3(ii)). If Y_1 and Y_2 are both open in X and U is an open subset of Z, then $h^{-1}(U) \cap Y_1 = f_1^{-1}(U)$ is open in Y_1 and $h^{-1}(U) \cap Y_2 = f_2^{-1}(U)$ is open in Y_2. So $h^{-1}(U) \cap Y_1$ and $h^{-1}(U) \cap Y_2$ are open in X (see Proposition 2.2.3(i)) and hence, $h^{-1}(U)$ is open in X.

5. Suppose f is a continuous real-valued function defined on $\mathbf{R}$ such that $\lim_{x \to \infty} f(x) = \lim_{x \to -\infty} f(x) = 0$. Prove that f is uniformly continuous.
   Hint: Let $\varepsilon > 0$ be given. By hypothesis, there exists $M > 0$ such that $|f(x)| < \varepsilon/2$ whenever $x \notin [-M, M]$. The function f is uniformly continuous on $[-M, M]$. Therefore, there exists $\delta_1 > 0$ such that $|f(x) - f(y)| < \varepsilon/2$ whenever $x, y \in [-M, M]$ and $|x - y| < \delta_1$. By continuity of f at $\pm M$, there exist δ_2 and δ_3 such that $|f(x) - f(M)| < \varepsilon/2$ whenever $|x - M| < \delta_2$, and $|f(x) - f(-M)| < \varepsilon/2$ whenever $|x - (-M)| < \delta_3$. Let $\delta = \min\{\delta_1, \delta_2, \delta_3\}$. If x and y are such that $x < M$ and $y > M$ while $|x - y| < \delta$, then

$$|f(x) - f(y)| \leq |f(x) - f(M)| + |f(M) - f(y)| < \varepsilon.$$

6. Let (X, d) be a separable metric space and f a continuous mapping of X into a metric space Y. Then $f(X)$ is separable.
   Hint: Let X_0 be a countable dense subset of X. By continuity, $f(X) = f(\bar{X}_0) \subseteq \overline{f(X_0)}$. Moreover, $f(X_0)$ is countable.

7. Show that the function $f: \mathbf{R} \to (-1, 1)$ defined by

$$f(x) = \frac{x}{1 + |x|}$$

is a homeomorphism. Also, show that f is uniformly continuous on $\mathbf{R}$. Hint:

$$f(x) = \begin{cases} \dfrac{x}{(1+x)} & x \geq 0, \\[2mm] \dfrac{x}{(1-x)} & x < 0. \end{cases}$$

The inverse function $g: (-1, 1) \to \mathbf{R}$ is given by

$$g(y) = \begin{cases} \dfrac{y}{(1-y)} & y \geq 0, \\[2mm] \dfrac{y}{(1+y)} & y < 0. \end{cases}$$

In fact, for $x \geq 0$, we have $x/(1+x) \geq 0$ and hence,

$$g \circ f(x) = g\left(\frac{x}{1+x}\right) = \frac{x/(1+x)}{1 - x/(1+x)} = x,$$

whereas, for $x < 0$, we have $x/(1-x) < 0$ and hence,

$$g \circ f(x) = g\left(\frac{x}{1-x}\right) = \frac{x/(1-x)}{1 + x/(1-x)} = x.$$

Similar computations for $f \circ g$ are left to the reader. Since the function f and its inverse g are both continuous, f is a homeomorphism.

We next check the uniform continuity of f. For $x, y > 0$,

$$|f(x) - f(y)| = \left|\frac{x}{1+x} - \frac{y}{1+y}\right| = \frac{|x-y|}{(1+x)(1+y)} < |x-y|.$$

Similarly, for $x < 0$ $y < 0$, we have $|f(x) - f(y)| < |x-y|$. For $x > 0$ and $y < 0$,

$$|f(x) - f(y)| = \left|\frac{x}{1+x} - \frac{y}{1-y}\right| = \frac{|x-y-2xy|}{(1+x)(1-y)} \leq \frac{x-y-2xy+x^2+y^2}{(1+x)(1-y)}$$

$$= \frac{(x-y)(1+x-y)}{(1+x)(1-y)} \leq x - y = |x - y|.$$

The case when x or y is 0 is left to the reader.

8. (a) Show that the image of a complete metric space under homeomorphism need not be complete.
   (b) Show that the image of a complete metric space under a one-to-one uniformly continuous mapping need not be complete.
   Hint: The function f in Exercise 7 is both a homeomorphism and uniformly continuous. $\mathbf{R}$ is complete but $f(\mathbf{R}) = (-1, 1)$ is not.

9. Let $X = [0, 2\pi] \subseteq \mathbf{R}$ and $Y = \{z \in \mathbf{C} : |z| = 1\}$, the unit circle in the complex plane. Consider the map $f: X \to Y$ defined by $f(t) = e^{it}, 0 \leq t < 2\pi$. Show that f is both one-to-one and onto. It is continuous but not a homeomorphism (see Example 3.1.13(ii)).

Hint: The inverse mapping $h: Y \to X$ is defined by

$$h(z) = h(e^{it}) = -i \log(e^{it}) = -i\{\log|e^{it}| + i \arg(e^{it})\} = t.$$

The function h is not continuous at $z=1$. In fact, if $z_n = e^{i(2\pi - 1/n)}$, $\lim_{n\to\infty} z_n = 1$, then

$$\lim_{n\to\infty} h(z_n) = \lim_{n\to\infty} (2\pi - 1/n) = 2\pi \neq 0 = h(1) = h(\lim_{n\to\infty} z_n).$$

10. Let $S_2 = \{x = (x_1, x_2, x_3) \in \mathbf{R}^3 : x_1^2 + x_2^2 + x_3^2 = 1\}$ and $a = (0,0,1)$. Show that $S_2 \backslash \{a\}$ and $\mathbf{R}^2$ are homeomorphic.
Hint: The mapping $\Pi : S_2 \backslash \{a\} \to \mathbf{R}^2$ defined by

$$\Pi(x_1, x_2, x_3) = \left(\frac{x_1}{1 - x_3}, \frac{x_2}{1 - x_3} \right)$$

and its inverse

$$\Pi^{-1}(x_1, x_2) = \left(\frac{2x_1}{1 + x_1^2 + x_2^2}, \frac{2x_2}{1 + x_1^2 + x_2^2}, \frac{-1 + x_1^2 + x_2^2}{1 + x_1^2 + x_2^2} \right)$$

are continuous since each of the component mappings defining Π and Π^{-1} is continuous. (See Example 3.1.13(i).)

11. A mapping $f: X_1 \to X_2$, where (X_1, d_1) and (X_2, d_2) are metric spaces is said to be an **open mapping** if $f(G)$ is open in X_2 for every open set G in X_1. It is said to be a **closed mapping** if $f(F)$ is closed in X_2 for every closed set F in X_1. Prove that a one-to-one function from X_1 to X_2 is a homeomorphism if and only if f is continuous and either open or closed.
Hint: A homeomorphism maps open (respectively closed) subsets of X_1 onto open (respectively closed) subsets of X_2. To see why, consider an open subset G of X_1 and a homeomorphism f. Then f^{-1} is a continuous map from X_2 onto X_1 and hence, $f(G) = (f^{-1})^{-1}(G)$ is open in X_2. Conversely, assume that f is both continuous and open from X_1 to X_2. (The argument for closed sets is left to the reader.) Let G be an open subset of X_1. Now $(f^{-1})^{-1}(G) = f(G)$, which is open because f has been assumed open. Hence, $f^{-1}: X_2 \to X_1$ is a continuous map.

12. (a) Give an example of a function f that is continuous and open, but is not a homeomorphism.
    (b) Give an example of a function f that is continuous and closed, but is not a homeomorphism.
    (c) Give an example of a function f that is continuous and closed, but is not open.
    (d) Give an example of a function f that is continuous and open, but is not closed.

Hint: (a)   The   map $f: \mathbf{R} \to S = \{(x,y) : x^2 + y^2 = 1\}$ defined   by $f(t) = (\cos(2\pi t), \sin(2\pi t))$ has continuous components and is, therefore, continuous. We shall show that it is also open. Let $(a,b) = I$ and write $\ell(I) = b - a$. If $\ell(I) > 1$, then $f(I) = S$, which is open in S. If $\ell(I) < 1$, then $f(I)$ is an arc without endpoints and so is open in S, being the intersection of S with an open disc with the missing endpoints on its edge. The function is not one-to-one and thus, cannot be a homeomorphism.

(b) The function $f : \mathbf{R}^2 \to \mathbf{R}$ defined by $f(x_1, x_2) = 0$ for all $(x_1, x_2) \in \mathbf{R}^2$ is continuous and closed, but is not a homeomorphism, as it is not one-to-one.

(c) See Example (b) above.

(d) The function $f : \mathbf{R}^2 \to \mathbf{R}$ defined by $f((x_1, x_2)) = x_1$ for all $(x_1, x_2) \in \mathbf{R}^2$ is continuous and open but not closed, as it maps the hyperbola $x_1 x_2 = 1$ onto the set $\mathbf{R} \backslash \{0\}$, which is not closed.

13. Let $X = \mathbf{R}^n$ and for any $x, y \in \mathbf{R}^n$, $d_p(x, y) = (\sum_{i=1}^{n} |x_i - y_i|^p)^{1/p}, 1 \leq p < \infty$. Show that $\lim_{p \to \infty} d_p(x, y) = d_\infty(x, y)$, where $d_\infty(x, y) = \max \{|x_i - y_i| : 1 \leq i \leq n\}$. Hence, or otherwise, prove that d_p and d_∞ are equivalent metrics on $\mathbf{R}^n$.
Hint: $\max \{|x_i - y_i| : 1 \leq i \leq n\} \leq d_p(x, y) \leq n^{1/p} \max \{|x_i - y_i| : 1 \leq i \leq n\}$.

14. Let d and d^* be metrics on $C^1[0, 1]$ defined by

$$d(x, y) = \sup\{|x(t) - y(t)| : t \in [0, 1]\} + \sup\{|x'(t) - y'(t)| : t \in [0, 1]\}$$

and

$$d^*(x, y) = \sup\{|x(t) - y(t)| : t \in [0, 1]\}.$$

Then d and d^* are not equivalent metrics in $C^1[0, 1]$.
Hint: The sequence $\{\frac{t^n}{n}\}_{n \geq 2}$ is such that

$$d^* \left(\frac{t^n}{n}, 0 \right) = \sup_t \left| \frac{t^n}{n} \right| \leq \frac{1}{n} \to 0 \text{ as } n \to \infty,$$

while

$$d \left(\frac{t^n}{n}, 0 \right) = \sup_t \left| \frac{t^n}{n} \right| + \sup_t |t^{n-1}| \leq \frac{1}{n} + 1 \to 1 \text{ as } n \to \infty.$$

15. Let f be a mapping of a complete metric space (X_1, d_1) onto (X_2, d_2). If f is an isometry, show that (X_2, d_2) is complete.
Hint: Let $\{y_n\}_{n \geq 1}$ be a Cauchy sequence in (X_2, d_2) and let $\varepsilon > 0$ be given. There exists n_0 such that

$$d_1(f^{-1}(y_n), f^{-1}(y_m)) = d_2(y_n, y_m) < \varepsilon \qquad \text{for } n, m \geq n_0.$$

So, $\{f^{-1}(y_n)\}_{n \geq 1}$ is a Cauhy sequence in (X_1, d_1), and since (X_1, d_1) is complete, we have $\lim_{n \to \infty} f^{-1}(y_n) = x$ for some $x \in X_1$ and hence $\lim_{n \to \infty} y_n = f(x)$.

16. Let $X = \ell_p$ and let $d(x, y) = (\sum_{i=1}^{\infty} |x_i - y_i|^p)^{1/p}, p \geq 1$, where $x = \{x_i\}_{i \geq 1}$, $y = \{y_i\}_{i \geq 1}$ are in ℓ_p. Define $\varphi : \ell_p \to \ell_p$ by $\varphi(x) = \varphi(x_1, x_2, \dots) = (0, x_1, x_2, \dots)$. Show that φ is an isometry.

Hint: $d(\varphi(x), \varphi(y)) = d((0, x_1, x_2, \dots), (0, y_1, y_2, \dots)) = \sum_{i=1}^{\infty} |x_i - y_i|^p = d(x, y)$.

17. Let X be a metric space and x_0 a point in X. The space of all real-valued, continuous bounded functions on X with the uniform metric $d_u(f, g) = \sup\{|f(x) - f(y)|: x \in X\}$ is denoted by $C_b(X, \mathbf{R})$. Show that

$$f_x(y) = d(y, x) - d(y, x_0) \text{ for } x, y \in X$$

defines an isometry of X into $C_b(X, \mathbf{R})$. Hence, give an alternate proof that every metric space has a unique completion.

Hint: The function $f_x: X \to C_b(X, \mathbf{R})$ is bounded and uniformly continuous:

$$|f_x(y)| = |d(y, x) - d(y, x_0)| \leq d(x, x_0)$$

and

$$|f_x(y_1) - f_x(y_2)| = |d(y_1, x) - d(y_1, x_0) - d(y_2, x) + d(y_2, x_0)|$$
$$\leq |d(y_1, x) - d(y_2, x)| + |d(y_1, x_0) - d(y_2, x_0)| \leq 2d(y_1, y_2).$$

The mapping $F: X \to C_b(X, \mathbf{R})$ defined by $F(x) = f_x$ is an isometry because

$$|f_{x_1}(y) - f_{x_2}(y)| = |d(y, x_1) - d(y, x_0) - d(y, x_2) + d(y, x_0)|$$
$$= |d(y, x_1) - d(y, x_2)| \leq d(x_1, x_2).$$

Moreover,

$$f_{x_1}(x_2) - f_{x_2}(x_2) = d(x_1, x_2).$$

Hence,

$$d_u(f_{x_1}, f_{x_2}) = d(x_1, x_2).$$

Define X^* as $\overline{F(X)}$ in $C_b(X, \mathbf{R})$. Being a closed subset of a complete metric space, X^* is complete; besides, it contains an isometric image of X. Thus, X^* is a completion of X.

If (Y_1, d_1) and (Y_2, d_2) are two completions of (X, d), then there exist two isometries

$$F_1: X \to Y_1 \quad \text{and} \quad F_2: X \to Y_2.$$

Then $h = F_2 \circ F_1^{-1}$ is an isometry from $F_1(X)$ onto $F_2(X)$. Since $F_1(X)$ is dense in Y_1, h is uniformly continuous and Y_2 is complete, it follows, using Theorem 3.4.9, that there exists a uniformly continuous extension h^* of h to all of Y_1. The mapping h^* is indeed an isometry. By Exercise 15, $h^*(Y_1)$ is complete; since it contains the dense subset $F_2(X)$, it can be easily shown that $h^*(Y_1) = Y_2$.

18. Let f be an isometry between metric spaces (X_1, d_1) and (X_2, d_2). Show that f is a homeomorphism between X_1 and X_2. Is the converse true? Justify your answer. Hint: For $x_1, y_1 \in X_1, f(x_1) = f(y_1)$ implies $d_1(x_1, y_1) = 0$, which, in turn, implies $x_1 = y_1$. So, f is one-to-one. Also, $d_1(f^{-1}(x_2), f^{-1}(y_2)) =$

$d_2(f(f^{-1}(x_2)),f(f^{-1}(y_2)))=d_2(x_2,y_2)$. Thus, f^{-1} is also an isometry. The homeomorphism of Exercise 7 above is not an isometry.

19. Let $\{f_n\}_{n\geq 1}$ be a sequence of continuous functions on a metric space X that converges uniformly to a function f. Prove that

$$\lim_{n\to\infty} f_n(x_n) = f(x)$$

for every sequence $\{x_n\}_{n\geq 1}$ of points in X such that $x_n \to x \in X$. Give an example of a sequence $\{f_n\}_{n\geq 1}$ of continuous functions on $[0,1]$ converging pointwise to a continuous function f and a sequence $\{x_n\}_{n\geq 1}$ of points in $[0,1]$ converging to a limit x such that $f_n(x_n)$ does not converge to $f(x)$.
Hint: By hypothesis, for every $\varepsilon > 0$, there exists $n_1 \in \mathbf{N}$ such that $n \geq n_1$ implies

$$|f_n(x) - f(x)| < \frac{\varepsilon}{2} \text{ for all } x \in X.$$

Observe that f is continuous by Theorem 3.6.5. Since $\lim_{n\to\infty} x_n = x$, there exists n_2 such that

$$|f(x_n) - f(x)| < \frac{\varepsilon}{2} \text{ for } n \geq n_2.$$

Let $n_0 = \max\{n_1, n_2\}$. For $n \geq n_0$ and $x \in X$, we have

$$|f_n(x_n) - f(x)| \leq |f_n(x_n) - f(x_n)| + |f(x_n) - f(x)| < \frac{\varepsilon}{2} + \frac{\varepsilon}{2}.$$

For the example, let

$$f_n(x) = \begin{cases} n^2 x & 0 \leq x \leq \dfrac{1}{n}, \\ -n^2 x + 2n & \dfrac{1}{n} \leq x \leq \dfrac{2}{n}, \\ 0 & \dfrac{2}{n} \leq x \leq 1. \end{cases}$$

Then $f_n \to f$ pointwise, where $f(x) = 0 \ \forall x \in [0,1]$. Note that $1/n \to 0$ as $n \to \infty$ but $f_n(1/n) = n \to \infty$ as $n \to \infty$.

20. Prove that the series $\sum_{k=1}^{\infty} (\frac{x}{1+x}\sin x)^k$ is uniformly convergent on $[0,1]$.
Hint: The function $f(x) = \frac{x}{1+x}\sin x$ vanishes at $x = 0$ and is monotonically increasing. In fact,

$$f'(x) = \frac{x(1+x)\cos x + \sin x}{(1+x)^2} \geq 0, \quad x \in [0,1].$$

So,

$$|f(x)| = f(x) \leq f(1) = \frac{1}{2}\sin 1.$$

21. Let

$$I(x) = \begin{cases} 0 & x \le 0, \\ 1 & x > 0, \end{cases}$$

and let $\{x_n\}_{n \ge 1}$ be a sequence of distinct points of (a, b). If $\sum_{n=1}^{\infty} |c_n|$ converges, prove that the series

$$f(x) = \sum_{n=1}^{\infty} c_n I(x - x_n) \quad a \le x \le b$$

converges uniformly and that f is continuous for every $x \ne x_n, n = 1, 2, \ldots$.
Hint: $|c_n I(x - x_n)| \le |c_n|$ and $\Sigma |c_n|$ is convergent; it follows by the Weierstrass M-test that the series defining $f(x)$ is uniformly and absolutely convergent. If $x \ne x_n, n = 1, 2, \ldots$, then every term $c_n I(x - x_n)$ is continuous at x and hence, so is the sum of the series at x, using the argument in the proof of Theorem 3.6.5.

22. If f is a continuous function on a closed set F in $\mathbf{R}$, then show that there exists a continuous extension f^* of f to $\mathbf{R}$. Moreover, if $|f(x)| \le M$ for all $x \in F$, then $|f^*(x)| \le M$ for all $x \in \mathbf{R}$.

Hint: $\mathbf{R} \backslash F$, being open, is a union of disjoint open intervals (a_n, b_n). Define

$$f^*(x) = f(a_n) + \frac{f(b_n) - f(a_n)}{b_n - a_n} (x - a_n), \quad x \in (a_n, b_n).$$

$n = 1, 2, \ldots$. The function g, which equals f on F and f^* on $\mathbf{R} \backslash F$, is continuous on $\mathbf{R}$. Clearly, g is continuous on F° and $\mathbf{R} \backslash F$. A neighbourhood of a boundary point contains points from F as well as $\mathbf{R} \backslash F$. Since the values of f at the points of F in this neighbourhood are close to the values at the boundary point and the values of f^* at points of $\mathbf{R} \backslash F$ in this neighbourhood are interpolated values, the latter are also close to the value at the boundary point.

23. Let

$$f_n(x) = \begin{cases} 0 & x < \dfrac{1}{n+1}, \\ \sin^2\left(\dfrac{\pi}{x}\right) & \dfrac{1}{n+1} \le x \le \dfrac{1}{n}, \\ 0 & \dfrac{1}{n} < x. \end{cases}$$

Show that $f_n \to 0$ pointwise, but not uniformly.
Hint: For positive $x \in \mathbf{R}$, choose n_0 so large that $1/n_0 < x$. Then for $n \ge n_0, f_n(x) = 0$. Therefore, $\lim_{n \to \infty} f_n(x) = 0$ if $x > 0$; the same is true when $x \le 0$, because $f_n(x) = 0$ in this case. Since $f_n(2/(2n+1)) = 1$, $n = 1, 2, \ldots$, $f_n(x)$ cannot be less than 1 for all large n and all $x \in \mathbf{R}$.

24. Prove that there exists no sequence $\{f_n\}_{n \geq 1}$ of functions continuous on $\mathbf{R}$ such that

$$\lim_{n \to \infty} f_n(x) = f(x),$$

where $f(x) = 1$ if x is rational, and 0, otherwise.

Hint: Suppose there does exist a sequence $\{f_n\}_{n \geq 1}$ of continuous functions such that $\lim_{n \to \infty} f_n(x) = f(x)$, $-\infty < x < \infty$. For each n, the set $E_n = \{x \in \mathbf{R}: f_n(x) \leq 1/2\} = f_n^{-1}((-\infty, 1/2])$ is closed, being the inverse image of the closed set $(-\infty, 1/2]$ under the continuous function f_n. Let $F_k = \bigcap_{n=k}^{\infty} E_n$. It is closed, being the intersection of closed sets. Let x be an irrational number; then $\lim_{n \to \infty} f_n(x) = f(x) = 0$. So, for $0 < \varepsilon < 1/2$, there exists k such that $n \geq k$ implies $|f_n(x)| < \varepsilon$; thus, $x \in F_k$. On the other hand, if $x \in F_k$ for some k, then x is irrational. In fact, $f_n(x) \leq 1/2$ for $n \geq k$ and hence, $f(x) \leq 1/2$. Thus, $\bigcup_{k=1}^{\infty} F_k$ is precisely the set of irrationals. Thus, the set of irrationals constitute an F_σ set. Moreover, each F_k is nowhere dense. This contradicts Corollary 2.4.4. (However, f can be represented as the double limit of continuous functions thus: $f(x) = \lim_{k \to \infty} \{\lim_{n \to \infty} \cos^{2n}(k!\pi x)\}$.)

25. Let K be defined and continuous on the closed square $[a, b] \times [a, b]$, g be continuous on $[a, b]$ and let λ be a real number. Prove that the integral equation

$$f(x) = \lambda \int_a^b K(x, y) f(y) dy + g(x), \quad x \in [a, b]$$

has a unique solution in $C[a, b]$ for sufficiently small λ.

Hint: Define a mapping $T: C[a, b] \to C[a, b]$ by

$$(T\varphi)(x) = \lambda \int_a^b K(x, y) \varphi(y) dy + g(x)$$

and show that T has a unique fixed point. In fact,

$$\begin{aligned}
d(T\varphi_1, T\varphi_2) &= \sup \left\{ \left| \lambda \int_a^b K(x, y)[\varphi_1(y) - \varphi_1(y)] dy \right| : x \in [a, b] \right\} \\
&\leq |\lambda| M(b - a) \sup \{ |\varphi_1(y) - \varphi_1(y)| : y \in [a, b] \} \\
&= |\lambda| M(b - a) d(\varphi_1, \varphi_2),
\end{aligned}$$

where $M = \sup \{ |K(x, y)| : x, y \in [a, b] \}$. For $|\lambda| < 1/M(b - a)$, the mapping T is a contraction.

26. **Theorem.** Let S denote the strip $[a, b] \times \mathbf{R}$ in the (x, y)-plane. Let f be a real-valued continuous function defined on S, satisfying

$$0 < m \leq \frac{f(x, y_2) - f(x, y_1)}{y_2 - y_1} \leq M \tag{3.24}$$

for all distinct y_1, y_2 in $\mathbf{R}$. Then there exists a unique, real-valued continuous function g defined on $[a, b]$ satisfying the relation

$$f(x, g(x)) = 0$$

for all $x \in [a, b]$.

Proof. Recall that $C[a, b]$ is a complete metric space with the uniform metric d; i.e., for $f, g \in C[a, b]$,

$$d(f, g) = \sup_x |f(x) - g(x)|.$$

Define a mapping $T: C[a, b] \to C[a, b]$ by setting, for any $g \in C[a, b]$,

$$(Tg)(x) = g(x) - \frac{2}{M + m} f(x, g(x)).$$

For $g, h \in C[a, b]$,

$$(Tg)(x) - (Th)(x) = g(x) - h(x) - \frac{2}{M + m}\{f(x, g(x)) - f(x, h(x))\}.$$

Elementary calculations using (3.24) show that

$$-\frac{M - m}{M + m} \leq 1 - \frac{2}{M + m}\frac{f(x, y_2) - f(x, y_1)}{y_2 - y_1} \leq \frac{M - m}{M + m}.$$

Thus,

$$\begin{aligned} d(Tg, Th) &= \sup_x |(Tg)(x) - (Th)(x)| \\ &\leq \sup_x \{|g(x) - h(x)|\}\frac{M - m}{M + m} \\ &= \frac{M - m}{M + m} d(g, h). \end{aligned}$$

This shows that T is a contraction on $C[a, b]$. Hence, there exists a unique element $g \in C[a, b]$ such that $Tg = g$, that is, $f(x, g(x)) = 0$ for all $x \in [a, b]$. $\square$

27. Let $\{f_n\}_{n \geq 1}$ be a sequence of real valued continuous functions defined on a metric space (X, d). Assume that for every $x \in X, f_n(x) \to f(x)$. The pointwise convergence of the sequence $\{f_n\}_{n \geq 1}$ does not guarantee the continuity of f. (Recall that $f_n(x) = x^n, n = 1, 2, \ldots, x \in [0, 1]$, converges pointwise to $f(x)$, which is 0 on $[0,1)$ and 1 when $x = 1$.) The limit function f, however, cannot be extremely discontinuous as the following theorem due to Baire shows.

Theorem. (Baire) Let $\{f_n\}_{n \geq 1}$ be a sequence of real-valued continuous functions defined on a metric space (X, d). Let

$$\lim_{n\to\infty} f_n(x) = f(x)$$

for each $x \in X$. Then the set of points of discontinuity of f constitute a set of category I.

Proof. Let $\varepsilon > 0$ be given. Put

$$U_m(\varepsilon) = \{x \in X : |f(x) - f_m(x)| \leq \varepsilon\}$$

and

$$U(\varepsilon) = \bigcup_{m=1}^{\infty} U_m^{\circ}(\varepsilon), \tag{3.25}$$

where $U_m{}^{\circ}(\varepsilon)$ denotes the interior of $U_m(\varepsilon)$. If Y denotes the set of all points of continuity of f, we shall show that

$$Y = \bigcap_{n=1}^{\infty} U\left(\frac{1}{n}\right).$$

Let $x_0 \in X$ be such that f is continuous at x_0. We claim that $x_0 \in \cap_{n=1}^{\infty} U(\frac{1}{n})$. Since $\lim_{n\to\infty} f_n(x_0) = f(x_0)$, there exists an integer $m > 0$ such that

$$|f(x_0) - f_m(x_0)| < \frac{1}{3}\varepsilon.$$

Since f and f_m are continuous at x_0, there exists $S(x_0, \delta)$ such that

$$|f(x) - f(x_0)| < \frac{1}{3}\varepsilon \quad \text{and} \quad |f_m(x) - f_m(x_0)| < \frac{1}{3}\varepsilon$$

for $x \in S(x_0, \delta)$. So, for $x \in S(x_0, \delta)$, we have

$$|f(x) - f_m(x)| \leq |f(x) - f(x_0)| + |f(x_0) - f_m(x_0)| + |f_m(x_0) - f_m(x)| < \varepsilon,$$

which proves that $x \in U_m(\varepsilon)$. Thus, $S(x_0, \delta) \subseteq U_m(\varepsilon)$. In particular, $x_0 \in (U_m(\varepsilon))^{\circ}$ and so $x_0 \in U(\varepsilon)$. Since $\varepsilon > 0$ is arbitrary, it follows that $x_0 \in \cap_{n=1}^{\infty} U(\frac{1}{n})$.

Conversely, suppose that $x_0 \in \cap_{n=1}^{\infty} U(1/n)$. Let $\varepsilon > 0$ be given. There exists an integer $n > 0$ such that $1/n < (1/3)\varepsilon$. Since $x_0 \in U(1/n)$, it follows that $x_0 \in U_m^{\circ}((1/3)\varepsilon)$ for some integer m. So there exists $S(x_0, \delta_1)$ such that for every $x \in S(x_0, \delta_1)$, we have

$$|f(x) - f_m(x)| < \frac{1}{3}\varepsilon.$$

Since f_m is continuous, it follows that there exists $\delta_2 > 0$ such that $|f_m(x_0) - f_m(x)| < (1/3)\varepsilon$ for all $x \in S(x_0, \delta_2)$. Let $\delta = \min\{\delta_1, \delta_2\}$. Now for $x \in S(x_0, \delta)$,

$$|f(x) - f(x_0)| \leq |f(x) - f_m(x)| + |f_m(x_0) - f_m(x)| + |f(x_0) - f_m(x_0)|$$
$$< \frac{1}{3}\varepsilon + \frac{1}{3}\varepsilon + \frac{1}{3}\varepsilon = \varepsilon,$$

using the continuity of f_m. Thus, x_0 is a point of continuity of f. This establishes (3.25).

It remains to show that $X \setminus Y$ is of category I. Put

$$F_m(\varepsilon) = \{x : |f_m(x) - f_{m+k}(x)| \leq \varepsilon \; (k = 0, 1, 2, \ldots)\}.$$

Since the functions f_m are continuous, $F_m(\varepsilon)$ is a closed set, being the intersection of the inverse images of $[-\varepsilon, \varepsilon]$ under the continuous map $f_m - f_{m+k}$. Since $\lim_{n \to \infty} f_n(x) = f(x)$, it follows not only that

$$X = \bigcup_{m=1}^{\infty} F_m(\varepsilon)$$

but also that

$$F_m(\varepsilon) \subseteq U_m(\varepsilon).$$

Thus, $F_m^\circ(\varepsilon) \subseteq U_m^\circ(\varepsilon)$ and so

$$\bigcup_{m=1}^{\infty} F_m^\circ(\varepsilon) \subseteq \bigcup_{m=1}^{\infty} U_m^\circ(\varepsilon) = U(\varepsilon).$$

For any closed set F, the difference $F \setminus F^\circ$ is a nowhere dense closed set. Thus,

$$X \setminus \bigcup_{m=1}^{\infty} F_m^\circ(\varepsilon) \subseteq \bigcup_{m=1}^{\infty} (F_m(\varepsilon) \setminus F_m^\circ(\varepsilon))$$

is a set of category I. So the set $X \setminus U(\varepsilon) = U(\varepsilon)^c$ is also a subset of category I. Consequently, the set $X \setminus Y$ of points of discontinuity of f, which is expressible in the form (see 3.25)

$$X \setminus \bigcap_{n=1}^{\infty} U(\frac{1}{n}) = \bigcup_{n=1}^{\infty} \left[U\left(\frac{1}{n}\right) \right]^c,$$

is a set of category I.

Remark. The result of Exercise 24 also follows from the above theorem.

28. Let X be any set (no metric required) and $T : X \to X$ a map such that T^n has a unique fixed point $x_0 \in X$. Then T has x_0 as its unique fixed point.
    Hint. Tx_0 satisfies $T^n(Tx_0) = T(T^n x_0) = Tx_0$ and is therefore a fixed point of T^n. Hence, by the uniqueness of the fixed point of T^n, we have $Tx_0 = x_0$, i.e., x_0 is a fixed point of T. Trivially, any fixed point of T is also a fixed point of T^n. Since the latter has only one fixed point, so does T.

29. Let (X, d_X) and (Y, d_Y) be metric spaces and $\mathcal{B}$ be a base for the open sets of Y (see Definition 2.3.4). Show that $f : X \to Y$ is continuous if and only if $f^{-1}(B)$ is open for every $B \in \mathcal{B}$.

In Example 3.7.15, the map T of a metric space into itself is not a contraction, but its square is. Exercise 30 shows that for any positive integer n, there exists a map T of some metric space into itself such that T^{n+1} is a contraction, but T^n is not.

30. For any $A > 0$ and the map $T: C[0, A] \to C[0, A]$ given by

$$(Tx)(t) = \int_0^t x(s)\,ds,$$

show that T^n is a contraction with reference to the uniform metric on $C[0,A]$ if and only if $A^n/n! < 1$. Hence, show that for a given positive integer n, it is possible to choose A so that T^{n+1} is a contraction, but T^n is not.
Hint: Let d denote the uniform metric on $C[0,A]$. It is easy to prove by induction that

$$|(T^n x)(t) - (T^n y)(t)| \le \frac{t^n}{n!} d(x, y) \qquad \text{for all } t \in [0, A].$$

It follows that

$$d(T^n x, T^n y) \le \frac{A^n}{n!} d(x, y).$$

In order to show that equality can hold in the preceding inequality, consider the particular elements x and y of $C[0,A]$, such that x is 1 everywhere and y is 0 everywhere. We have $(T^n x)(t) = t^n/n!$ and $(T^n y)(t) = 0$ everywhere. Therefore,

$$d(T^n x, T^n y) = \sup_{0 \le t \le A} |\frac{t^n}{n!} - 0| = \frac{A^n}{n!} = \frac{A^n}{n!} d(x, y),$$

because $d(x, y) = 1$. Thus, T^n is a contraction if and only if $A^n/n! < 1$.
   Now for any positive integer n, we can show by induction that $(n!)^{1/n} < ((n+1)!)^{1/(n+1)}$ or, equivalently, that $(n!)^{n+1} < ((n+1)!)^n$. The case $n = 1$ is trivial. Assuming it true for $n = k$, we have

$$\begin{aligned}((k+2)!)^{k+1} &= (k+2)^{k+1}((k+1)!)^{k+1} = (k+2)^{k+1}((k+1)!)^k(k+1)! \\ &> (k+2)^{k+1}(k!)^{k+1}(k+1)! > (k+1)^{k+1}(k!)^{k+1}(k+1)! \\ &= ((k+1)!)^{k+1}(k+1)! = ((k+1)!)^{k+2},\end{aligned}$$

completing the induction.
   The inequality just established shows that for a given positive integer n, there exists A such that $(n!)^{1/n} < A < ((n+1)!)^{1/(n+1)}$, so that $A^n/n! > 1$ but $A^{n+1}/(n+1)! < 1$.

4 Connected Spaces

In this chapter, we discuss connectedness, local connectedness and arcwise connect-edness. Geometrically, it is evident that an interval cannot be written as a disjoint union of two nonempty relatively open *subintervals*. That it cannot even be written as a disjoint union of two nonempty relatively open *subsets* will be proved in this chapter. An attempt to generalise the above property leads to the concept of connected metric spaces. An interval also has the property that any two points in it can be joined by an "arc" without exiting the interval. The generalised version of this property in metric spaces is called "arcwise connectedness" and it plays a vital role in analytic function theory. The two concepts are, in general, not equivalent; however, they coincide in open subsets of the complex plane (more generally, in open subsets of a Euclidean space). The local version of connectedness is also discussed in this chapter and shown not to be equivalent to connectedness.

4.1. Connectedness

The intuitive notion of a connected set is that the entire set is in one piece, that is, it should not consist of two or more parts that are "separated" from each other. For example, the intervals on the real line are connected subsets and, in fact, they are the only connected subsets of $\mathbf{R}$, as we shall soon see. It will then follow that the set

$$(a, b) \cup (c, d) = \{x \in \mathbf{R} : a < x < b \text{ or } c < x < d\}, \qquad \text{where } b < c,$$

is not connected. This is in perfect agreement with our general understanding of connectedness. It is this idea which we make precise below. We shall then establish the fundamental facts of the theory of connectedness.

We begin with the definition of a disconnected space.

Definition 4.1.1. A metric space (X, d) is said to be **disconnected** if there exist two nonempty subsets A and B of X such that

(i) $X = A \cup B$;
(ii) $A \cap \overline{B} = \emptyset$ and $\overline{A} \cap B = \emptyset$.

156

That is, the subsets must be nonempty, together they must constitute the whole space and neither may contain a point of the closure of the other. If no such subsets exist, then (X, d) is said to be **connected**; this means that if we do split X into two nonempty parts A and B having no points in common, then at least one of the subsets contains a limit point of the other.

A nonempty subset Y of a metric space (X, d) is said to be connected if the subspace $(Y, d|_Y)$ with the metric induced from X is connected.

Examples 4.1.2. (i) Let $X = \{x \in \mathbf{R}: -1 \le x < 0 \text{ or } 0 < x \le 1\}$ with the metric induced from the real line. Then X is disconnected, as can be seen by taking $A = [-1, 0)$ and $B = (0, 1]$. (See Figure 4.1.) So is the space $X = \{z \in \mathbf{C}: |z| \le 1\}$ $\cup \{z \in \mathbf{C}: |z - 3| \le 1\}$ with the metric induced from $\mathbf{C}$, as illustrated in Figure 4.2.
(ii) The rationals $\mathbf{Q} \subseteq \mathbf{R}$ with the metric induced from $\mathbf{R}$ are not connected. In fact, if

$$A = \{x \in \mathbf{Q}: x < \sqrt{2}\} \text{ and } B = \{x \in \mathbf{Q}: x > \sqrt{2}\},$$

then A and B are disjoint subsets and $A \cap \overline{B} = \varnothing, \overline{A} \cap B = \varnothing, A \cup B = \mathbf{Q}$.
(iii) A singleton set $\{x\}$ in any metric space is always connected.

We now establish some equivalent versions of disconnectedness.

Theorem 4.1.3. Let (X, d) be a metric space. Then the following statements are equivalent:

(i) (X, d) is disconnected;
(ii) there exist two nonempty disjoint subsets A and B, both open in X, such that $X = A \cup B$;
(iii) there exist two nonempty disjoint subsets A and B, both closed in X, such that $X = A \cup B$;
(iv) there exists a proper subset of X that is both open and closed in X.

Proof. (i)$\Rightarrow$(ii). Let $X = A \cup B$, where A and B are nonempty and $A \cap \overline{B} = \varnothing$, $\overline{A} \cap B = \varnothing$. Then $A = X \backslash \overline{B}$. In fact, $A \subseteq X \backslash \overline{B} \subseteq X \backslash B = A$. So A is

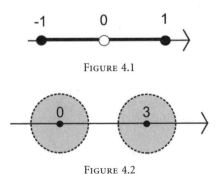

FIGURE 4.1

FIGURE 4.2

open in X. Similarly, B is open in X. Since A and $\overline{B}$ are disjoint, a fortiori, A and B are disjoint, which proves (ii).

That (ii) and (iii) are equivalent is trivial.

(iii)$\Rightarrow$(iv) Since $A = X \backslash B$, A is open. Thus A is both a closed and open proper subset of X, and so (iv) is proved.

(iv)$\Rightarrow$(i) Let A be a proper subset of X that is both open and closed in X and let $B = X \backslash A$. Then $X = A \cup B$, $A \cap B = \emptyset$. Since $A = \overline{A}$ (A being closed), it follows that $\overline{A} \cap B = \emptyset$. Similarly, $A \cap \overline{B} = \emptyset$. This completes the proof. $\qquad\square$

The following theorem characterises connected subsets of real numbers.

Theorem 4.1.4. Let $(\mathbf{R}, d)$ be the space of real numbers with the usual metric. A subset $I \subseteq \mathbf{R}$ is connected if and only if I is an interval, i.e., I is of one of the following forms:

$$(a, b), \ [a, b), \ (a, b], \ [a, b], \ (-\infty, b), \ (-\infty, b], \ (a, \infty), \ [a, \infty), \ (-\infty, \infty).$$

Proof. Let I be a connected subset of real numbers and suppose, if possible, that I is not an interval. Then there exist real numbers x, y, z with $x < z < y$ and $x, y \in I$ but $z \notin I$. Then I may expressed as $I = A \cup B$, where

$$A = (-\infty, z) \cap I \qquad \text{and} \qquad B = (z, \infty) \cap I.$$

Since $x \in A$ and $y \in B$, therefore, A and B are nonempty; also, they are clearly disjoint and open in I. Thus, I is disconnected.

To prove the converse, suppose I is an interval but is not connected. Then there are nonempty subsets A and B such that

$$I = A \cup B, \qquad A \cap \overline{B} = \emptyset, \qquad \overline{A} \cap B = \emptyset.$$

Pick $x \in A$ and $y \in B$ and assume (without loss of generality) that $x < y$. Observe that $[x, y] \subseteq I$, for I is an interval. Define

$$z = \sup\,(A \cap [x, y]).$$

The supremum exists since $A \cap [x, y]$ is bounded above by y and it is nonempty, as x is in it. Then $z \in \overline{A}$. (We shall show that if $z \notin A$, then z is a limit point of A. Let $\varepsilon > 0$ be arbitrary. By the definition of supremum, there exists some element $a \in A$ such that $z - \varepsilon < a \le z$, i.e., every neighbourhood of z contains a point of A.) Hence, $z \notin B$, for $\overline{A} \cap B = \emptyset$; in particular, $x \le z < y$.

If $z \notin A$, then $x < z < y$ and $z \notin I$. This contradicts the fact that $[x, y] \subseteq I$.

If $z \in A$, then $z \notin \overline{B}$, for $A \cap \overline{B} = \emptyset$. So there exists a $\delta > 0$ such that $(z - \delta, z + \delta) \cap B = \emptyset$. This implies that there exists $z_1 \notin B$ satisfying the inequality $z < z_1 < y$. Then $x \le z < z_1 < y$ and $z_1 \notin I$, for z_1 being greater than $\sup\,(A \cap [x, y])$ is not in A. This contradicts the fact that $[x, y] \subseteq I$. $\qquad\square$

Remark 4.1.5. It follows as a special case of Theorem 4.1.4 that the entire real line $\mathbf{R}$ is a connected set. It now follows from Theorem 4.1.3(iv) that the only subsets of $\mathbf{R}$ that are both open and closed are the empty set and $\mathbf{R}$ itself.

Let $X_0 = \{0, 1\}$ and let d_0 denote the discrete metric on X_0. We shall call (X_0, d_0) the **discrete two point space**. Definition 4.1.1 can be reformulated in the following handier fashion:

Theorem 4.1.6. Let (X, d) be a metric space. Then the following statements are equivalent:

(i) (X, d) is disconnected;
(ii) there exists a continuous mapping of (X, d) onto the discrete two element space (X_0, d_0).

Proof. (i)$\Rightarrow$(ii). Let $X = A \cup B$, where A and B are disjoint nonempty open subsets (see Theorem 4.1.3(ii)). Define a mapping $f: X \rightarrow X_0$ by

$$f(x) = \begin{cases} 0 & \text{if } x \in A, \\ 1 & \text{if } x \in B; \end{cases}$$

the mapping f is clearly onto. It remains to show that f is continuous from (X, d) to (X_0, d_0). The open subsets of the discrete metric space are precisely $\emptyset, \{0\}, \{1\}$ and $\{0,1\}$. Observe that $f^{-1}(\emptyset) = \emptyset$, $f^{-1}(\{0, 1\}) = X$ and the subsets $\emptyset, X$ are open in (X, d). Moreover, $f^{-1}(\{0\}) = A$, $f^{-1}(\{1\}) = B$, which are open subsets of (X, d). Hence, f is continuous and thus (ii) is proved.

(ii)$\Rightarrow$(i) Let $f: (X, d) \rightarrow (X_0, d_0)$ be continuous and onto. Let $A = f^{-1}(\{0\})$ and $B = f^{-1}(\{1\})$. Then A and B are nonempty disjoint subsets of X, both open and such that $X = A \cup B$. It follows upon using Theorem 4.1.3(ii) that X is disconnected.   □

A reformulation of Theorem 4.1.6 is the following:

Corollary 4.1.7. Let (X, d) be a metric space. Then the following statements are equivalent:

(i) (X, d) is connected;
(ii) the only continuous mappings of (X, d) into (X_0, d_0) are the constant mappings, namely, the mappings $f(x) = 1$ for every $x \in X$, $g(x) = 0$ for every $x \in X$.

The continuous image of a connected space is connected. More precisely, we have the following theorem.

Theorem 4.1.8. Let (X, d_X) be a connected metric space and $f : (X, d_X) \rightarrow (Y, d_Y)$ be a continuous mapping. Then the space $f(X)$ with the metric induced from Y is connected.

Proof. The map $f: X \rightarrow f(X)$ is continuous. If $f(X)$ were not connected, then there would be, by Theorem 4.1.6, a continuous mapping, g say, of $f(X)$ onto the discrete two element space (X_0, d_0). Then $g \circ f \colon X \to X_0$ would also be a continuous mapping of X onto X_0, contradicting the connectedness of X. □

The intermediate value theorem of real analysis (see Proposition 0.5.3) is a special case of Theorem 4.1.8.

Theorem 4.1.9. (Intermediate Value Theorem) If $f \colon [a, b] \to \mathbf{R}$ is continuous over $[a, b]$, then for every y such that $f(a) \leq y \leq f(b)$ or $f(b) \leq y \leq f(a)$ there exists $x \in [a, b]$ for which $f(x) = y$.

Proof. We need consider only the case when $f(a) \neq y \neq f(b)$. Let y be any real number such that $f(a) < y < f(b)$. By Theorem 4.1.4, $[a, b]$ is a connected subset of $\mathbf{R}$. Hence, $f([a, b])$ is an interval by Theorems 4.1.8 and 4.1.4. Therefore, there exists an $x \in [a, b]$ such that $f(x) = y$. The case where $f(b) < y < f(a)$ is dealt with in a similar way. □

The following converse of the intermediate value theorem also holds.

Theorem 4.1.10. Let (X, d_X) be a metric space. If every continuous function $f \colon (X, d_X) \to (\mathbf{R}, d)$ has the intermediate value property (i.e., if $y_1, y_2 \in f(X)$ and y is a real number between y_1 and y_2, then there exists an $x \in X$ such that $f(x) = y$), then (X, d_X) is a connected metric space.

Proof. Suppose, if possible, (X, d_X) is not connected. Then, by Theorem 4.1.6, there exists a continuous map $g \colon (X, d_X) \to (X_0, d_0)$ that is onto. Define a map $h \colon (X_0, d_0) \to (\mathbf{R}, d)$ by $h(0) = 0$ and $h(1) = 1$. Then h is continuous. Consider the map $h \circ g \colon (X, d_X) \to (\mathbf{R}, d)$. Clearly, $h \circ g$ is continuous, being the composition of continuous maps h and g. Besides, $\{0, 1\} \subseteq h \circ g(X)$. However, there exists no $x \in X$ such that $h \circ g(x) = 1/2$. In fact, $(h \circ g)^{-1}(\{1/2\}) = g^{-1} \circ h^{-1}(\{1/2\}) = g^{-1}(\varnothing) = \varnothing$. □

An interesting application of the Weierstrass intermediate value theorem is the following "fixed point theorem":

Theorem 4.1.11. Let $I = [-1, \ 1]$ and let $f \colon I \to I$ be continuous. Then there exists a point $c \in I$ such that $f(c) = c$.

Proof. If $f(-1) = -1$ or $f(1) = 1$, the required conclusion follows; hence, we can assume that $f(-1) > -1$ and $f(1) < 1$. Define

$$g(x) = f(x) - x, \ x \in I.$$

Note that g is continuous, being the difference of continuous functions, and that it satisfies the inequalities $g(-1) = f(-1) + 1 > 0$ and $g(1) = f(1) - 1 < 0$. Hence, by the Weierstrass intermediate value theorem, there exists $c \in (-1, 1)$ such that $g(c) = 0$, that is, $f(c) = c$. □

Remarks 4.1.12. (i) The interval $[-1, 1]$ cannot be replaced by $[-1, 1)$ or $[-1, \infty)$. In the former case, the function $f(t) = (t + 1)/2, \ -1 \leq t < 1$ is continuous, maps $[-1,1)$ into itself and yet has no fixed point. Indeed, $f(t) = t$ implies $t = 1$. In the latter case, $f(t) = t + 1$, $-1 \leq t < \infty$, is continuous, maps $[-1, \infty)$ into itself and yet has no fixed point, for $f(t) = t$ implies $1 = 0$.

(ii) The foregoing theorem is possibly the simplest case of the famous fixed point theorem of L.E.J. Brouwer, according to which every continuous mapping of the closed unit ball in the Euclidean space $\mathbf{R}^n$ into itself has a fixed point. The proofs for the cases $n \geq 2$ are not easy and are beyond the scope of the present text.

Theorem 4.1.13. If Y is a connected set in a metric space (X, d) then any set Z such that $Y \subseteq Z \subseteq \bar{Y}$ is connected.

Proof. Suppose A and B are two nonempty open sets in Z such that $A \cup B = Z$ and $A \cap B = \varnothing$; as Y is dense in Z, $Y \cap A$ and $Y \cap B$ are nonempty open sets in Y and we have

$$Y = (Y \cap A) \cup (Y \cap B), \ (Y \cap A) \cap (Y \cap B) = Y \cap (A \cap B) = \varnothing,$$

a contradiction. $\qquad\square$

Remark 4.1.14. Since $Y \subseteq \bar{Y} \subseteq \bar{Y}$, it follows that $\bar{Y}$ is connected if Y is connected in (X, d).

Example 4.1.15. Since $Y = \{(x, y): y = \sin(1/x), \ 0 < x \leq 1\} \subseteq \mathbf{R}^2$ is a continuous image of $(0,1]$, it follows that $\bar{Y} = Y \cup \{(0, y): -1 \leq y \leq 1\}$ is connected. Observe that with the omission of any subset of $\{(0, y): -1 \leq y \leq 1\}$, the resulting set is still connected.

Theorem 4.1.16. Let (X, d) be a metric space and let $\{Y_\lambda : \lambda \in \Lambda\}$ be a family of connected sets in (X, d) having a nonempty intersection. Then $Y = \bigcup_{\lambda \in \Lambda} Y_\lambda$ is connected.

Proof. Suppose that Y is disconnected, i.e., $Y = A \cup B$, where A and B are nonempty open sets in Y such that $A \cap B = \varnothing$. Let $y \in \bigcap_{\lambda \in \Lambda} Y_\lambda$. Without loss of generality, assume that $y \in A$. By assumption, there is at least one λ such that $B \cap Y_\lambda \neq \varnothing$; then, as $A \cap Y_\lambda \neq \varnothing$, the sets $A \cap Y_\lambda$ and $B \cap Y_\lambda$ are open in Y_λ and satisfy

$$(A \cap Y_\lambda) \cup (B \cap Y_\lambda) = Y_\lambda, \ (A \cap Y_\lambda) \cap (B \cap Y_\lambda) = \varnothing,$$

a contradiction. $\qquad\square$

Corollary 4.1.17. Let $\{Y_i\}_{1 \leq i \leq n}$ be a sequence of connected sets such that $Y_i \cap Y_{i+1} \neq \varnothing$ for $1 \leq i \leq n - 1$; then $\bigcup_{i=1}^n Y_i$ is connected.

Proof. This follows at once from Theorem 4.1.16 by induction. $\qquad\square$

A disconnected metric space can be decomposed uniquely into connected "components"; the number of components gives a rough estimate of the disconnectedness of the space.

Let (X, d) be a metric space and $x \in X$. The union $C(x)$ of all connected subsets of (X, d) that contain x is a connected subset of (X, d) by Theorem 4.1.16.

Definition 4.1.18. The union $C(x)$ of all connected subsets containing the point x is called the **connected component of** x in (X, d).
Clearly, $C(x)$ is a maximal connected subset of X.

Examples 4.1.19. (i) Let **Q** be the set of rationals in $(\mathbf{R}, d)$. The component of each $x \in \mathbf{Q}$ is the set consisting of x alone. In other words, any subset A of **Q** containing more than one point is disconnected. Indeed, if $x, y \in A$, $x < y$, then $(-\infty, \alpha) \cap A$ and $(\alpha, \infty) \cap A$ provide a disconnection of A, when $x < \alpha < y$ and α is irrational.

(ii) Let $Y \subseteq \mathbf{R}^2$ be the subspace consisting of the segments joining the origin to the points $\{(1, 1/n) : n \in \mathbf{N}\}$ together with the segment $(1/2, 1]$. The line joining $(0,0)$ and $(1, 1/n)$ is the image of the connected set $[0,1]$ under the continuous map $y = x/n$ and, therefore, connected. If Z denotes the union of these lines, then Z is connected since the origin is common to all the line segments. Finally, Y is such that

$$Z \subset Y \subset \bar{Z},$$

where $\bar{Z} = Z \cup (0, 1]$, and so Y is connected, by Theorem 4.1.13 and Theorem 4.1.16. (See Figure 4.3.) However, $Y \backslash \{(0, 0)\}$ is not connected. In $Y \backslash \{(0, 0)\}$, the component of each point is the segment containing it.

Theorem 4.1.20. Let (X, d) be a metric space. Then

(i) each connected subset of (X, d) is contained in exactly one component;
(ii) each nonempty connected subset of (X, d) that is both open and closed in (X, d) is a component of (X, d);
(iii) each component of (X, d) is closed.

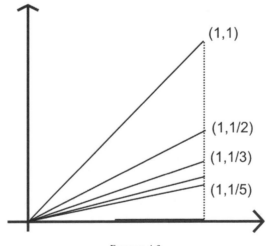

FIGURE 4.3

Proof. (i) Observe that if $C(x) \cap C(x') \neq \emptyset$, then $C(x) \cup C(x')$ is connected (see Theorem 4.1.16). This contradicts the maximality of $C(x)$ unless $C(x) = C(x')$. Thus, any two distinct connected components are disjoint. Now, let A be a connected subset of X containing x. By the maximality of $C(x)$, it follows that $A \subseteq C(x)$. Since any two distinct components are disjoint, the statement (i) follows.

(ii) Let A be a connected subset of (X, d) that is both open and closed in (X, d). Let $x \in A$, so that $A \subseteq C(x)$. Then A is both open and closed in $(C(x), d|_{C(x)})$ by Theorem 2.2.2 and consequently, $A = C(x)$ (see Theorem 4.1.3(d)).

(iii) Since $C(x)$ is connected, so also is $\overline{C(x)}$ (see Theorem 4.1.13); but the maximality of $C(x)$ implies $C(x) \supseteq \overline{C(x)}$. Hence, $C(x)$ is closed. $\square$

Remark 4.1.21. Components need not be open sets in (X, d). Let

$$X = \{0\} \cup \left\{ \frac{1}{n} : n \in \mathbf{N} \right\}$$

with the induced metric from $(\mathbf{R}, d)$; then $\{0\}$ is a component of (X, d) but is not open in (X, d). In fact, if $C(0)$, the component of 0 in X, contains another point of X, say, $1/m$, then there exists an irrational number α such that $0 < \alpha < 1/m$. The sets $(-\infty, \alpha) \cap C(0), (\alpha, \infty) \cap C(0)$ are nonempty, disjoint and open in $C(0)$, whose union is $C(0)$, a contradiction.

4.2. Local Connectedness

Many of the important spaces occurring in analysis are not connected but enjoy a property that is a local version of connectedness. In this section, we briefly discuss this local property.

Definition 4.2.1. A metric space (X, d) is said to be **locally connected** if, for $x \in X$, there is a base of connected neighbourhoods of x. Thus, X is locally connected if and only if the family of all open connected sets is a base for the open subsets of X.

Examples 4.2.2. (i) Let $X = \{x \in \mathbf{R} : a < x < b \text{ or } c < x < d\}$, where $b < c$, with the induced metric from $(\mathbf{R}, d)$. Then X is locally connected but not connected.

(ii) Consider the discrete two-point space (X_0, d_0) defined after Remark 4.1.5. Here $X_0 = \{0, 1\}$ and d_0 is the discrete metric on X_0. It is not connected since $\{0\}$ and $\{1\}$ are both open in X_0 and $X_0 = \{0\} \cup \{1\}$. But (X_0, d_0) is locally connected, for $\{ \{0\}, \{1\} \}$ is a base of connected subsets for the open subsets of X_0.

(iii) An example of a connected subset Y of the plane that is not locally connected may be constructed as follows (see Example 4.1.19(ii)):

$$Y = \left[\frac{1}{2}, 1 \right] \cup \{ (x, y) : y = \frac{x}{n}, 0 \leq x \leq 1 \}.$$

It follows from Theorems 4.1.13 and 4.1.16 that Y is connected. But this space is not locally connected since an open set containing $\alpha = (3/4, 0)$ will not contain a connected open set containing α. In fact, $S(\alpha, r) \cap Y$ must intersect a ray joining $(0,0)$ to some $(1, 1/n)$ and this intersection is both open and closed in $S(\alpha, r) \cap Y$, because this intersection is a union of disjoint open segments.

It was observed in Remark 4.1.21 that components need not be open sets in the space. However, if the space is locally connected, then the following holds.

Theorem 4.2.3. Let (X, d) be a metric space. X is locally connected if and only if the connected components of the open sets in X are open in X. (For any subset A of X, the connected components of the points of the subspace A are called the connected components of A).

Proof. Let G be an open subset of X, and C be a component of G, and $\{U_\lambda : \lambda \in \Lambda\}$ be a basis consisting of open connected sets for the open sets of X. Let $x \in C$. Since $x \in G$, there is a U_λ such that $x \in U_\lambda \subseteq G$; but since C is the component of x and U_λ is connected, $x \in U_\lambda \subseteq C$. We have thus shown that C is a union of open subsets of X and is, therefore, open.

On the other hand, if O is any open set containing a point $x \in X$, the connected component of x in the subspace O is a connected neighbourhood of x contained in O; hence, X is locally connected. $\square$

We close this section with the following characterisation of open subsets of $\mathbf{R}$, based on Theorems 4.1.4 and 4.2.3 (see Theorem 2.1.11 of Chapter 2):

Theorem 4.2.4. Let O be a nonempty open subset of $(\mathbf{R}, d)$. Then O is a union of an at most countable family of open intervals, no two of which have common points.

Proof. It follows from Theorems 4.1.4 and 4.2.3 that the connected components of O are intervals and that they are open subsets of O and, therefore, of $\mathbf{R}$ as well. Thus, they are open intervals. Note that $O \cap \mathbf{Q}$, where $\mathbf{Q}$ denotes the set of rationals in $\mathbf{R}$, is countable and each component of O contains a point of $O \cap \mathbf{Q}$. The mapping $r \to C(r)$ of $O \cap \mathbf{Q}$ into the set of connected components of O is onto and, therefore, the set of connected components of O is countable. $\square$

Remark 4.2.5. The property of being locally connected is not preserved by continuous maps.

Let $X_1 = \{0, 1, \ldots\}$ and $X_2 = \{0\} \cup \{1/n : n \in \mathbf{N}\}$, equipped with metrics induced from $(\mathbf{R}, d)$. The mapping $f: X_1 \to X_2$ defined by

$$f(x) = \begin{cases} 0 & \text{if } x = 0, \\ 1/x & \text{if } x = n. \end{cases}$$

Then X_1 is locally connected and f is a continuous map of X_1 onto X_2; but X_2 is not locally connected (see Remark 4.1.21 and Theorem 4.2.3).

4.3. Arcwise Connectedness

A metric space (X, d) is regarded as "arcwise connected" if any two of its points can be joined by what is called a "path" (see definition below). Arcwise connectedness is particularly important in the study of functions of a complex variable. Although a "region" is defined to be an open connected subset of the plane, most of the proofs use the arcwise connectedness of a region rather than its connectedness. That these two properties are equivalent for an open subset of the plane will be proved in this section.

Definition 4.3.1. Let (X, d) be a metric space, $Y \subseteq X$ and $I = [0, 1]$. A **path** (or **arc**) in Y is a continuous mapping $f: I \to Y$. If $f: I \to Y$ is a path in Y, we call $f(0) \in Y$ its **initial point**, $f(1) \in Y$ its **final point** and say that f **joins** $f(0)$ and $f(1)$ or that f **runs from** $f(0)$ **to** $f(1)$. A subset Y of X is said to be **arcwise** (or **pathwise**) **connected** if, for any two points in Y, there exists a path running from one to the other, i.e. the points can be "joined by" a path in Y.

It should be noted that if $f: I \to Y$ is a path, then the mapping $t \to f(1 - t), t \in I$, is a path that runs from $f(1)$ to $f(0)$.

Examples 4.3.2. (i) The subset $Y = A \cup B$ of $\mathbf{R}^2$, where

$$A = \{(0, y) : -1 \le y \le 1\}$$
$$B = \{(x, y) : y = \sin 1/x, 0 < x \le 1\}$$

is a connected subset of $\mathbf{R}^2$ (see Example 4.1.15). However, Y is not arcwise connected; in fact, there exists no path from a point in A to a point in B.

(ii) The set $S(0, 1) = \{z \in \mathbf{C}: |z| < 1\}$ is arcwise connected. In fact,

$$f(t) = tz_2 + (1 - t)z_1, 0 \le t \le 1$$

is a continuous path joining z_1 and z_2 in $S(0,1)$.

A useful reformulation of arcwise connectedness is given in the following proposition.

Proposition 4.3.3. Let (X, d) be a metric space and $x_0 \in X$ be any element. X is arcwise connected if and only if each $x \in X$ can be joined to x_0 by a path.

Proof. If X is arcwise connected, the condition trivially holds. Conversely, assume that the condition is satisfied and let x_1, x_2 be any two points in X. Let $f: I \to X$ run from x_1 to x_0 and $g: I \to X$ run from x_0 to x_2; then

$$h(t) = \begin{cases} f(2t) & \text{if } 0 \le t \le \dfrac{1}{2}, \\ g(2t - 1) & \text{if } \dfrac{1}{2} \le t \le 1 \end{cases}$$

is continuous (because at $t = 1/2$, we have $f(1) = x_0 = g(0)$) and is a path running from x_1 to x_2. $\square$

The general relation between connectedness and arcwise connectedness is given in the following theorem.

Theorem 4.3.4. Let (X, d) be an metric space. If X is arcwise connected, then it is connected.

Proof. Let $x \in X$; then for each $y \in X$, there exists a continuous function $\varphi_y : [0, 1] \to X$ such that $\varphi_y(0) = x$ and $\varphi_y(1) = y$. Since $[0, 1]$ is connected and φ_y is continuous, $\varphi_y([0, 1])$ is connected by Theorem 4.1.8; moreover,

$$x \in \bigcap_{y \in X} \varphi_y([0, 1])$$

and, hence,

$$\bigcup_{y \in X} \varphi_y([0, 1])$$

is connected in X by Theorem 4.1.16 and therefore X is connected.   □

Alternative proof. If X were not connected, then there would exist nonempty disjoint subsets A and B such that $X = A \cup B$ and $\bar{A} \cap B = \varnothing = A \cap \bar{B}$. Choose $a \in A, b \in B$ and a path $p : I \to X$ joining a and b. Since $\overline{p^{-1}(A)} \subseteq p^{-1}(\bar{A})$ by Theorem 3.1.12 (ii)), $I = p^{-1}(A) \cup p^{-1}(B)$ would constitute a disconnection of I, contradicting the fact that I is connected (see Theorem 4.1.4). The converse of the above theorem is false, that is, a connected metric space need not be arcwise connected (see Example 4.3.2(i)).

We next show that arcwise connectedness is invariant under a continuous map.

Theorem 4.3.5. Let (X, d) be a metric space and $Y \subseteq X$. If Y is arcwise connected and f is a continuous map from the metric space (X, d_X) to the metric space (Z, d_Z), then $f(Y)$ is arcwise connected.

Proof. Let $z_1, z_2 \in f(Y)$. Then there exist $z_1^*, z_2^* \in Y$ such that $f(z_i^*) = z_i, i = 1, 2$. But Y is arcwise connected and so there exists a path $g : I \to Y$ such that $g(0) = z_1^*, g(1) = z_2^*$. Now the composition of continuous functions is continuous (see Theorem 3.1.11) and so $f \circ g : I \to Z$ is continuous. Moreover,

$$f \circ g(0) = f(z_1^*) = z_1, \qquad f \circ g(1) = f(z_2^*) = z_2,$$
$$f \circ g(I) = f[g(I)] \subseteq f(Y).$$

Thus, $f(Y)$ is arcwise connected.   □

Theorem 4.3.6. Let $\mathcal{A}$ be a class of arcwise connected subsets of a metric space (X, d) with nonempty intersection. Then

$$B = \bigcup \{A; A \in \mathcal{A}\}$$

is arcwise connected.

Proof. Let $a, b \in B$. Then there exist A_a and $A_b \in \mathcal{A}$ such that $a \in A_a$ and $b \in A_b$. Now $\mathcal{A}$ has a nonempty intersection; so there exists $p \in \bigcap \{A: A \in \mathcal{A}\}$. Then $p \in A_a$ and, since A_a is arcwise connected, there is a path $f: I \to A_a \subseteq B$ from a to p. Similarly, there is a path $g: I \to A_b \subseteq B$ from p to b. Then $h: I \to B$ defined by

$$
h(t) = \begin{cases} f(2t) & \text{if } 0 \le t \le \dfrac{1}{2}, \\ g(2t - 1) & \text{if } \dfrac{1}{2} \le t \le 1 \end{cases}
$$

is continuous (because at $t = 1/2$, we have $f(1) = p = g(0)$), and it is a path running from a to b contained in B. Hence, B is arcwise connected. $\square$

Our final result in this section says that, for open subsets of the complex plane, the properties of being connected and arcwise connected are equivalent.

Theorem 4.3.7. Let Ω be a nonempty open connected subset of the complex plane. Then Ω is arcwise connected.

Proof. Let $z \in \Omega$ and let G consist of those points in Ω which can be joined to z by a path in Ω. By Proposition 4.3.3, it is sufficient to show that $G = \Omega$.

First of all, we claim that G is open. Let $\omega \in G \subseteq \Omega$. Now Ω is open and so there exists an open ball $S(\omega, r)$ such that $\omega \in S(\omega, r) \subseteq \Omega$. But $S(\omega, r)$ is arcwise connected (see Example 4.3.2(ii)); hence, each point $\xi \in S(\omega, r)$ can be joined to ω, which in turn, can be joined to z. Hence each point $\xi \in S(\omega, r)$ can be joined to z and so $\omega \in S(\omega, r) \subseteq G$. Accordingly, G is open. Let $H = \Omega \backslash G$; i.e., H consists of those points in Ω which cannot be joined to z by a path in Ω. We claim that H is open. Let $\omega^* \in H \subseteq \Omega$. Since Ω is open, there exists an open disc $S(\omega^*, r^*)$ such that $\omega^* \in S(\omega^*, r^*) \subseteq \Omega$. Since $S(\omega^*, r^*)$ is arcwise connected, no $\xi^* \in S(\omega^*, r^*)$ can be joined to z by a path in Ω, and so $\xi^* \in S(\omega^*, r^*) \subseteq H$. Hence, H is open.

But Ω is connected and therefore cannot be the union of two nonempty disjoint open subsets. Thus, H is empty and so $\Omega = G$. As already noted, this completes the proof that Ω is arcwise connected. $\square$

Remark 4.3.8. The requirement that Ω be open cannot be dropped (see Example 4.3.2(i)).

4.4. Exercises

1. Show that the subset $A \subseteq \mathbf{R}^2$, where

$$
A = \{(x, y): x^2 - y^2 \ge 4\}
$$

is disconnected.

Hint: If $G = \{(x, y): x < -1\}$ and $H = \{(x, y): x > 1\}$, then $G \cap A$ and $H \cap A$ are nonempty open disjoint sets in A whose union is A.

2. If C is a connected subset of a disconnected metric space $X = A \cup B$, where A, B are nonempty and $\overline{A} \cap B = \emptyset = A \cap \overline{B}$, then either $C \subseteq A$ or $C \subseteq B$.
   Hint: Clearly, $C = C \cap X = C \cap (A \cup B) = (C \cap A) \cup (C \cap B)$. Also,

   $$[(C \cap A) \cap \overline{(C \cap B)}] \cup [\overline{(C \cap A)} \cap (C \cap B)] \subseteq (A \cap \overline{B}) \cup (\overline{A} \cap B) = \emptyset.$$

   If both $(C \cap A)$ and $(C \cap B)$ are nonempty, then they form a disconnection of C. So, either $C \cap A = \emptyset$, in which case $C \subseteq B$, or $C \cap B = \emptyset$, in which case $C \subseteq A$.

3. If every two points in a metric space X are contained in some connected subset of X, then X is connected.
   Hint: If not, $X = A \cup B$, where A, B are nonempty disjoint and $\overline{A} \cap B = \emptyset = A \cap \overline{B}$. Since A and B are nonempty, choose $a \in A$ and $b \in B$. By hypothesis, a and b are contained in some connected subset $C \subseteq X$. In view of Exercise 2 above, $C \subseteq A$ or $C \subseteq B$. Since $A \cap B = \emptyset$, this leads to a contradiction.

4. A metric space X is said to be **totally disconnected** if for each pair of points $x, y \in X$ there exist A and B, $A \neq \emptyset, B \neq \emptyset, X = A \cup B, \overline{A} \cap B = \emptyset = A \cap \overline{B}$ such that $x \in A$ and $y \in B$. Show that the set $\mathbf{Q}$ of rationals with the usual metric is totally disconnected.
   Hint: Let $x, y \in \mathbf{Q}$. Without loss of generality, assume that $x < y$. There exists an irrational number z such that $x < z < y$. Set

   $$G = \{u \in \mathbf{Q} : u < z\} \text{ and } H = \{u \in \mathbf{Q} : u > z\}.$$

   Then G and H constitute a disconnection of $\mathbf{Q}$ and $x \in G, y \in H$.

5. The components of a totally disconnected space X are singleton subsets of X. (In view of Exercises 4 and 5, it follows that the connected components of $\mathbf{Q}$ are singletons (see Example 4.1.19(i)).)
   Hint: Let C be a component of X and suppose $x, y \in C$ with $x \neq y$. Since X is totally disconnected, there exists a disconnection $G \cup H$ of X such that $x \in G$ and $y \in H$. Since $C \cap G$ and $C \cap H$ are nonempty,

   $$(C \cap G) \cap \overline{(C \cap H)} \subseteq (C \cap G) \cap (\overline{C} \cap \overline{H}) = C \cap (G \cap \overline{H}) = C \cap \emptyset = \emptyset$$

   and

   $$\overline{(C \cap G)} \cap (C \cap H) = \emptyset,$$

   therefore, $C = (C \cap G) \cup (C \cap H)$ is a disconnection of C. But C, being a component, must be connected.

6. Let $X = C[0, 1]$ with the metric $d(x, y) = \sup\{|x(t) - y(t)| : 0 \leq t \leq 1\}$. Show that $(C[0, 1], d)$ is a connected metric space.
   Hint: Any two elements $x, y \in X$ can be joined by an arc $tx + (1 - t)y$. Therefore, the space is arcwise connected and, hence, connected.

7. Show that the connected components of any metric space X form a partition of X. Hint: Certainly, $\cup\{C(x): x \in X\} = X$. Distinct components are disjoint, as observed in the proof of Theorem 4.1.20(i).

8. Prove the following:

(a) $\mathbf{R}^2\backslash\{0\}$ is arcwise connected and, hence, connected.

(b) The set $S^1 = \{z \in \mathbf{C}: |z| = 1\}$ is connected.

(c) Let $Y = \left\{ \begin{pmatrix} \cos\theta & \sin\theta \\ -\sin\theta & \cos\theta \end{pmatrix} : \theta \in \mathbf{R} \right\}$ with the metric induced from $\mathbf{R}^4$.

Then Y is connected.

Hint: (a) Indeed any two points can be joined to a third point (infinitely many possibilities) by line segments not passing through zero.

(b) The mapping $z \rightarrow z/|z|$ from $\mathbf{R}^2\backslash\{(0,0)\}$ to S^1 is both continuous and onto.

(c) The mapping that carries $\begin{pmatrix} \cos\theta & \sin\theta \\ -\sin\theta & \cos\theta \end{pmatrix}$ to $\cos\theta + i \sin\theta$ is a homeomorphism.

5 Compact Spaces

One of the distinguishing properties of a bounded closed interval $[a, b]$ is that every sequence in it has a subsequence converging to a limit in the interval. This need not happen with an unbounded interval such as $[0, \infty)$ or a bounded nonclosed interval such as $(0,1]$; the former contains the sequence $\{n\}_{n \geq 1}$, which has no convergent subsequence, and the latter contains the sequence $\{1/n\}_{n \geq 1}$, which has no subsequence converging to a limit belonging to the interval. In fact, it is true of any bounded closed subset of $\mathbf{R}$ that any sequence in it has a subsequence converging to a limit belonging to the subset. To see why, we first note that any sequence in a bounded subset must, by the Bolzano-Weierstrass theorem (Proposition 0.4.2), have a convergent subsequence with limit in $\mathbf{R}$; this limit must then be in the closed subset by the definition of a closed subset.

Another characterisation of bounded closed subsets of $\mathbf{R}$, closely related to the Bolzano-Weierstrass theorem, is the Heine-Borel theorem (Proposition 0.7.1.). This asserts that any open cover of a bounded closed subset F of $\mathbf{R}$ contains a finite subcover. That is, if $\{G_\lambda : \lambda \in \Lambda\}$ is a class of open subsets of $\mathbf{R}$ such that

$$\bigcup_\lambda G_\lambda \supseteq F,$$

then there exists a finite number of open subsets $G_{\lambda_i}, i = 1, 2, \ldots, n$, say, such that

$$G_{\lambda_1} \cup G_{\lambda_2} \cup \ldots \cup G_{\lambda_n} \supseteq F.$$

In fact, this is true of any bounded closed subset of $\mathbf{R}^n$ (see Section 5.5(d) below). How these results translate to arbitrary metric spaces will be investigated.

We shall also introduce the related concepts of local compactness and countable compactness. It will be shown that countable compactness of a metric space is equivalent to its compactness. However, local compactness is not equivalent to compactness but only a consequence of the latter. Certain important spaces in analysis are locally compact but not compact, notably $\mathbf{R}^n$. The chapter ends with a description of compact subsets of some special metric spaces.

5.1. Bounded sets and Compactness

Let (X, d) be a metric space and $Y \subseteq X$. Let $\mathcal{G}$ be a collection of open sets in X with the property that $Y \subseteq \cup\{G : G \in \mathcal{G}\}$; equivalently, for each $x \in Y$, there is a $G \in \mathcal{G}$ such that $x \in G$. Then $\mathcal{G}$ is called an "open cover" or an "open covering" of Y. A finite subcollection of $\mathcal{G}$ which is itself a cover is called a finite subcover or a finite subcovering of Y (see Definition 2.2.8).

Definition 5.1.1. A metric space (X, d) is said to be **compact** if every open covering $\mathcal{G}$ of X has a finite subcovering, that is, there is a finite subcollection $\{G_1, G_2, \ldots, G_n\} \subseteq \mathcal{G}$ such that

$$X = \bigcup_{i=1}^{n} G_i.$$

A nonempty subset Y of X is said to be compact if it is a compact metric space with the metric induced on it by d. In view of the definition of relatively open subsets of a metric space (see Definition 2.1.4), this is equivalent to saying that a nonempty subset Y is compact if every covering $\mathcal{G}$ of Y by relatively open sets of Y has a finite subcovering.

Observe that if $\mathcal{G}$ is an open covering of X, then the collection $\mathcal{F}$ of complements of sets in $\mathcal{G}$ is a collection of closed sets whose intersection is empty, and, conversely. Thus, a metric space (X, d) is compact if and only if every collection of closed sets with empty intersection has a finite subcollection with empty intersection.

Examples 5.1.2. (i) The interval $(0, 1)$ in the metric space $(\mathbf{R}, d)$, where d denotes the usual metric, is not compact. In order to prove the assertion, it suffices to exhibit an open covering from which no finite subcovering can be selected. The open covering $\{(1/n, 1) : n = 2, 3, \ldots\}$ is one such covering of $(0, 1)$ from which no finite subcovering can be selected. More generally, the open ball $S(0, 1)$ in $(\mathbf{R}^n, d)$ (see Example 1.2.2(iii)) is not compact. Indeed,

$$\bigcup_{n=2}^{\infty} S(0, 1 - 1/n) \supseteq S(0, 1).$$

However, no finite subcollection of $\{S(0, 1 - 1/n) : n = 2\ 3, \ldots\}$ covers the open ball $S(0, 1)$.

(ii) Let $\mathbf{Z}$ denote the set of integers in the metric space $(\mathbf{R}, d)$, where d denotes the usual metric. Observe that $\{n\} = \mathbf{Z} \cap (n - 1/2, n + 1/2)$, and is, therefore, a relatively open subset of $\mathbf{Z}$. Now, the collection $\{\{n\} : n \in \mathbf{Z}\}$ is an open cover of $\mathbf{Z}$ that admits no finite subcover. Thus, $\mathbf{Z}$ is not compact.

(iii) Let Y be a finite subset of a metric space (X, d). Then Y is compact.

Definition 5.1.3. A collection $\mathcal{F}$ of sets in X is said to have the **finite intersection property** if every finite subcollection of $\mathcal{F}$ has a nonempty intersection.

The following proposition now holds.

Proposition 5.1.4. Let (X, d) be a metric space. The following statements are equivalent:

(i) (X, d) is compact;
(ii) every collection of closed sets in (X, d) with empty intersection has a finite subcollection with empty intersection;
(iii) every collection of closed sets in (X, d) with the finite intersection property has nonempty intersection.

Proof. That (i) is equivalent to (ii) has been proved in the paragraph preceding Example 5.1.2. The statements (ii) and (iii) are equivalent; in fact, each is the contrapositive of the other.   □

The reader will have noticed that the set considered in Example 5.1.2 (i) was not closed and the one considered in (ii) was not bounded. This is not a coincidence. In fact, if a subset Y of a metric space (X, d) is compact, then it is both closed and bounded.

Theorem 5.1.5. Let (X, d) be a metric space and Y a subset of X. If Y is a compact subset of (X, d), then Y is closed and bounded.

Proof. Let Y be a compact subset of (X, d) and $y \in Y, x \in Y^c$. For some real number $\varepsilon(y)$ (choose $\varepsilon(y) < (1/2)d(x, y)$), there exist open balls $S(y, \varepsilon(y))$ and $S(x, \varepsilon(y))$ with centres at y and x, respectively, such that

$$S(y, \varepsilon(y)) \cap S(x, \varepsilon(y)) = \varnothing.$$

Clearly,

$$Y \subseteq \bigcup_{y \in Y} S(y, \varepsilon(y)).$$

Since Y is compact, there exist $y_1, y_2, \ldots, y_n$ such that

$$Y \subseteq \bigcup_{i=1}^{n} S(y_i, \varepsilon(y_i)).$$

For each of the $y_i, i = 1, 2, \ldots, n$, the open balls $S(x, \varepsilon(y_i))$ satisfy

$$S(y_i, \varepsilon(y_i)) \cap S(x, \varepsilon(y_i)) = \varnothing.$$

Let $Z = \bigcap_{i=1}^{n} S(x, \varepsilon(y_i))$. Then Z is an open subset of X containing $x \in Y^c$. We next show that $Y \cap Z = \varnothing$. If $t \in Y \cap Z$, then $t \in S(y_j, \varepsilon(y_j))$ for some j in the set $\{1, 2, \ldots, n\}$ and $t \in S(x, \varepsilon(y_j))$. Therefore, $S(y_j, \varepsilon(y_j)) \cap S(x, \varepsilon(y_j)) \neq \varnothing$ and this contradicts the way $\varepsilon(y_j)$ were chosen. So, no point of Y^c can be a limit point of Y. Hence, all the limit points of Y belong to Y, that is, Y is closed.

We next show that Y is bounded. If Y were not bounded, there would exist x and y in Y such that for any preassigned positive M, $d(x, y) > M$. Consider the open balls centred at the points of Y, each of radius 1. Clearly,

$$Y \subseteq \bigcup_{y \in Y} S(y, 1).$$

In view of the compactness of Y, there exist $y_1, y_2, \ldots, y_n$ such that

$$Y \subseteq \bigcup_{i=1}^{n} S(y_i, 1).$$

Now, let $k = \max \{d(y_i, y_j): i, j = 1, 2, \ldots, n\}$. There exist x and y in Y such that

$$d(x, y) > k + 2.$$

Since x and y are in Y, there exist y_i and y_j such that

$$x \in S(y_i, 1) \text{ and } y \in S(y_j, 1).$$

So

$$d(x, y) \leq d(x, y_i) + d(y_i, y_j) + d(y_j, y) < k + 2.$$

This contradicts the way x and y were chosen. Consequently, Y is bounded. $\qquad\square$

Remark 5.1.6. The converse of the above, Theorem 5.1.5, is, however, false. In fact, let X be an infinite set with the discrete metric (see Example 1.2.2 (v)). Each subset of X is both closed and bounded, because the open ball $S(x, 1/2)$ is the set $\{x\}$ consisting of x alone and $d(x, y) \leq 1, x, y \in X$. The open cover $\{\{x\}: x \in X\}$ has no finite subcover.

Another example that illustrates the same phenomenon is the following: Consider the metric space (ℓ_2, d) (see Example 1.2.2(vii)). Let $Y = \{e_1, e_2, \ldots, e_n, \ldots\}$, where e_n denotes the sequence all of whose terms are 0 except the nth term, which is 1. Also, let $e_0 = (0, 0, 0, \ldots)$; then $e_n \in \bar{S}(e_0, 2)$ for all n; so Y is bounded. Moreover, $d(e_n, e_m) = \sqrt{2}$ for all $n, m(n \neq m)$. So, Y has no limit points and, hence, is closed. Thus, Y is both a closed and bounded subset of (ℓ_2, d); it is, however, not compact. In fact, the open cover $\{S(e_n, 1): n = 1, 2, \ldots\}$ of Y contains no finite subcover of Y.

The above argument without substantial change will yield that $Y \subseteq (\ell_p, d), p \geq 1$, is not compact.

We now proceed to describe other ways of characterising compactness.

Definition 5.1.7. Let (X, d) be a metric space and ε be an arbitrary positive number. Then a subset $A \subseteq X$ is said to be an **ε-net** for X if, given any $x \in X$, there exists a point $y \in A$ such that $d(x, y) < \varepsilon$. In other words, A is an ε-net for X if

$$X = \bigcup \{S(y, \varepsilon): y \in A\}.$$

A finite subset of X that is an ε-net for X is called a **finite ε-net** for X.

Thus, for example, the set $A = \{(m, n) : m, n \in \mathbf{Z}\}$ is a $(2^{-1/2} + \delta)$-net $(\delta > 0)$ for $(\mathbf{R}^2, d)$.

Definition 5.1.8. The metric space (X, d) is said to be **totally bounded** if, for any $\varepsilon > 0$, there exists a finite ε-net for (X, d). A nonempty subset Y of X is said to be totally bounded if the subspace Y is totally bounded.

Examples 5.1.9. (i) A bounded interval in $\mathbf{R}$ is a totally bounded metric space. Let the endpoints of the interval be a and b $(a < b)$ and ε be an arbitrary positive number. Take an integer $n > (b - a)/\varepsilon$ and divide the interval into n equal sub-intervals each of length $(b - a)/n$. The points

$$\{a + (j - 1)\frac{b - a}{n} : j = 2, \ldots, n\}$$

constitute the required ε-net for the interval with endpoints a and b. In fact, consider any x in the interval. Then $a \leq x \leq b$. There exists an integer $k \in \{1, 2, \ldots, n\}$ such that

$$a + (k - 1)\frac{b - a}{n} \leq x \leq a + k\frac{b - a}{n}.$$

Consequently, the distance of x from each of the endpoints of the interval

$$\left[a + (k - 1)\frac{b - a}{n}, a + k\frac{b - a}{n}\right]$$

is less than or equal to $(b - a)/n$, which is strictly less than ε in view of the way in which n has been selected. It follows that any set containing at least one endpoint of each of the aforementioned subintervals, $k = 1, 2, \ldots, n$, forms an ε-net; the points in question constitute such a set.

(ii) Let (X, d) denote a discrete metric space with infinitely many points (see Example 1.2.2 (v)). Since

$$S\left(x, \frac{1}{2}\right) = \{x\}, \qquad x \in X,$$

it follows that X contains no finite $1/2$-net. So X is not totally bounded.

(iii) In the space (ℓ_2, d) (see Example 1.2.2(vii)), consider the set

$$Y = \{e_1, e_2, \ldots, e_n, \ldots\},$$

where e_n denotes the sequence all of whose terms are zero except the nth, which is equal to 1. It was observed in Remark 5.1.6 that Y is bounded. We now show that Y is not totally bounded. In fact, Y has no finite $1/\sqrt{2}$-net; for if it has one, say A, then for each positive integer n, there exists some $a_n \in A$ such that $d(e_n, a_n) < 1/\sqrt{2}$. Clearly, $m \neq n$ implies $a_n \neq a_m$, because, otherwise,

$$\sqrt{2} = d(e_n, e_m) \leq d(e_n, a_n) + d(a_n, a_m) + d(a_m, e_m) < \frac{1}{\sqrt{2}} + 0 + \frac{1}{\sqrt{2}} = \sqrt{2},$$

a contradiction. But this implies that the set $\{a_n : n = 1, 2, \ldots\}$ is infinite, which is impossible because it is a subset of the finite set A.

It is clear that every nonempty subset of a totally bounded metric space is totally bounded. (Let Y be a nonempty subset of a totally bounded metric space X and let $\{x_1, x_2, \ldots, x_n\}$ be a finite $(1/2)\varepsilon$-net in X. Since $Y \neq \varnothing$, $Y \cap S(x_j, (1/2)\varepsilon) \neq \varnothing$ for at least one $j, 1 \leq j \leq n$. Assume for $j = 1, 2, \ldots, m$, $Y \cap S(x_j, (1/2)\varepsilon) \neq \varnothing$ (x_j reindexed, if necessary) and m is the largest such positive integer. Choose $y_j \in Y \cap S(x_j, (1/2)\varepsilon)$, $j = 1, 2, \ldots, m$. If $y \in Y$, then $y \in S(x_j, (1/2)\varepsilon)$ for some j, $1 \leq j \leq m$ and hence $d(y, y_j) \leq d(y, x_j) + d(x_j, y_j) < \varepsilon$.) Since it was shown in Example 5.1.9 (iii) that (ℓ_2, d) contains a nonempty subset Y that is not totally bounded, it now follows that (ℓ_2, d) is not totally bounded.

"Totally bounded" is a stronger restriction than "bounded" on a metric space.

Proposition 5.1.10. A totally bounded metric space is bounded.

Proof. Let (X, d) be totally bounded and suppose $\varepsilon > 0$ has been given. Then there exists a finite ε-net for X, say A. Since A is a finite set of points, $d(A) = \sup\{d(y, z) : y, z \in A\} < \infty$. Now, let x_1 and x_2 be any two points of X. There exist points y and z in A such that

$$d(x_1, y) < \varepsilon \text{ and } d(x_2, z) < \varepsilon.$$

It follows, using the triangle inequality, that

$$d(x_1, x_2) \leq d(x_1, y) + d(y, z) + d(z, x_2)$$
$$\leq d(A) + 2\varepsilon.$$

So,

$$d(X) = \sup\{d(x_1, x_2) : x_1, x_2 \in X\} \leq d(A) + 2\varepsilon$$

and, hence, X is bounded. $\qquad\square$

Remark 5.1.11. In the case of $(\mathbf{R}^n, d)$ (see Example 1.2.2 (ii)), where d is the Euclidean metric, it can be shown that a subset $Y \subseteq \mathbf{R}^n$ is bounded if and only if it is totally bounded. In fact, a bounded subset in $\mathbf{R}^n$ can be enclosed in the interior of some sufficiently large n-rectangle. Divide the large n-rectangle into smaller rectangles, each with diagonal of length less than $\varepsilon/n^{1/2}$. Then the vertices of these smaller n-rectangles will form a finite ε-net in the large n-rectangle. So, the large n-rectangle is totally bounded. In view of the observation preceding Proposition 5.1.10, it follows that Y is totally bounded.

The following important criterion of total boundedness helps in determining when a set is totally bounded.

Theorem 5.1.12. Let Y be a subset of the metric space (X, d). Then Y is totally bounded if and only if every sequence in Y contains a Cauchy subsequence.

Proof. Suppose Y is totally bounded. Let $\{y_n\}_{n \geq 1}$ be a sequence in Y whose range may be assumed to be infinite. Choose a finite 1/2-net in Y. Then one of the balls of radius 1/2 with centre in the net contains infinitely many elements of the range of the sequence. We shall denote the subsequence formed by these elements by $\{y_n^{(1)}\}_{n \geq 1}$. Choose a finite 1/4-net in Y. Then one of the balls of radius 1/4 with centre in the finite 1/4-net contains infinitely many elements of the range of $\{y_n^{(1)}\}_{n \geq 1}$. We shall denote the subsequence formed as $\{y_n^{(2)}\}_{n \geq 1}$. Proceeding in this way, we obtain a sequence of sequences, each a subsequence of the preceding one, so that at the kth stage, the terms $\{y_n^{(k)}\}_{n \geq 1}$ lie in the ball of radius $1/2^k$ with centre in the $1/2^k$-net. Now $\{y_n^{(n)}\}_{n \geq 1}$ is a subsequence of $\{y_n\}_{n \geq 1}$. Let $\varepsilon > 0$ be given. Choose n_0 so large that $1/2^{n_0-2} < \varepsilon$. Then, for $m > n > n_0$, we have

$$d\left(y_n^{(n)}, y_m^{(m)}\right) \leq d\left(y_n^{(n)}, y_{n+1}^{(n+1)}\right) + \ldots + d\left(y_{m-1}^{(m-1)}, y_m^{(m)}\right)$$
$$< \frac{2}{2^n} + \frac{2}{2^{n+1}} + \ldots + \frac{2}{2^{m-1}} < \frac{1}{2^{n-2}} < \frac{1}{2^{n_0-2}} < \varepsilon,$$

so that the sequence $\{y_n^{(n)}\}_{n \geq 1}$ is a Cauchy sequence.

Conversely, suppose that every sequence in Y has a Cauchy subsequence. We shall show that Y is totally bounded. Let ε be a positive real number and let $y_1 \in Y$. If $Y \setminus S(y_1, \varepsilon) = \varnothing$, we have found an ε-net, namely, the set $\{y_1\}$. Otherwise choose $y_2 \in Y \setminus S(y_1, \varepsilon)$. If $Y \setminus [S(y_1, \varepsilon) \cup S(y_2, \varepsilon)] = \varnothing$, we have found an ε-net, namely, the set $\{y_1, y_2\}$. It is enough to show that this process terminates after a finite number of steps. If it does not terminate, we shall obtain an infinite sequence $\{y_n\}_{n \geq 1}$ with the property that $d(y_n, y_m) \geq \varepsilon$, $n \neq m$. Consequently, the sequence $\{y_n\}_{n \geq 1}$ would have no Cauchy subsequence, contrary to hypothesis.   □

We give below a characterisation of compact metric spaces.

Proposition 5.1.13. Let (X, d) be a compact metric space. Then (X, d) is totally bounded.

Proof. For any given $\varepsilon > 0$, the collection of all balls $S(x, \varepsilon)$ for $x \in X$ is an open cover of X. The compactness of X implies that this open cover contains a finite subcover. Hence, for $\varepsilon > 0$, X is covered by a finite number of open balls of radius ε, i.e., the centres of the balls in the finite subcover form a finite ε-net for X. So, X is totally bounded.   □

Corollary 5.1.14. Every compact metric space X is separable.

Proof. For each positive integer n, X has a finite $1/n$-net, say A_n. Then A_n is a finite set and

$$X = \bigcup_{a \in A_n} S\left(a, \frac{1}{n}\right).$$

Let $A = \cup_{n=1}^{\infty} A_n$. Then A is a countable subset of X, being a countable union of finite sets. We show that A is dense in X. For this, let $S(x, \varepsilon)$ be an open ball centred at x.

Choose a positive integer n such that $1/n < \varepsilon$. Since A_n is a finite $1/n$-net, therefore $x \in \bigcup_{a \in A_n} S(a, 1/n)$. So there exists an $a \in A_n$ such that $d(x, a) < 1/n$. Consequently, $a \in S(x, \varepsilon)$. This completes the proof. $\qquad\square$

Proposition 5.1.15. Let (X, d) be a compact metric space. Then (X, d) is complete.

Proof. Suppose, if possible, that (X, d) is a compact metric space that is not complete. Then there exists a Cauchy sequence $\{x_n\}_{n \geq 1}$ in (X, d) not having a limit in X. Let $y \in X$; since $\{x_n\}_{n \geq 1}$ does not converge to y, there exists an $\varepsilon_0 > 0$ such that

$$d(x_n, y) \geq 2\varepsilon_0 \qquad\qquad (5.1)$$

for infinitely many values of n. Since $\{x_n\}_{n \geq 1}$ is Cauchy, there exists an integer n_0 such that $n, m \geq n_0$ implies

$$d(x_n, x_m) < \varepsilon_0.$$

Choose $k > n_0$ for which $d(x_k, y) \geq 2\varepsilon_0$ (this is possible since the inequality (5.1) is satisfied for infinitely many values of n). Then

$$d(x_k, y) \leq d(x_k, x_m) + d(x_m, y),$$

which implies

$$d(x_m, y) \geq d(x_k, y) - d(x_k, x_m)$$
$$> 2\varepsilon_0 - \varepsilon_0 = \varepsilon_0$$

for all $m \geq n_0$. So, the open ball $S(y, \varepsilon_0)$ contains x_n for only finitely many values of n. In this manner, we can associate with each $y \in X$ a ball $S(y, \varepsilon_0(y))$, where $\varepsilon_0(y)$ is a positive number that depends on y, and the ball $S(y, \varepsilon_0(y))$ contains x_n for only finitely many values of n. Observe that

$$X = \bigcup \{S(y, \varepsilon_0(y)) : y \in X\},$$

which means that $\{S(y, \varepsilon_0(y)) : y \in X\}$ is a covering of X. Since X is compact, there exists a finite subcovering $S(y_i, \varepsilon_0(y_i))$, $i = 1, 2, \ldots, n$, of X. So

$$X = \bigcup_{i=1}^{n} S(y_i, \varepsilon_0(y_i)).$$

Since each ball contains x_n for only a finite number of values of n, therefore the balls in the finite subcovering, and hence, also X, must contain x_n for only a finite number of values of n. This, however, is impossible. Hence, (X, d) must be complete. $\qquad\square$

We have so far proved that if a metric space is compact, then it is totally bounded and complete. The converse, namely that a totally bounded and complete metric space is compact, is true as well.

Theorem 5.1.16. Let (X, d) be a totally bounded and complete metric space. Then (X, d) is compact.

Proof. Suppose, if possible, that (X, d) is totally bounded and complete but is not compact. Then there exists an open covering $\{G_\lambda\}_{\lambda \in \Lambda}$ of X that does not admit a finite subcovering.

Since (X, d) is totally bounded, it is bounded; hence, for some real number $r > 0$ and some $x_0 \in X$, we have $X \subseteq S(x_0, r)$. Observe that $X \subseteq S(x_0, r)$ implies $X = S(x_0, r)$. Let $\varepsilon_n = r/2^n$.

We know that X, being totally bounded, can be covered by finitely many balls of radius ε_1. By our hypothesis, at least one of these balls, say $S(x_1, \varepsilon_1)$, cannot be covered by a finite number of sets G_λ (for if each had a finite subcovering, the same would be true for X). Because $S(x_1, \varepsilon_1)$ is itself totally bounded (any nonempty subset of a totally bounded set is totally bounded, as shown above), we can find an $x_2 \in S(x_1, \varepsilon_1)$ such that $S(x_2, \varepsilon_2)$ cannot be covered by a finite number of sets G_λ.

In this way, a sequence $\{x_n\}_{n \geq 1}$ may be defined with the property that

for each n, $S(x_n, \varepsilon_n)$ cannot be covered by a finite number of sets G_λ (5.2)

and $x_{n+1} \in S(x_n, \varepsilon_n)$.

We next show that the sequence $\{x_n\}_{n \geq 1}$ is convergent. Since $x_{n+1} \in S(x_n, \varepsilon_n)$, it follows that $d(x_n, x_{n+1}) < \varepsilon_n$ and hence,

$$d(x_n, x_{n+p}) \leq d(x_n, x_{n+1}) + d(x_{n+1}, x_{n+2}) + \ldots + d(x_{n+p-1}, x_{n+p})$$
$$< \varepsilon_n + \varepsilon_{n+1} + \ldots + \varepsilon_{n+p-1}$$
$$< \frac{r}{2^{n-1}}.$$

So $\{x_n\}_{n \geq 1}$ is a Cauchy sequence in X, and since X is complete, it converges to $y \in X$, say. Since $y \in X$, there exists $\lambda_0 \in \Lambda$ such that $y \in G_{\lambda_0}$. Because G_{λ_0} is open, it contains $S(y, \delta)$ for some $\delta > 0$. Choose n so large that $d(x_n, y) < \delta/2$ and $\varepsilon_n < \delta/2$. Then, for any $x \in X$ such that $d(x, x_n) < \varepsilon_n$, it follows that

$$d(x, y) \leq d(x, x_n) + d(x_n, y)$$
$$< \frac{1}{2}\delta + \frac{1}{2}\delta = \delta,$$

so that $S(x_n, \varepsilon_n) \subseteq S(y, \delta)$. Therefore, $S(x_n, \varepsilon_n)$ admits a finite subcovering, namely by the set G_{λ_0}. Since this contradicts (5.2), the proof is complete. $\square$

We sum up the results of this section with the following theorem:

Theorem 5.1.17. A metric space is compact if and only if it is complete and totally bounded.

5.2. Other Characterisations of Compactness

The Bolzano-Weierstrass property of real numbers states that every bounded point set in **R** has at least one limit point. An equivalent formulation of the Bolzano-Weierstrass property is "every bounded sequence in **R** contains a convergent subsequence". In this section, we seek a reformulation of compactness that generalises the above property of real numbers. We begin by establishing an analogue of this property for arbitrary metric spaces.

Proposition 5.2.1. Let (X, d) be a metric space. Then the following statements are equivalent:

(i) every infinite set in (X, d) has at least one limit point in X;
(ii) every infinite sequence in (X, d) contains a convergent subsequence.

Proof. (i)$\Rightarrow$(ii). Let $\{x_n\}_{n \geq 1}$ be a sequence in X. If the set $\{x_1, x_2, x_3, \dots\}$ is finite, then one of the points, say x_{i_0}, satisfies $x_{i_0} = x_j$ for infinitely many $j \in \mathbf{N}$. Hence, the constant sequence $\{x_{i_0}\}$ is a subsequence of $\{x_n\}_{n \geq 1}$, which converges to the point x_{i_0}.

Suppose that the set $\{x_1, x_2, x_3, \dots\}$ is infinite. In view of (i), the infinite set $\{x_1, x_2, x_3, \dots\}$ has at least one limit point $x \in X$. A subsequence of $\{x_n\}_{n \geq 1}$ that converges to x may be obtained as follows: Let n_1 be any integer such that $d(x_{n_1}, x) < 1$. Having defined n_k, let n_{k+1} be the smallest integer such that $n_{k+1} > n_k$ and $d(x_{n_{k+1}}, x) < 1/(k+1)$. Then the sequence $\{x_{n_k}\}_{k \geq 1}$ converges to x.

(ii)$\Rightarrow$(i). Let Y be an infinite subset of X. Then there exists a sequence $\{y_n\}_{n \geq 1}$ in X of distinct terms. In view of (ii), $\{y_n\}_{n \geq 1}$ contains a subsequence $\{y_{n_i}\}_{i \geq 1}$ of distinct terms that converges to $y \in X$. Hence every open ball with centre y contains an infinite number of terms of the convergent subsequence $\{y_{n_i}\}_{i \geq 1}$. But the terms are distinct; hence, every open ball centred at y contains an infinite number of points of Y. Accordingly, $y \in X$ is a limit point of Y. $\square$

Theorem 5.2.2. The metric space (X, d) is compact if and only if every sequence of points in X has a subsequence converging to a point in X.

Proof. Suppose first that X is compact (equivalently, totally bounded and complete; see Theorem 5.1.17) and that $\{x_n\}_{n \geq 1}$ is any sequence of points in X. Since X is totally bounded, it follows, using Theorem 5.1.12, that $\{x_n\}_{n \geq 1}$ contains a Cauchy subsequence $\{x_{n_i}\}_{i \geq 1}$. But $\{x_{n_i}\}_{i \geq 1}$ converges to a point $x \in X$ because X is complete. Thus, if X is compact, then every sequence in X contains a convergent subsequence.

Conversely, suppose every sequence in X has a convergent subsequence. It follows in view of the fact that every convergent sequence is Cauchy and Theorem 5.1.12 that X is totally bounded. It remains to show that X is complete. To this end, let $\{x_n\}_{n \geq 1}$ be a Cauchy sequence in X. By assumption, $\{x_n\}_{n \geq 1}$ has subsequence $\{x_{n_i}\}_{i \geq 1}$ that converges to a point $x \in X$. We shall show that $\lim_n x_n = x$. Let ε be an arbitrary positive number. Since $\lim_i x_{n_i} = x$, there exists i_0 such that $i \geq i_0$ implies

$$d(x_{n_i}, x) < \frac{1}{2}\varepsilon. \tag{5.3}$$

Since the sequence $\{x_n\}_{n \geq 1}$ is Cauchy, there exists n_0 such that $n, m \geq n_0$ implies

$$d(x_n, x_m) < \frac{1}{2}\varepsilon. \tag{5.4}$$

If i is such that $i \geq i_0$ and $n_i \geq n_0$, then using (5.3) and (5.4), we have

$$d(x_n, x) \leq d(x_n, x_{n_i}) + d(x_{n_i}, x) < \frac{1}{2}\varepsilon + \frac{1}{2}\varepsilon = \varepsilon,$$

whenever $n \geq n_0$. This completes the proof. $\qquad\square$

The results of Sections 5.1 and 5.2 may be summed up as follows:

Theorem 5.2.3. Let (X, d) be a metric space. The following statements are equivalent:

(i) (X, d) is compact;
(ii) (X, d) is complete and totally bounded;
(iii) every infinite set in X has at least one limit point;
(iv) every sequence in X contains a convergent subsequence.

Here is a useful corollary.

Corollary 5.2.4. Let Y be a closed subset of the compact metric space (X, d). Then (Y, d_Y) is compact.

Proof. Let $\{y_n\}_{n \geq 1}$ be a sequence of points in Y. Then $\{y_n\}_{n \geq 1}$ considered as a sequence of points in (X, d) has a subsequence converging to a point $x \in X$ (see Theorem 5.2.3). But then $x \in Y$ since Y is closed, by Proposition 2.1.28. Thus, any sequence in Y has a subsequence converging to a point in Y. By Theorem 5.2.3, Y is compact. $\qquad\square$

In the other direction, we have the following.

Theorem 5.2.5. Let Y be a subset of the metric space (X, d). If (Y, d_Y) is compact, then Y is a closed subset of (X, d).

Proof. Let $x \in X$ be a limit point of Y. Then there is a sequence $\{y_n\}_{n \geq 1}$ in Y converging to x (see Proposition 2.1.20). But then $\{y_n\}_{n \geq 1}$ is a Cauchy sequence in Y. Since Y is complete, $\{y_n\}_{n \geq 1}$ converges to a point y in $Y \subseteq X$. This point y must be x and so $x \in Y$. Thus, Y contains all its limit points and is therefore closed. $\qquad\square$

Remark 5.2.6. The reader will note that the proof employs only the completeness of Y. However, the result will be useful in the context of compactness rather than completeness.

Another characterisation of compactness is considered below.

Definition 5.2.7. A metric space (X, d) is said to be **countably compact** if every countable open covering of X has a finite subcovering.

Since every space satisfying the second axiom of countability (see Definition 2.3.6) has the property that every open covering of it contains a countable subcovering (see Proposition 2.3.11), it follows that countable compactness is equivalent to compactness in the presence of the second axiom of countability. That the concepts of countable compactness and compactness for metric spaces are equivalent without the second axiom of countability will be proved below.

Theorem 5.2.8. A metric space (X, d) is compact if and only if it is countably compact.

Proof. It is enough to prove that countable compactness implies compactness, for if (X, d) is compact then it is trivial that it is also countably compact.

   Assume that X is countably compact. Suppose, if possible, that there is an infinite subset A of X having no limit point. Choose a countable set F of distinct points of A, i.e., let $F = \{x_1, x_2, x_3, \dots\}$, where $m \neq n$ implies $x_m \neq x_n$. Clearly, F has no limit point since A has no limit point, and hence F is closed. In particular, no $x_n (n = 1, 2, 3, \dots)$ is a limit point of F. So, for each $x_n \in F$, there exists $r(x_n) > 0$ such that $S(x_n, r(x_n))$ contains no other point of F. The collection

$$G = \{X \backslash F\} \bigcup \{S(x_n, r(x_n)) : n = 1, 2, \dots\}$$

consisting of open sets is a countable open covering of X, which clearly has no finite subcovering. In fact, if we remove even a single ball $S(x_n, r(x_n))$, say, from the collection, it will fail to cover x_n at least. This contradicts the hypothesis that X is countably compact. Therefore, there cannot be an infinite subset of X having no limit point. In view of Theorem 5.2.3, (X, d) must be compact.   □

   We shall use the next theorem in proving an analogue in metric spaces of the property that continuous real-valued functions defined on closed bounded intervals are uniformly continuous.

Definition 5.2.9. Let $\{G_\lambda : \lambda \in \Lambda\}$ be an open covering of a metric space (X, d). Any number $\delta > 0$ such that, for each $x \in X$ there exists $\lambda \in \Lambda$ (dependent on x) for which

$$S(x, \delta) \subseteq G_\lambda,$$

is called a **Lebesgue number** of the covering $\{G_\lambda : \lambda \in \Lambda\}$.
   Open coverings of compact metric spaces possess Lebesgue numbers.

Theorem 5.2.10. Let (X, d) be a compact metric space and $\{G_\lambda : \lambda \in \Lambda\}$ be an open covering of X. Then there exists a positive number δ such that each ball $S(x, \delta)$ is contained in at least one G_λ.

Proof. Each $x \in X$ belongs to an open set G_λ in the cover $\{G_\lambda : \lambda \in \Lambda\}$. Hence, there exists an open ball $S(x, r(x))$ with centre x and radius $r(x)$ such that $x \in S(x, r(x)) \subseteq G_\lambda$. Clearly,

$$X = \bigcup \{S\left(x, \frac{r(x)}{2}\right) : x \in X\}.$$

Since X is compact, there exist $x_1, x_2, \dots, x_n$ such that

$$X = \bigcup_{i=1}^{n} S\left(x_i, \frac{r(x_i)}{2}\right).$$

Let

$$\delta = \min\left\{\frac{r(x_1)}{2}, \frac{r(x_2)}{2}, \ldots, \frac{r(x_n)}{2}\right\};$$

then $\delta > 0$ is a Lebesgue number. For, given any $S(x, \delta)$, we have $x \in S(x_i, r(x_i)/2)$ for some i and so for any $z \in S(x, \delta)$,

$$d(z, x_i) \leq d(z, x) + d(x, x_i) < \delta + \frac{r(x_i)}{2} \leq r(x_i);$$

that is,

$$S(x, \delta) \subseteq S(x_i, r(x_i)).$$

Since the latter set lies is some G_λ, the results follows. $\square$

5.3. Continuous Functions on Compact Spaces

Some fundamental theorems concerning real-valued functions on closed bounded intervals of **R** possess natural generalisations when the domain of the function is replaced by a compact metric space. We begin with a theorem that is of tremendous importance in analysis.

Theorem 5.3.1. Let f be a continuous function from a compact metric space (X, d_X) into a metric space (Y, d_Y). Then the range $f(X)$ of f is also compact.

Proof. Let $\{G_\lambda : \lambda \in \Lambda\}$ be an open covering of $f(X)$. Since f is continuous, $f^{-1}(G_\lambda)$ is open in X (see Theorem 3.1.9). Moreover,

$$\{f^{-1}(G_\lambda) : \lambda \in \Lambda\}$$

is an open covering of X. Since X is compact, there exist $\lambda_1, \lambda_2, \ldots, \lambda_n$ in Λ such that

$$\bigcup_{i=1}^{n} f^{-1}(G_{\lambda_i}) = X.$$

Now

$$f(X) = f\left(\bigcup_{i=1}^{n} f^{-1}(G_{\lambda_i})\right) = \bigcup_{i=1}^{n} f(f^{-1}(G_{\lambda_i})) \subseteq \bigcup_{i=1}^{n} G_{\lambda_i};$$

so, $\{G_{\lambda_i} : i = 1, \ldots, n\}$ is a finite subcovering of $f(X)$. Consequently, $f(X)$ is compact. $\square$

Corollary 5.3.2. Let f be a homeomorphism of a metric space (X, d_X) onto a metric space (Y, d_Y). Then X is compact if and only if Y is compact.

Corollary 5.3.3. Let f be a continuous function from a compact metric space (X, d_X) onto a metric space (Y, d_Y). Then the range $f(X)$ of f is a bounded and closed subset of Y (see Theorem 5.1.5).

When $Y = \mathbf{R}$ and $X = [a, b]$, both having their usual metrics, Corollary 5.3.3 implies the result that "a real-valued continuous function f defined on a closed bounded interval in $\mathbf{R}$ is bounded".

Proposition 5.3.4. Let $K \subseteq \mathbf{R}$ be both closed and bounded and let $M = \sup K$, $m = \inf K$. Then M and m are in K.

Proof. Suppose $M \notin K$. For any $\varepsilon > 0$, there exists $k \in K$ such that $M - \varepsilon < k < M$. Thus, M is a limit point of K but $M \notin K$. Since K is closed, we have a contradiction.

Similarly, $m \in K$.   □

Theorem 5.3.5. If f is a continuous real-valued function on a compact metric space (X, d_X), then f is bounded and attains its bounds, i.e., if $M = \sup f(X)$, $m = \inf f(X)$, there exist x and y in X such that $f(x) = M$ and $f(y) = m$.

Proof. By Corollary 5.3.3, the range of the function f must be a bounded and closed subset of $\mathbf{R}$. Hence, $f(X)$ possesses a supremum M and an infimum m. By Proposition 5.3.4, $M, m \in f(X)$. So, there exist x and y in X such that $f(x) = M$ and $f(y) = m$.   □

Remark 5.3.6. The theorem is not true if X is not compact, as is shown by the following examples:

(i) The function $f: (0, 1) \to \mathbf{R}$ defined by $f(x) = 1/x$ is continuous but not bounded on $(0,1)$.
(ii) The function $f(x) = (x - 1)/x$, $1 < x < \infty$, is continuous and bounded. However, it does not attain its bounds, which are 0 and 1.

The domains of functions in both the above examples fail to be compact.

Corollary 5.3.7. If the real-valued function f is continuous on the closed and bounded interval $[a, b]$, then f is bounded and takes a maximum and a minimum value at points of $[a, b]$.

Theorem 5.3.8. If f is a one-to-one continuous mapping of a compact metric space (X, d_X) onto a metric space (Y, d_Y), then f^{-1} is continuous on Y and, hence, f is a homeomorphism of (X, d_X) onto (Y, d_Y).

Proof. Suppose $f: X \to Y$ is one-to-one and onto. Its inverse $f^{-1}: Y \to X$ is well defined. Let F be a closed subset of X. By Corollary 5.2.4, F is a compact metric space. By Theorem 5.3.1, $f(F)$ is compact and, hence, a closed subset of Y by Theorem 5.2.5. But $f(F) = (f^{-1})^{-1}(F)$ and so $(f^{-1})^{-1}(F)$ is closed in Y. Hence, by Theorem 3.1.10, f^{-1} is continuous.   □

Remark 5.3.9. The assumption "X is compact" is essential for the validity of theorem 5.3.8. In fact, if $X = [0, 1)$ with the usual metric of $\mathbf{R}$ and $Y = \{z \in \mathbf{C}: |z| = 1\}$ with the metric induced from $\mathbf{C}$, the function $f: [0, 1) \to Y$ defined by

$$f(x) = \exp \ (2\pi i x), \qquad x \in [0, 1),$$

is continuous. However, f^{-1} is not continuous at the point $f(0)$, i.e., at 1: Consider the sequence $\{x_n\}_{n \geq 1}$, $x_n = 1 - 1/n$. Then $f(x_n) = \exp (-2\pi i/n)$. So $f(x_n) \to 1$ as $n \to \infty$. Since $f(0) = 1$, it follows that $f(x_n) \to f(0)$. However, the corresponding preimage sequence $\{1 - 1/n\}_{n \geq 1}$ does not converge in X.

We end this Section with an application of Theorem 5.2.10.

Theorem 5.3.10. Let (X, d_X) be a compact metric space, (Y, d_Y) be an arbitrary metric space and $f: X \to Y$ be continuous. Then for each $\varepsilon > 0$, there exists a $\delta > 0$ (δ depending on ε only) such that $f(S(x, \delta)) \subseteq S(f(x), \varepsilon)$ for every $x \in X$. That is, f is uniformly continuous on X.

Proof. The collection of balls $\{S(y, \varepsilon/2): y \in Y\}$ constitute an open cover of Y. The sets $\{f^{-1}S(y, \varepsilon/2)): y \in Y\}$ therefore form an open cover of the compact metric space X. Let δ be a Lebesgue number of this open cover of X. Since each open ball $S(x, \delta)$ lies in one of these sets,

$$f(S(x, \delta)) \subseteq S\left(y, \frac{\varepsilon}{2}\right)$$

for some $y \in Y$. Because $f(x) \in S(y, \varepsilon/2)$, we find for any $z \in S(x, \delta)$ that

$$d(f(z), f(x)) \leq d(f(z), y) + d(y, f(x))$$

$$< \frac{\varepsilon}{2} + \frac{\varepsilon}{2} = \varepsilon,$$

i.e.,

$$f(S(x, \delta)) \subseteq S(f(x), \varepsilon).$$

This completes the proof.   □

Corollary 5.3.11. If the real-valued function f is continuous on the closed bounded interval $[a, b]$, then f is uniformly continuous on $[a, b]$.

5.4. Locally Compact Spaces

Many of the important spaces occurring in analysis are not compact but enjoy a property that is a local version of compactness.

Definition 5.4.1. A metric space (X, d) is said to be **locally compact** if each point $x \in X$ has a neighbourhood $S(x, r)$ such that the closure $\overline{S(x, r)}$ is compact.

Examples 5.4.2. (i) Consider the real line **R** with the usual metric. Observe that each point $x \in \mathbf{R}$ is in the interval $(x - \varepsilon, x + \varepsilon)$ whose closure $[x - \varepsilon, x + \varepsilon]$, being a closed bounded subset of **R**, is compact. On the other hand, **R** is not compact, since the collection

$$\{\ldots, (- 3, - 1), (- 2, 0), (- 1, 1), (0, 2), (1, 3), \ldots\}$$

is an open cover of **R** but contains no finite subcover.

(ii) An infinite discrete metric space X is locally compact but not compact. The reader may recall that the open cover consisting of singleton subsets contains no finite subcover. On the other hand, $S(x, 1/2)$, where $x \in X$, is a neighbourhood such that its closure $\overline{S(x, 1/2)} = S(x, 1/2) = \{x\}$ is compact, being a finite subset of X.

Thus we see from the above examples that a locally compact space need not be compact. However, any compact space is locally compact, since the full space is a neighbourhood of each of its points.

For the next theorem we shall need the following.

Proposition 5.4.3. Let (X, d) be a metric space and $A \subseteq X$ be nonempty. For $r > 0$, let $V_r(A) = \{x \in X : d(x, A) < r\}$. Then $V_r(A)$ is an open set in X containing A, called an **open neighbourhood** of A.

Proof. Clearly, $A \subseteq V_r(A)$. We need only show that $V_r(A)$ is open. Let $x \in V_r(A)$ be arbitrary. Then $d(x, A) < r$. If $y \in S(x, r - d(x, A))$, then by the definition of $d(y, A)$, we get

$$\begin{aligned}
d(y, A) &= \inf \{d(y, u) : u \in A\} \\
&\leq d(y, z), \qquad z \in A \\
&\leq d(y, x) + d(x, z) \\
&< r - d(x, A) + d(x, A) = r.
\end{aligned}$$

Hence, $V_r(A)$ contains an open ball around each of its points and is, therefore, open in X.   □

Theorem 5.4.4. Let K be a compact subset of a locally compact metric space (X, d). Then there exists an $r > 0$ such that $V_r(K)$ has compact closure, i.e., $\overline{V_r(K)}$ is compact in X.

Proof. For each $x \in K$, there exists $r(x) > 0$ (r depending on x) such that $S(x, r(x))$ has compact closure. The collection

$$\{S(x, r(x)) : x \in K\}$$

is an open covering of K. Since K is compact, there exists a finite subset $\{x_1, x_2, \ldots, x_n\}$ of K such that

$$K \subseteq \bigcup_{i=1}^{n} S(x_i, r(x_i)).$$

The set

$$U = \bigcup_{i=1}^{n} S(x_i, r(x_i))$$

is open, being a finite union of open sets. Also,

$$\bar{U} = \overline{\bigcup_{i=1}^{n} S(x_i, r(x_i))} = \bigcup_{i=1}^{n} \overline{S(x_i, r(x_i))},$$

being a finite union of compact sets, is itself compact. Observe that the function

$$x \to d(x, X \backslash U)$$

defined on K is a strictly positive continuous function (see Example 3.4.2 (ii) and Proposition 3.4.3). By Theorem 5.3.5, there exists x_0 in K such that

$$d(x_0, X \backslash U) = \inf \{d(x, X \backslash U) : x \in K\}.$$

But $d(x_0, X \backslash U) = r > 0$; hence, $V_r(K) \subseteq U$. In fact, if $z \notin U$, then $d(z, K) = \inf \{d(x, z) : x \in K\} \geq \inf \{d(x, X \backslash U) : x \in K\} = d(x_0, X \backslash U) = r$; so $z \notin V_r(K)$. Moreover, $\overline{V_r(K)}$, being a closed subset of the compact set $\bar{U}$, is compact (see Corollary 5.2.4). $\qquad \square$

Theorem 5.4.5. Let (X, d) be a locally compact metric space. The following properties are equivalent:

(i) There exists an increasing sequence $\{G_n\}_{n \geq 1}$ of open sets with compact closures in X such that $\overline{G_n} \subseteq G_{n+1}$ for $n = 1, 2, \ldots$, and $X = \bigcup_n G_n$;

(ii) X is a countable union of compact sets;

(iii) X is separable.

Proof. (i)$\Rightarrow$(ii). If $\{G_n\}_{n \geq 1}$ is an increasing sequence of open sets satisfying the hypothesis in (i), then

$$X = \bigcup_n G_n \subseteq \bigcup_n \overline{G_n} \subseteq \bigcup_n G_{n+1} \subseteq X.$$

So, $X = \bigcup_n \overline{G_n}$, where each $\overline{G_n}$ is compact in X.

(ii)$\Rightarrow$(iii). If X is the union of a sequence $\{K_n\}_{n \geq 1}$ of compact sets, each subspace K_n is separable (see Corollary 5.1.14). If D_n is an at most denumerable subset in K_n, dense with respect to K_n, then $D = \bigcup_n D_n$ is at most denumerable. Moreover, D is dense in X.

In fact,

$$X = \bigcup_n K_n \subseteq \bigcup_n \overline{D_n} \subseteq \bar{D}.$$

(iii)$\Rightarrow$(i). Assume that X is separable and let $\{V_n\}_{n \geq 1}$ be an at most denumerable basis for the open subsets of X (see Theorem 2.3.16 and Definition 2.3.6). For every $x \in X$, there exists a ball $S(x, r)$ such that $\overline{S(x, r)}$ is compact. Since $\{V_n\}_{n \geq 1}$ is a basis for the open sets of X, there is an index $n(x)$ such that

$$x \in V_{n(x)} \subseteq S(x, r);$$

in particular, $\overline{V_{n(x)}} \subseteq \overline{S(x, r)}$ is compact (see Corollary 5.2.4). It follows that those of the V_n having compact closures constitute a basis for the open sets of X. We may therefore assume that all the V_n have compact closures. Define G_n inductively as follows:

$$G_1 = V_1, \ G_{n+1} = V_{n+1} \cup V_{r_n}(\overline{G_n}), \ n = 1, 2, \ldots,$$

where r_n has been chosen so that $V_{r_n}(\overline{G_n})$ has compact closure (see Theorem 5.4.4). The G_n so constructed satisfy (i). This completes the proof. $\qquad \square$

Theorem 5.4.6. An open or closed subset of a locally compact metric space is locally compact.

Proof. Suppose A is open and let $a \in A$. Then $a \in X$ and by local compactness of X, there exists an $r > 0$ such that $\overline{S(a, r)}$ is compact. Let $r' \leq r$ be positive and such that $S(a, r') \subseteq A$. This is possible since A is open. Moreover,

$$\overline{S(a, r')} \subseteq \overline{S(a, r)}.$$

Hence, $\overline{S(a, r')}$, being a closed subset of a compact subset, is itself compact (see Corollary 5.2.4). We have thus shown that, for each $a \in A$, there is an open ball centred at a with compact closure. So A is locally compact.

Now suppose A is closed in X, and let $a \in A$. Then $a \in X$ and by local compactness of X, there exists an $r > 0$ such that $\overline{S(a, r)}$ is compact. Now,

$$S(a, r) \cap A$$

is a neighbourhood of $a \in A$. Note that the A-closure of a subset of A is the same as its X-closure because A is closed. Moreover,

$$\overline{S(a, r) \cap A} \subseteq \overline{S(a, r)} \cap A$$

is compact since the right hand side of the above inclusion relation is compact and the left hand side is closed (see Corollary 5.2.4). $\qquad \square$

5.5. Compact Sets in Special Metric Spaces

The application of compactness criteria in Section 5.3 to individual special metric spaces is not always simple. For such instances, special criteria can be given that are more suitable for verification. We examine these criteria for several different metric spaces in (a) through (d) below.

(a) In analysis, one important metric space is the space $C[0,1]$ with the uniform metric. For subsets of this space, a frequently used criterion of compactness is given by the Arzelà-Ascoli theorem. In order to formulate this theorem, we introduce the following:

Definition 5.5.1. A nonempty subset K of $C[0,1]$ is said to be **equicontinuous** if for each $\varepsilon > 0$ there exists $\delta > 0$ such that, for every $f \in K$,

$$|x - y| < \delta \qquad \text{implies} \quad |f(x) - f(y)| < \varepsilon.$$

Clearly, each $f \in K$ is uniformly continuous. The δ in the definition of uniform continuity depends upon ε and the function f under consideration. The condition that K is equicontinuous requires that a single $\delta > 0$ can be chosen independently of $f \in K$.

Examples 5.5.2. (i) Each finite subset of $C[0,1]$ is equicontinuous on $[0,1]$.

(ii) Let S denote the closed unit ball in $C[0,1]$, i.e.,

$$S = \{f \in C[0, 1]: \sup_{x} |f(x)| \leq 1\}.$$

For each $f \in S$, define

$$g(x) = \int_0^x f(t)dt, \qquad 0 \leq x \leq 1,$$

and denote the set of all such g by S_1. Then for every $g \in S_1$,

$$\begin{aligned}
|g(x) - g(y)| &= \left| \int_0^x f(t)dt - \int_0^y f(t)dt \right| \\
&\quad \left| \int_y^x f(t)dt \right| \\
&\leq \sup_{t} |f(t)| \cdot |x - y| \leq |x - y|.
\end{aligned}$$

The right hand side of the above inequality is independent of g. Thus, for $\varepsilon > 0$, if we choose $\delta = \varepsilon$, then

$$|x - y| < \delta \qquad \text{implies} \quad |g(x) - g(y)| < \varepsilon.$$

The set S_1 is, therefore, equicontinuous.

(iii) Let $f_n(x) = \frac{nx}{1+n^2x^2}, 0 \leq x \leq 1$. The set $\{f_n\}_{n \geq 1}$ is not equicontinuous. The function f_n assumes its maximum value $1/2$ at $x = n^{-1}, n = 1, 2, \ldots$. If $\{f_n\}_{n \geq 1}$ were to be equicontinuous, then for $\varepsilon = 1/2$, there would exist a positive δ, independent of n, such that $|f_n(x) - f_n(0)| = |f_n(x)| < \varepsilon$ whenever $0 \leq x < \delta$. Since this δ would be independent of n, we could choose $n > 1/\delta$ so that $x = 1/n$ would satisfy $0 \leq x < \delta$ and hence, $|f_n(x)| < \varepsilon$, i.e., $|f_n(1/n)| < 1/2$. But $f_n(1/n) = 1/2$; the contradiction shows that δ cannot be chosen independently of n.

Definition 5.5.3. A nonempty subset K of $C[0, 1]$ is said to be **uniformly bounded** if there exists an $M > 0$ such that

$$|f(x)| \leq M \qquad \text{for all } x \in [0, 1] \text{ and all } f \in K.$$

Examples 5.5.4 (i) Each finite subset of $C[0, 1]$ is uniformly bounded.

(ii) The set $S_1 = \{g \in C[0, 1]: g(x) = \int_0^x f(t)dt, f \in C[0, 1], \sup |f(x)| \leq 1\}$ is uniformly bounded. To see why, note that $|g(x)| \leq \sup_t |f(t)| \cdot |x| \leq 1$ for all $x \in [0, 1]$ and $g \in S_1$.

Theorem 5.5.5. (Arzelà-Ascoli) Let K be a closed subset of $C[0, 1]$. Then the following are equivalent:

(i) K is compact;
(ii) K is uniformly bounded and equicontinuous.

Proof. Suppose K is compact and hence a bounded (see Theorem 5.1.5) subset of $C[0,1]$. Thus, K is uniformly bounded as a set of functions. It remains to show that K is equicontinuous.

Let $\varepsilon > 0$ be given and let $f_1, f_2, \ldots, f_n$ be an $(\varepsilon/3)$-net in K (such a net exists since K is compact; see Proposition 5.1.13). Now, let $f \in K$. For each $i = 1, 2, \ldots, n$,

$$|f(x) - f(y)| \leq |f(x) - f_i(x)| + |f_i(x) - f_i(y)| + |(f_i(y) - f(y)|.$$

Choose j so that

$$\sup \{|f(x) - f_j(x)|: x \in [0, 1]\} < \frac{\varepsilon}{3}.$$

Then

$$|f(x) - f(y)| \leq |f_j(x) - f_j(y)| + \frac{2\varepsilon}{3}.$$

Since the interval $[0,1]$ is compact, the functions $f_1, f_2, \ldots, f_n$ are uniformly continuous (see Theorem 5.3.10). Therefore, there exists a $\delta > 0$ such that

$$|x - y| < \delta \qquad \text{implies } |f_j(x) - f_j(y)| < \frac{\varepsilon}{3}, \qquad j = 1, 2, \ldots, n.$$

Consequently,

$$x, y \in [0,1], |x - y| < \delta \qquad \text{implies } |f(x) - f(y)| < \varepsilon$$

for each $f \in K$. Thus, K is equicontinuous.

Conversely, suppose that K is uniformly bounded and equicontinuous. Since $[0,1]$ with the induced metric is a compact metric space, it is separable (see Corollary 5.1.14). Let $\{x_1, x_2, \ldots\}$ be a countable dense subset of $[0,1]$. Now, let $\{f_n\}_{n \geq 1}$ be an arbitrary sequence in K. We shall show that it has a convergent subsequence. By hypothesis, K is a bounded subset of $C[0,1]$. So, there exists an $M > 0$ such that $|f(x)| \leq M$ for all $x \in [0,1]$ and all $f \in K$. Consider the sequence $\{f_n(x_1)\}_{n \geq 1}$. Since this sequence is bounded, by the Bolzano-Weierstrass theorem (see Proposition 0.4.2), we may choose a subsequence $\{f_{n,1}\}_{n \geq 1}$ of $\{f_n\}_{n \geq 1}$ such that $\{f_{n,1}(x_1)\}_{n \geq 1}$ converges. We next consider the sequence $\{f_{n,1}(x_2)\}_{n \geq 1}$. Again by the Bolzano-Weierstrass theorem, we may select a subsequence $\{f_{n,2}\}_{n \geq 1}$ of $\{f_{n,1}\}_{n \geq 1}$ such that $\{f_{n,2}(x_2)\}_{n \geq 1}$ converges. Continuing this process indefinitely, we obtain sequences $\{f_{n,m}\}_{n \geq 1}$, one for each m, such that $\{f_{n,m+1}\}_{n \geq 1}$ is a subsequence of $\{f_{n,m}\}_{n \geq 1}$ and $\{f_{n,m+1}(x_{m+1})\}_{n \geq 1}$ converges. Finally, we consider the diagonal sequence $\{f_{n,n}\}_{n \geq 1}$. For fixed k, $\{f_{n,n}(x_k)\}_{n \geq k}$ is a subsequence of $\{f_{n,k}(x_k)\}_{n \geq k}$ and hence converges. Therefore, $\{f_{n,n}\}_{n \geq 1}$ converges at every point of the dense subset $\{x_1, x_2, \ldots\}$. It remains to show that $\{f_{n,n}\}_{n \geq 1}$, as a sequence of functions in $C[0,1]$, is a Cauchy sequence. The completeness of $C[0,1]$ (see Proposition 1.4.13) will then ensure that the diagonal sequence converges.

Let $\varepsilon > 0$ be given. Since K is equicontinuous, there exists a $\delta > 0$ such that

$$|x - y| < \delta \qquad \text{implies } |f_{n,n}(x) - f_{n,n}(y)| < \frac{\varepsilon}{3} \qquad (5.5)$$

for all members of the sequence $\{f_{n,n}\}_{n \geq 1}$. Consider the collection of open intervals

$$\{(x_i - \delta, x_i + \delta) : i = 1, 2, \ldots\},$$

Since $\{x_1, x_2, \ldots\}$ is a dense subset of $[0,1]$,

$$\bigcup_{i=1}^{\infty} (x_i - \delta, x_i - \delta) \supseteq [0,1].$$

The compactness of $[0,1]$ then ensures the existence of finitely many points $\{\xi_1, \xi_2, \ldots, \xi_j\}$ in $\{x_1, x_2, \ldots\}$ such that

$$\bigcup_{n=1}^{j} (\xi_n - \delta, \xi_n + \delta) \supseteq [0,1].$$

The sequence $\{f_{n,n}\}_{n \geq 1}$ converges at each of the points $\{\xi_1, \xi_2, \ldots, \xi_j\}$ and, a fortiori, satisfies the Cauchy criterion at these points. So there exists a positive integer n_0 such that

$$m \geq n_0 \text{ and } n \geq n_0 \qquad \text{implies } |f_{m,m}(\xi_k) - f_{n,n}(\xi_k)| < \frac{\varepsilon}{3} \qquad (5.6)$$

for $k = 1, 2, \ldots, j$. Let $x \in [0, 1]$. There exists a $k \in \{1, 2, \ldots, j\}$ such that $|x - \xi_k| < \delta$, and therefore, for $m \geq n_0$ and $n \geq n_0$,

$$|f_{m,m}(x) - f_{n,n}(x)| \leq |f_{m,m}(x) - f_{m,m}(\xi_k)| + |f_{m,m}(\xi_k) - f_{n,n}(\xi_k)| + |f_{n,n}(\xi_k) - f_{n,n}(x)|$$
$$< \frac{\varepsilon}{3} + \frac{\varepsilon}{3} + \frac{\varepsilon}{3} = \varepsilon,$$

using (5.5) and (5.6) above. We have thus shown that every sequence in K has a convergent subsequence; so K is compact (see Theorem 5.2.2). $\qquad\square$

(b) Let $X = \ell_p$ with metric d defined by

$$d(x, y) = \left(\sum_{i=1}^{\infty} |x_i - y_i|^p \right)^{1/p}, \qquad p \geq 1,$$

where $x = \{x_i\}_{i \geq 1}$ and $y = \{y_i\}_{i \geq 1}$ are in ℓ_p (see Example 1.2.2(vii)). The space (ℓ_p, d) is complete (see Proposition 1.4.10) but not compact. If fact, if $Y = \{e_1, e_2, \ldots, e_n, \ldots\}$, where e_n denotes the sequence all of whose terms are equal to zero except the nth term, which is equal to 1 then Y is both closed and bounded but is not compact (see Remark 5.1.6). It therefore follows that (ℓ_p, d) is not compact. That the space ℓ_p is not compact also follows from the fact that it is not even bounded. Indeed, if $M > 0$, choose n larger than M^p; then the element y with 1 in each of the first n places and zero elsewhere is such that $d(y, e_0) = n^{1/p} > M$.

The following criterion describes totally bounded subsets of (ℓ_p, d).

Theorem 5.5.6. Let Y be a nonempty subset of ℓ_p. Then Y is a totally bounded subset of ℓ_p if and only if Y satisfies the following conditions:

(i) Y is bounded, that is, there exists an $M > 0$ such that

$$\left(\sum_{i=1}^{\infty} |y_i|^p \right)^{1/p} \leq M$$

for all $y = \{y_i\}_{i \geq 1}$ in Y;

(ii) given $\varepsilon > 0$, there exists an integer N such that

$$\left(\sum_{i=N}^{\infty} |y_i|^p \right)^{1/p} < \varepsilon$$

for all $y = \{y_i\}_{i \geq 1}$ in Y.

Proof. Let Y be totally bounded. Then Y is bounded (see Proposition 5.1.10), and, therefore, there exists an $M_1 > 0$ such that $d(Y) \leq M_1$. Let $y^o \in Y$ be fixed. It then follows that $d(y, y^o) \leq M_1$ for all $y \in Y$. By Minkowski's inequality (see Theorem 1.1.5),

$$\left(\sum_{i=1}^{\infty}|y_i|^p\right)^{1/p} = \left(\sum_{i=1}^{\infty}|y_i - y_i^o + y_i^o|^p\right)^{1/p}$$

$$\leq \left(\sum_{i=1}^{\infty}|y_i - y_i^o|^p\right)^{1/p} + \left(\sum_{i=1}^{\infty}|y_i^o|^p\right)^{1/p}$$

$$\leq M_1 + \left(\sum_{i=1}^{\infty}|y_i^o|^p\right)^{1/p}$$

for $y \in Y$. If $M \geq M_1 + (\sum_{i=1}^{\infty}|y_i^o|^p)^{1/p}$, then

$$\left(\sum_{i=1}^{\infty}|y_i|^p\right)^{1/p} \leq M$$

for $y \in Y$. This proves (i) and we proceed to prove (ii).

Using the total boundedness of Y again, we get a finite $(\varepsilon/2)$-net $\{y^{(1)}, y^{(2)}, \ldots, y^{(m)}\}$, say. Since

$$\left(\sum_{i=1}^{\infty}|y_i^{(n)}|^p\right)^{1/p} < \infty$$

for $n = 1, 2, \ldots, m$, there exist integers i_n such that

$$\left(\sum_{i=i_n}^{\infty}|y_i^{(n)}|^p\right)^{1/p} < \frac{\varepsilon}{2} \qquad \text{for } n = 1, 2, \ldots, m.$$

Let $N = \max\{i_1, i_2, \ldots, i_m\}$. Then

$$\left(\sum_{i=N}^{\infty}|y_i^{(n)}|^p\right)^{1/p} < \frac{\varepsilon}{2} \qquad \text{for } n = 1, 2, \ldots, m. \qquad (5.7)$$

Now, let $y \in Y$. Using the $(\varepsilon/2)$-net $\{y^{(1)}, y^{(2)}, \ldots, y^{(m)}\}$, choose $y^{(n)}$ such that

$$\left(\sum_{i=1}^{\infty}|y_i - y_i^{(n)}|^p\right)^{1/p} < \frac{\varepsilon}{2}$$

and, in particular,

$$\left(\sum_{i=N}^{\infty}|y_i - y_i^{(n)}|^p\right)^{1/p} < \frac{\varepsilon}{2} \qquad (5.8)$$

Using Minkowski's inequality (see Theorem 1.1.5) with (5.7) and (5.8), we have

$$\left(\sum_{i=N}^{\infty}|y_i|^p\right)^{1/p} \leq \left(\sum_{i=N}^{\infty}|y_i - y_i^{(n)}|^p\right)^{1/p} + \left(\sum_{i=N}^{\infty}|y_i^{(n)}|^p\right)^{1/p} < \frac{\varepsilon}{2} + \frac{\varepsilon}{2} = \varepsilon$$

for $y \in Y$. Thus, (ii) holds.

For the converse, suppose that Y is a subset of ℓ_p which satisfies (i) and (ii). Let $\varepsilon > 0$ be arbitrary. By hypothesis (ii), there exists a positive integer N such that

$$\sum_{i=N+1}^{\infty} |y_i|^p < \frac{1}{2}\varepsilon^p$$

for all $y \in Y$. Associate with each $y \in Y$ the element $(y_1, y_2, \ldots, y_N, 0, 0, \ldots)$ and call the set of all such elements as Y_ε. Let $\tilde{Y}_\varepsilon$ be the subset of $\mathbf{R}^N$ defined by

$$\tilde{Y}_\varepsilon = \{(y_1, y_2, \ldots, y_N) : y \in Y\}$$

equipped with the metric

$$d_p(u, v) = \left(\sum_{i=1}^{N} |u_i - v_i|^p\right)^{1/p}.$$

$(\mathbf{R}^N, d_p)$ is a metric space (see Example 1.2.2 (iii)). Moreover,

$$\left(\sum_{i=1}^{N} |y_i|^p\right)^{1/p} \leq M$$

for all elements $(y_1, y_2, \ldots, y_N) \in \tilde{Y}_\varepsilon$, using (i) of the hypothesis. So $\tilde{Y}_\varepsilon$ is a bounded subset of $\mathbf{R}^N$ and, hence, a totally bounded subset (see Remark 5.1.11). Let $\{u^{(1)}, u^{(2)}, \ldots, u^{(m)}\}$ be an $(\varepsilon/2^{1/p})$-net for $\tilde{Y}_\varepsilon$, where $u^{(j)} = (u_1^{(j)}, u_2^{(j)}, \ldots, u_N^{(j)})$, $j = 1, 2, \ldots, m$. Then

$$\{y^{(j)} = (u_1^{(j)}, u_2^{(j)}, \ldots, u_N^{(j)}, 0, 0, \ldots) : j = 1, 2, \ldots, m\}$$

is an ε-net for Y. In fact, for $y \in Y, y = (y_1, y_2, \ldots)$,

$$\sum_{i=1}^{N} |y_i - u_i^{(j)}|^p < \frac{\varepsilon^p}{2}$$

for some j. So, for this j,

$$\sum_{i=1}^{\infty} \left|y_i - y_i^{(j)}\right|^p = \sum_{i=1}^{N} \left|y_i - u_i^{(j)}\right|^p + \sum_{i=N+1}^{\infty} |y_i|^p$$

$$< \frac{\varepsilon^p}{2} + \frac{\varepsilon^p}{2} = \varepsilon^p,$$

which implies

$$\left(\sum_{i=1}^{\infty} |y_i - y_i^{(j)}|^p\right)^{1/p} < \varepsilon.$$

We have thus proved that Y is totally bounded.   □

(c) Let (X, d) be a discrete metric space (see Example 1.2.2 (v)) A subset $Y \subseteq X$ is compact if and only if it is finite.

Let $Y = \{y_1, y_2, \ldots, y_n\}$ and let $\{G_\lambda : x \in \Lambda\}$ be an open cover of Y. Then there exist G_{λ_i}, $i = 1, 2, \ldots, n$, such that $y_i \in G_{\lambda_i}$. Hence, $\{G_{\lambda_i} : i = 1, 2, \ldots, n\}$ is a finite subcovering of (Y, d_Y), where d_Y denotes the induced metric. So, (Y, d_Y) is compact.

On the other hand, suppose that Y is a compact subset of (X, d). Since every subset of X is open in (X, d), the collection $\{\{y\}: y \in Y\}$ is an open covering of Y. This open covering of Y contains a finite subcovering $\{\{y_1\}, \{y_2\}, \ldots, \{y_n\}\}$. So,

$$\{y_1\} \cup \{y_2\} \cup \ldots \cup \{y_n\} = Y,$$

that is, Y is a finite subset of X.

(d) Let $X = \mathbf{R}^n$ with the **Euclidean metric** d_2. It follows from Proposition 2.2.6, Remark 5.1.11 and Theorem 5.2.3 that a closed bounded subset of $\mathbf{R}^n$ is compact. The converse follows from Proposition 5.1.10 and Theorem 5.2.5. Thus, a subset of $\mathbf{R}^n$ is compact if and only if it is both closed and bounded. Recall that the metrics d_1, d_2 and d_∞ are equivalent (see Example 3.5.7(i)). Also, the identity map $i: (X, d) \to (X, d')$, where d and d' are equivalent metrics on the underlying set, is a homeomorphism. By Remark 3.5.5, a subset $Y \subseteq X$ is compact in (X, d) if and only if it is compact in (X, d'). So, a subset of $\mathbf{R}^n$ is compact in one of the three metrics d_i, $i = 1, 2, \infty$, if and only if it is both closed and bounded with respect to any one of the three.

5.6 Exercises

1. Let (X, d) be a compact metric space and $\{f_n\}_{n \geq 1}$ be a sequence of real-valued continuous functions on X that converges pointwise to a continuous function f on X. If the sequence $\{f_n\}_{n \geq 1}$ is monotonic, then show that $f_n \to f$ uniformly on X. Give an example to show that none of the conditions–namely, X is compact, the sequence $\{f_n\}_{n \geq 1}$ is monotonic, f is continuous–can be dropped in the hypotheses of the above result (known as Dini's theorem).

   Hint: Assume that $f_n(x) \geq f_{n+1}(x)$ for every $x \in X$ and $n = 1, 2, \ldots$. Set $g_n(x) = f_n(x) - f(x)$. Then $g_n \to 0$ pointwise and $g_n(x) \geq g_{n+1}(x)$ on X. We shall show that $g_n \to 0$ uniformly on X. Let $\varepsilon > 0$ be given. For each $x \in X$ there exists an integer n_x such that $0 \leq g_{n_x}(x) < \varepsilon$. By continuity of g_n and by monotonicity of the sequence $\{g_n\}_{n \geq 1}$, there exists an open ball $S(x, r_x)$ such that for every $y \in S(x, r_x)$ and $n \geq n_x$, $0 \leq g_n(y) < \varepsilon$. Using compactness of X, we obtain $x_1, x_2, \ldots, x_m$ such that

$$X \subseteq \bigcup_{i=1}^{m} S(x_i, r_{x_i}).$$

   If $n_0 = \max\{n_{x_1}, \ldots, n_{x_m}\}$, we obtain

$$0 \leq g_n(x) < \varepsilon$$

   for all $x \in X$ and $n \geq n_0$.

(i) For $x \in (0,1), f_n(x) = 1/(1 + nx) \to 0$ monotonically on the noncompact space (0,1). However, the convergence fails to be uniform, as $f_n(1/n) = 1/2$.

(ii) For $x \in [0,1], f_n(x) = \sin(n\pi \cdot \min(x, 1/n)) \to 0$ on the compact space [0,1] but not monotonically. However, the convergence fails to be uniform, as $f_n(1/2n) = 1$.

(iii) For $x \in [0,1], f_n(x) = 1/(1 + nx) \to f$ monotonically, where $f(0) = 1$ and $f(x) = 0$ when $0 < x \le 1$. The limit f is not continuous and the convergence fails to be uniform.

2. Show that $S(0,1) = \{z \in \mathbf{C}: |z| < 1\}$ and the closed ball $S[0,1] = \{z \in \mathbf{C}: |z| \le 1\}$ are not homeomorphic.
   Hint: The open ball $S(0,1)$ is not compact (see Example 5.2.2 (i)). The closed ball $S[0,1]$, being a closed bounded subset of the complex plane, is compact (see [19], Theorem 2.41). It is also a consequence of the last paragraph of Section 5.6.

3. Let A be a compact subset of a metric space (X, d). Show that for any $B \subseteq X$, there is a point $p \in A$ such that $d(p, B) = d(A, B)$.
   Hint: Let $\alpha = d(A, B) = \inf\{d(a, b): a \in A, b \in B\}$. For every positive integer n, there exists $a_n \in A$ and $b_n \in B$ such that $\alpha \le d(a_n, b_n) < \alpha + 1/n$.
   Now A is compact. So the sequence $\{a_n\}_{n \ge 1}$ has a subsequence converging to a point $p \in A$. We claim that $d(p, B) = \alpha = d(A, B)$.
   Suppose $d(p, B) > \alpha$, say $d(p, B) = \alpha + \delta$, $\delta > 0$. Since a subsequence of $\{a_n\}_{n \ge 1}$ converges to p, there exists $n_0 \in \mathbf{N}$ such that $d(p, a_{n_0}) < \delta/2$ and $d(a_{n_0}, b_{n_0}) < \alpha + 1/n_0 < \alpha + \delta/2$. Thus,

$$d(p, a_{n_0}) + d(a_{n_0}, b_{n_0}) < \frac{1}{2}\delta + \alpha + \frac{1}{2}\delta = \alpha + \delta = d(p, B) \le d(p, b_{n_0}).$$

   But this contradicts the triangle inequality. Hence, $d(p, B) = d(A, B)$.

4. Let A be a compact subset of a metric space (X, d) and B be a closed subset of X such that $A \cap B = \varnothing$. Show that $d(A, B) > 0$.
   Hint: Suppose $d(A, B) = 0$. Then by Exercise 3 above, there exists $p \in A$ such that $d(p, B) = d(A, B) = 0$. Since B is closed and $d(p, B) = 0$, it follows that $p \in B$ and so $A \cap B \ne \varnothing$. This contradicts the hypothesis; so $d(A, B) > 0$.

5. Show that the Hilbert cube, namely, the subset of ℓ_2 of points $x = \{x_n\}_{n \ge 1}$ such that

$$|x_n| \le \frac{1}{n}, \qquad n = 1, 2, \ldots,$$

   is compact.
   Hint: Let $\{x^{(p)}\}_{p \ge 1}$, where $x^p = \{x_n^{(p)}\}_{n \ge 1}$, be a sequence of points in the Hilbert cube. The sequence of first coordinates, namely, $\{x_1^{(1)}, x_1^{(2)}, x_1^{(3)}, \ldots\}$ are such that $|x_1^{(p)}| \le 1$, and so the original sequence has a subsequence $\{x^{p(1)}\}$, the first coordinates of which converge to a number x_1 satisfying $|x_1| \le 1$. In the same way, this subsequence must itself have a subsequence $\{x^{p(2)}\}$ whose second

coordinates converge to a number x_2 satisfying $|x_2| \leq 1/2$, while the first coordinates still converge to x_1. We may continue this process by induction. Finally, the diagonal sequence $\{x^{p(p)}\}$ of these subsequences will converge coordinatewise to $(x_1, x_2, \ldots)$, which also belongs to the Hilbert cube.

6. The closed unit ball in $C[0,1]$, i.e.,

$$\bar{S}(0, 1) = \{x \in C[0, 1] : \sup_t |x(t)| \leq 1\}$$

is not compact.
Hint: The sequence

$$f_n(t) = \begin{cases} \dfrac{-1}{2} & \text{if } 0 \leq t \leq \dfrac{1}{n+1}, \\ \dfrac{1}{2}[2n(n+1)t - (2n+1)] & \text{if } \dfrac{1}{n+1} \leq t \leq \dfrac{1}{n}, \\ \dfrac{1}{2} & \text{if } \dfrac{1}{n} \leq t \leq 1 \end{cases}$$

is in $\bar{S}(0, 1)$ and $d(f_n, f_m) = 1$ whenever $n \neq m$. Thus, we have a sequence of points in $\bar{S}(0,1)$ that has no convergent subsequence.

7. Show that the metric space $C[0, 1]$ is not locally compact.
Hint: Modify the Example in Exercise 6.

8. Show by using the finite intersection property that $(\mathbf{R}, d)$ is not compact.
Hint: The sets $F_n = \{x : x \geq n\}$ are closed and each finite family has a nonempty intersection; yet $\bigcap_{n=1}^{\infty} F_n = \varnothing$.

9. Let $\mathcal{F}$ be the family of all continuous maps $[0, 1] \to [0, 1]$ such that

$$|f(s) - f(t)| \leq |s - t|, \qquad s, t \in [0, 1].$$

Define

$$d(f, g) = \sup \{|f(s) - f(t)| : 0 \leq t \leq 1\}.$$

Prove that $\mathcal{F}$ is compact.
Hint: Arzelà-Ascoli theorem (see Theorem 5.6.5).

10. Let f be a continuous mapping of a compact metric space (X, d) into itself such that $f(X)$ is everywhere dense in X. Prove that $f(X) = X$.
Hint: The subset $f(X)$, being a continuous image of the compact space X, must be compact and, hence, a closed subset of X. Being dense, it must be equal to X.

11. Let $\{x_n\}_{n \geq 1}$ be a sequence in a compact metric space. If every convergent subsequence of it has the same limit x, show that $\lim_n x_n = x$. Give an example to show that this may not happen in a noncompact metric space.

12. If E is a nonempty compact subset of a metric space (X, d), then there exist points x and y in E such that $\operatorname{diam}(E) = d(x, y)$.

    Hint: Suppose, if possible, that there are no such points. Then $d(u, v) < \operatorname{diam}(E)$ whenever $u, v \in E$. By definition of diameter, there exist x_1 and y_1 in E such that $\operatorname{diam}(E) - 1 < d(x_1, y_1)$. If x_n and y_n are in E and $\operatorname{diam}(E) - 1/n < d(x_n, y_n)$, then since $\operatorname{diam}(E) - d(x_n, y_n) > 0$ by supposition, $\min\{1/n, \operatorname{diam}(E) - d(x_n, y_n)\}$ is a positive number. Again by the definition of diameter, there exist x_{n+1} and y_{n+1} in E such that $\operatorname{diam}(E) - \min\{1/(n+1), \operatorname{diam}(E) - d(x_n, y_n)\} < d(x_{n+1}, y_{n+1})$. Consequently,

    $$\operatorname{diam}(E) - \frac{1}{(n+1)} < d(x_{n+1}, y_{n+1}),$$

    as well as

    $$\operatorname{diam}(E) - (\operatorname{diam}(E) - d(x_n, y_n)) < d(x_{n+1}, y_{n+1}).$$

    The first one of these inequalities ensures that $\operatorname{diam}(E) - 1/n < d(x_n, y_n)$ for all n and the second one ensures that $\{d(x_n, y_n)\}_{n \geq 1}$ is an increasing sequence. Since E is compact, the sequences $\{x_n\}_{n \geq 1}$ and $\{y_n\}_{n \geq 1}$ possess convergent subsequences $\{x_{n_k}\}_{k \geq 1}$ and $\{y_{n_k}\}_{k \geq 1}$ with limits x and y, say, belonging to E. Then $d(x, y) \geq d(x_{n_k}, y_{n_k}) > \operatorname{diam}(E) - 1/n_k$ for all k. Hence, $d(x, y) \geq \operatorname{diam}(E)$. On the other hand, $d(x, y) \leq \operatorname{diam}(E)$ because x and y belong to E.

13. Let (X, d) be a metric space. A mapping $T \colon X \to X$ is said to be **contractive** if $d(Tx, Ty) < d(x, y)$ for every $x, y \in X$, $x \neq y$. Show that a contractive map T from a compact metric space X into itself has a unique fixed point.

    Hint: The function $f \colon X \to \mathbf{R}$ defined by $f(x) = d(x, Tx)$ is uniformly continuous. Since X is compact, there exists $x \in X$ such that $f(x) \leq f(y) \forall y \in X$. We claim that x is a fixed point of T. Suppose $Tx \neq x$. Then $f(x) \leq f(Tx)$, i.e., $d(x, Tx) \leq d(Tx, T(Tx)) < d(x, Tx)$, which is a contradiction.

14. Let T be a contractive map of a compact metric space X into itself and $x_0 \in X$. Then show that $Tx_0, T^2 x_0, T^3 x_0, \ldots$ converge to the unique fixed point of T (see Exercise 13).

    Hint: Let x denote the unique fixed point of T. Then

    $$d(T^{n+1} x_0, x) = d(T(T^n x_0), Tx) < d(T^n x_0, x), \qquad n = 1, 2, \ldots.$$

    Thus, $\{d(T^n x_0, x)\}_{n \geq 1}$ is a decreasing sequence of nonnegative numbers and, hence, converges to λ, say, $\lim_n d(T^n x_0, x) = \lambda$. The sequence $\{T^n x_0\}$, being in the compact metric space, has a convergent subsequence $\{T^{n_k} x_0\}_{k \geq 1}$. Let $\lim_k T^{n_k} x_0 = y$. Then

    $$d(y, x) = \lim_k d(T^{n_k} x_0, x) = \lambda.$$

If $\lambda \neq 0$, then $y \neq x$ and so

$$\lambda = d(y,x) > d(Tx, Ty)$$
$$= \lim_k d(Tx, T(T^{n_k}x_0))$$
$$\geq \limsup_k d(T^2x, T^2(T^{n_k}x_0)) \geq \ldots \geq$$
$$\geq \limsup_k d((T^{n_{k+1}-n_k})x, T^{n_{k+1}}x_0)$$
$$= \limsup_k d(x, T^{n_{k+1}}x_0) = \lambda.$$

This is a contradiction. So $\lambda = 0$ and hence, $y = x$. Thus, x is the only sub-sequential limit of $\{T^n x_0\}$. The result now follows from Exercise 11.

15. Let A be a compact subset of a metric space (X, d). Then show that the derived set A' of A is compact.
Hint: A' is a closed subset of X (see Proposition 2.1.23). Moreover, $A' \subseteq A$ since A is closed. So A' is a closed subset of A (see Theorem 2.2.2(ii)) and is, therefore, compact because A is compact.

16. A sequence $\{x_n\}_{n\geq 1}$ in a metric space X is convergent and $x = \lim_n x_n$. Prove that the union $\{x\} \cup \{x_n : n = 1, 2, \ldots\}$ is a compact subset of X.
Hint: Let $\{U_\alpha\}_{\alpha \in \Lambda}$ be an open cover of $\{x\} \cup \{x_n : n = 1, 2, \ldots\}$ and let $x \in U_{\alpha_0}$. There exists n_0 such that $n \geq n_0 \Rightarrow x_n \in U_{\alpha_0}$. If $x_i \in U_{\alpha_i}$ for $i = 1, 2, \ldots, n_0 - 1$, then $\cup_{i=0}^{n_0-1} U_{\alpha_i}$ contains $\{x\} \cup \{x_n : n = 1, 2, \ldots\}$.

17. Let f be a mapping of a metric space X into a metric space Y. Prove that if $f|_A$ is continuous on A for all compact subsets A of X, then f is continuous on X.
Hint: Let $\{x_n\}_{n\geq 1}$ be a sequence in X and let $x = \lim_n x_n$. Then $\{x\} \cup \{x_n : n = 1, 2, \ldots\}$ is compact by Exercise 16. Since f restricted to this compact set is continuous, it follows that $f(x) = \lim_n f(x_n)$. Thus, f is continuous.

18. Let f be a continuous mapping of a compact metric space X into a metric space. Prove that

$$f(\bar{A}) = \overline{f(A)}.$$

Hint: $\bar{A}$ is a compact subset of X. So $f(\bar{A})$ is compact (and hence, closed), f being continuous. Now, $f(A) \subseteq f(\bar{A})$ and hence, $\overline{f(A)} \subseteq f(\bar{A})$. The opposite inclusion is a consequence of the fact that f is continuous.

19. Let K be a subset of $C[0,1]$, the space of continuous real-valued functions on $[0, 1]$. Suppose that each $f \in K$ is differentiable and that there exists an $M > 0$ such that $\sup_t |f'(t)| \leq M$ for all $f \in K$. Prove that K is equicontinuous.
Hint: For $x, y \in [0, 1]$, $y \neq x$ and $f \in K$, we have

$$\left|\frac{f(x) - f(y)}{x - y}\right| = |f'(\xi)| \leq M, \qquad x < \xi < y.$$

So, $|f(x) - f(y)| \leq M|x - y|$ for $x, y \in [0, 1]$ and $f \in K$.

20. If $\{f_n\}_{n \geq 1}$ is a uniformly convergent sequence of continuous functions on $[0,1]$, then show that $\{f_n\}_{n \geq 1}$ is an equicontinuous family on $[0,1]$.

    Hint: Let $\varepsilon > 0$ be given. By hypothesis, there exists n_0 such that $n, m \geq n_0$ implies $|f_n(x) - f_m(x)| < \varepsilon/3$ for all $x \in [0,1]$. Since continuous functions are uniformly continuous on compact sets, there exists $\delta > 0$ such that

$$|f_k(x) - f_k(y)| < \frac{\varepsilon}{3} \qquad \text{for } 1 \leq k \leq n_0 \text{ and } |x - y| < \delta. \tag{5.9}$$

    Thus, if $x, y \in [0,1]$, $|x - y| < \delta$ and $n \geq n_0$, we have

$$|f_n(x) - f_n(y)| \leq |f_n(x) - f_{n_0}(x)| + |f_{n_0}(x) - f_{n_0}(y)| + |f_{n_0}(y) - f_n(y)|$$
$$< \frac{\varepsilon}{3} + \frac{\varepsilon}{3} + \frac{\varepsilon}{3} = \varepsilon.$$

    This, together with (5.9), proves the result. It is also a consequence of Exercise 16 in light of the Arzela-Ascoli theorem.

21. Let $\{f_n\}_{n \geq 1}$ be a sequence of real-valued functions defined on a set E of real numbers. We say that $\{f_n\}_{n \geq 1}$ is **pointwise bounded** on E if, for all $x \in E$, there exists an $M_x > 0$ such that

$$|f_n(x)| \leq M_x, \qquad n = 1, 2, \ldots$$

    If $\{f_n\}_{n \geq 1}$ is pointwise bounded and equicontinuous on $[0,1]$, then $\{f_n\}_{n \geq 1}$ is uniformly bounded on $[0,1]$.

    Hint: Set $g(x) = \sup\{|f_n(x)| : n = 1, 2, \ldots\}$. Let $\varepsilon > 0$ be given. By hypothesis, there exists $\delta > 0$ such that $x, y \in [0,1]$, $|x - y| < \delta$ implies $|f_n(x) - f_n(y)| < \varepsilon$ for $n = 1, 2, \ldots$. Fix x and y. Then

$$|f_n(x)| < |f_n(y)| + \varepsilon \text{ and } |f_n(y)| < |f_n(x)| + \varepsilon \text{ for } n = 1, 2, \ldots$$

    so that

$$g(x) \leq g(y) + \varepsilon \text{ and } g(y) \leq g(x) + \varepsilon.$$

    It follows that $|g(y) - g(x)| \leq \varepsilon$ whenever $|x - y| < \delta$. So, g is continuous on $[0,1]$ and hence, bounded.

22. Prove that the family $\{\sin nx\}_{n \geq 1}$ is uniformly bounded but not equicontinuous on $[-\pi, \pi]$.

    Hint: In fact, for any $\delta > 0$, there exists n large enough so that $\pi/n < \delta$. If $x = -\pi/2n$ and $y = \pi/2n$, then $|x - y| = \pi/n < \delta$ but $|\sin(nx) - \sin(ny)| = |-1 - 1| = 2$.

23. Let $f_n(x) = \frac{x^2}{x^2 + (1 - nx)^2}$, $0 \leq x \leq 1$, $n = 1, 2, \ldots$. Prove that $\{f_n\}_{n \geq 1}$ is uniformly bounded on $[0,1]$ but is not equicontinuous.

Hint: Clearly, $|f_n(x)| \leq 1$, so that $\{f_n\}_{n \geq 1}$ is uniformly bounded on $[0,1]$. Also $f_n(1/n) = 1$, $n = 1, 2, \ldots$. For any $\delta > 0$, choose n so large that $1/n < \delta$. If $x = 1/n$, $y = 0$, then $|x - y| = 1/n < \delta$ but $|f_n(x) - f_n(0)| = f_n(1/n) = 1$.

24. Let $\{f_n\}_{n \geq 1}$ be a sequence of twice differentiable functions on $[0,1]$ such that $f_n(0) = f_n'(0) = 0$ for all n. Suppose also that $|f_n''(x)| \leq 1$ for all n and all $x \in [0, 1]$. Prove that there is a subsequence of $\{f_n\}_{n \geq 1}$ that converges uniformly on $[0,1]$.
Hint: $f_n(x) = f_n(0) + f_n'(0) \cdot x + \frac{f_n''(\xi)}{2} x^2 = \frac{f_n''(\xi)}{2} x^2$ for some $\xi \in (0, 1)$.
So,

$$|f_n(x)| \leq \frac{1}{2} \qquad \text{for all } x \in [0, 1].$$

Also, $|f_n'(x)| = |f_n'(x) - f_n'(0)| \leq |f_n''(\xi) \cdot (x - 0)| \leq 1$ for all $x \in [0, 1]$. Therefore, $|f_n(x) - f_n(y)| \leq |x - y|$ for all $x, y \in [0, 1]$. Apply the Arzelà-Ascoli theorem.

25. Let $f: X \to Y$ be a continuous open mapping of a locally compact metric space X into Y. Then $f(X)$ is locally compact.
Hint: Let $y \in f(X)$. Then there exists an $x \in X$ such that $y = f(x)$. Let $S(x, r)$ be an open ball centred at x such that $\overline{S(x, r)}$ is compact. Since f is an open map, $f(S(x, r))$ is open and contains y. Also, $f(\overline{S(x, r)})$ is compact and hence, closed. Now

$$\overline{f(S(x, r))} \subseteq \overline{f(\overline{S(x, r)})} = f(\overline{S(x, r)});$$

so $\overline{f(S(x, r))}$ is compact, being a closed subset of the compact set $f(\overline{S(x, r)})$.

6 Product Spaces

There are two main techniques for constructing new metric spaces out of given ones. The first of these, and the simplest, is to form a subspace from a given space. The second is to "multiply together" a number of given spaces. The genesis of this study lies in the study of functions of several variables. Our purpose here is to describe the way in which this is carried out. Infinite metric products are discussed in Section 6.3, which also contains the famous Tichonov Theorem, namely, that an infinite product of compact metric spaces is compact and conversely. A representation of Cantor's set as an infinite product, together with some of its properties, are discussed in the last section.

6.1. Finite and Infinite Products of Sets

We begin by recalling the definition of the Cartesian product of a finite collection of sets.

Definition 6.1.1. The **Cartesian product of a finite** collection of sets $X_1, X_2, \ldots, X_n$, denoted by $X_1 \times X_2 \times \ldots \times X_n$, or also by $\prod_{i=1}^{n} X_i$, is the collection of all ordered n-tuples $(x_1, x_2, \ldots, x_n)$, where $x_i \in X_i$ for each $i = 1, 2, \ldots, n$. Note that an n-tuple may be considered as a mapping from the indexing set $\{1, 2, \ldots, n\}$ into $\cup_i X_i$ such that the value of the mapping at i, written x_i, is in the set X_i.

Generalising the notion of a product of a finite number of sets to a countable number of sets, we shall let the **product $\prod_{i=1}^{\infty} X_i$ of a countable number of sets** $\{X_i\}_{i \geq 1}$ be the family of all sequences $x = \{x_i\}_{i \geq 1}$, where each $x_i \in X_i$. It is the collection of all the mappings of the set $\mathbf{N}$ of natural numbers into $\bigcup_{i=1}^{\infty} X_i$ such that the value of the mapping at $i \in \mathbf{N}$ is in the set X_i. The set X_n is called the **nth factor** of $\prod_{i=1}^{\infty} X_i$; for each $n \in \mathbf{N}$, the map

$$p_n : \prod_{i=1}^{\infty} X_i \to X_n$$

given by $p_n(\{x_i\}_{i \geq 1}) = x_n$ is the **projection onto the nth factor.**

Examples 6.1.2. (i) $\mathbf{R}^n = \prod_{i=1}^{n} X_i$, where each $X_i = \mathbf{R}$. The projection p_k maps $(x_1, x_2, \ldots, x_n) \in \mathbf{R}^n$ into its kth coordinate x_k.

(ii) If each X_i has exactly one element, then $\prod_{i=1}^{\infty} X_i$ consists of a single element. If one X_i is empty, then $\prod_{i=1}^{\infty} X_i$ is empty.

(iii) Let $X_i = \{0, 2\}$ for $i \in \mathbf{N}$; then $\prod_{i=1}^{\infty} X_i$ is the set of all sequences of 0s and 2s:

$$\{\{x_i\} : x_i = 0 \text{ or } 2, \ i = 1, 2, \ldots\}.$$

The map $f : \prod_{i=1}^{\infty} X_i \rightarrow [0, 1]$ defined by

$$f(\{x_i\}_{i \geq 1}) = \sum_{i=1}^{\infty} \frac{x_i}{3^i}$$

is easily seen to be one-to-one; the image is called the **Cantor set**.

The following property of the Cartesian product may be verified easily: If A_n and B_n are subsets of X_n, then

$$\prod_n A_n \cap \prod_n B_n = \prod_n (A_n \cap B_n) \text{ and } \prod_n A_n \cup \prod_n B_n \subseteq \prod_n (A_n \cup B_n).$$

For a given $n \in \mathbf{N}$ and $A_n \subseteq X_n$, we denote $p_n^{-1}(A_n)$ by $<A_n>$; i.e., this is the Cartesian product ΠY_i in which $Y_n = A_n$, while $Y_i = X_i$ for $i \neq n$. In symbols,

$$p_n^{-1}(A_n) = <A_n> = A_n \times \Pi\{X_i : i \in \mathbf{N}, \ i \neq n\}.$$

It is, of course, understood that the factor A_n appears at the nth place in the product on the right hand side. Similarly, for finitely many indices $i_1, i_2, \ldots, i_n$ and sets

$$A_{i_1} \subseteq X_{i_1}, \ldots, A_{i_n} \subseteq X_{i_n},$$

the subset

$$<A_{i_1}> \cap \ldots \cap <A_{i_n}> = p_{i_1}^{-1}(A_{i_1}) \cap \ldots \cap p_{i_n}^{-1}(A_{i_n})$$

$$= \prod_{j=1}^{n} (A_{i_j} \times \prod \{X_i : i \in \mathbf{N}, i \neq i_1, i_2, \ldots, i_n\})$$

is denoted by

$$<A_{i_1}, A_{i_2}, \ldots, A_{i_n}>.$$

If $A_n \subseteq X_n$ for $n = 1, 2, \ldots$, then the following may also be easily verified:

(i) $\Pi_n A_n = \bigcap_n <A_n>$
(ii) $<A_n>^c = <A_n^c>$ and $(\Pi A_n)^c = \bigcup_n <A_n^c>$.

6.2. Finite Metric Products

Let $(X_1, d_1), (X_2, d_2), \ldots, (X_n, d_n)$ be metric spaces and let $x = (x_1, x_2, \ldots, x_n)$ and $y = (y_1, y_2, \ldots, y_n)$ be arbitrary points in the product $X = \prod_{i=1}^{n} X_i$. Define

$$d(x, y) = \max\{d_i(x_i, y_i) : 1 \leq i \leq n\}.$$

Proposition 6.2.1. (X, d) is a metric space.

Proof. Clearly, $d(x, y) \geq 0$ and $d(x, y) = 0$ if and only if $d_i(x_i, y_i) = 0$ for $1 \leq i \leq n$, which is the case if and only if $x_i = y_i$ for $1 \leq i \leq n$, i.e., if and only if $x = y$. It is equally clear that $d(x, y) = d(y, x)$. It remains to verify the triangle inequality.

Observe that

$$d_i(x_i, z_i) \leq d_i(x_i, y_i) + d_i(y_i, z_i),$$

where $x = (x_1, x_2, \ldots, x_n)$, $y = (y_1, y_2, \ldots, y_n)$ and $z = (z_1, z_2, \ldots, z_n)$, are points in X. This implies

$$d_k(x_k, z_k) \leq \max\{d_i(x_i, y_i) : 1 \leq i \leq n\} + \max\{d_i(y_i, z_i) : 1 \leq i \leq n\}$$

for $k = 1, 2, \ldots, n$. So

$$d(x, z) = \max\{d_k(x_k, z_k) : 1 \leq k \leq n\} \leq d(x, y) + d(y, z).$$

Thus, the function d satisfies (MS1)–(MS4) in Definition 1.2.1 and hence, (X, d) is a metric space. ◻

Definition 6.2.2. The metric space obtained by taking

$$d(x, y) = \max\{d_i(x_i, y_i) : 1 \leq i \leq n\}$$

as the distance on $X = \prod_{i=1}^{n} X_i$ is called the **product of the metric spaces** $(X_1, d_1), (X_2, d_2), \ldots, (X_n, d_n)$.

Remark 6.2.3. (i) The functions

$$d'(x, y) = \sum_{i=1}^{n} d_i(x_i, y_i),$$

$$d''(x, y) = \left[\sum_{i=1}^{n} (d_i(x_i, y_i))^2\right]^{1/2},$$

where $x = (x_1, x_2, \ldots, x_n)$ and $y = (y_1, y_2, \ldots, y_n)$ belong to X, are also metrics on X. The proof of the statement that d' is a metric is almost trivial. In fact, if $x = (x_1, x_2, \ldots, x_n)$, $y = (y_1, y_2, \ldots, y_n)$ and $z = (z_1, z_2, \ldots, z_n)$ are in X, then

$$d'(x, z) = \sum_{i=1}^{n} d_i(x_i, z_i)$$

$$\leq \sum_{i=1}^{n} (d_i(x_i, y_i) + d_i(y_i, z_i))$$

$$= \sum_{i=1}^{n} d_i(x_i, y_i) + \sum_{i=1}^{n} d_i(y_i, z_i)$$

$$= d'(x, y) + d'(y, z).$$

The proof of the statement that d'' is a metric is no different from the one given in Example 1.2.2(ii) and is, therefore, not included.

(ii) The metrics d' and d'' are equivalent to d. Indeed,

$$d(x, y) \leq d''(x, y) \leq d'(x, y) \leq nd(x, y).$$

(See Corollary 1.4.9 and Definition 3.5.4.)

(iii) For all questions dealing with metric properties or Cauchy sequences or uniformly continuous functions, we may consider on X any one of the metrics d, d', d''. In what follows, we shall consider on X the metric d. Open (respectively, closed) balls for the distances will be written S, S', S'' (respectively, $\bar{S}$, $\bar{S}'$, $\bar{S}''$).

Proposition 6.2.4. The open ball $S(x, r)$, $x = (x_1, x_2, \ldots, x_n)$ and $r > 0$, in X is the product of the open balls $S_1(x_1, r)$, $S_2(x_2, r), \ldots, S_n(x_n, r)$. That is,

$$S(x, r) = \prod_{i=1}^{n} S_i(x_i, r),$$

where $S_i(x_i, r)$ is the open ball centred at $x_i \in X_i$ and radius $r > 0$ in X_i.

Proof. We have $y \in S(x, r)$ if and only if $\max \{d_i(y_i, x_i) : 1 \leq i \leq n\} < r$ if and only if $d_i(y_i, x_i) < r$, $1 \leq i \leq n$. So, $y \in S(x, r)$ if and only if $y_i \in S_i(x_i, r)$, $1 \leq i \leq n$, that is, if and only if $y \in S_1(x_1, r) \times S_2(x_2, r) \times \ldots \times S_n(x_n, r)$.   □

Corollary 6.2.5. The collection $\{ \prod_{i=1}^{n} S_i(x_i, r) : x_i \in X_i, \ 1 \leq i \leq n \text{ and } r > 0\}$ is a base for the open subsets of the metric space (X, d).

Proof. See Proposition 2.3.5.   □

Proposition 6.2.6. If $G_i \subseteq X_i$, $1 \leq i \leq n$ are open subsets in X_i, then $\prod_{i=1}^{n} G_i$ is open in X.

Proof. (See Figure 6.1.) If $(x_1, x_2, \ldots, x_n) \in \prod_{i=1}^{n} G_i$, then there exist positive $r_1, r_2, \ldots, r_n$ such that $S_i(x_i, r_i) \subseteq G_i$, $1 \leq i \leq n$. Let $r = \min \{r_1, r_2, \ldots, r_n\}$. Then by Proposition 6.2.4, we have $S(x, r) = \prod_{i=1}^{n} S_i(x_i, r) \subseteq \prod_{i=1}^{n} G_i$.   □

Definition 6.2.7. A function $f : (X, d_X) \to (Y, d_Y)$ is said to be an **open map** if $f(G)$ is open in Y for every open set G in X, that is, f is an open map if the image of every open set in X is an open set in Y. Similarly, a function $f : (X, d_X) \to (Y, d_Y)$ is said to be a **closed map** if $f(F)$ is closed in Y for every closed set F in X, that is, f is a closed map if the image of every closed set in X is a closed set in Y. [See Exercise 11 of Chapter 3.]

Proposition 6.2.8. The projection map $p_i : \prod_{j=1}^{n} X_j \to X_i$, $1 \leq i \leq n$, is both continuous and open.

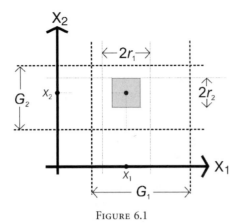

FIGURE 6.1

Proof. Observe that

$$p_i^{-1}(S_i(x_i, r)) = S_i(x_i, r) \times \prod_{j \neq i} X_j.$$

Let $G_i \subseteq X_i$ be open. Then $G_i = \bigcup \{S_i(x_i, r_i) : x_i \in G_i$ and r_i (depending on x_i) $> 0\}$, where the union is taken over *all* $x_i \in G_i$ and *one* suitable r_i for each x_i (see Proposition 2.3.5). So

$$p_i^{-1}(G_i) = p_i^{-1} \left(\bigcup S_i(x_i, r_i) \right)$$

$$= \bigcup p_i^{-1}(S_i(x_i, r_i))$$

$$= \bigcup \left(S_i(x_i, r_i) \times \prod_{j \neq i} X_j \right)$$

$$= \bigcup S_i(x_i, r_i) \times \prod_{j \neq i} X_j$$

$$= G_i \times \prod_{j \neq i} X_j.$$

Since $G_i \times \prod_{j \neq i} X_j$ is open in X by Proposition 6.2.6, it follows that p_i is a continuous function (see Theorem 3.1.9).

Let $G \subseteq X$ be open. Then

$$G = \bigcup \{S(x, r) : x \in G \text{ and } r \text{ (depending on } x) > 0\},$$

where (by Proposition 6.2.4)

$$S(x, r) = \prod_{i=1}^{n} S_i(x_i, r), \qquad x = (x_1, x_2, \ldots, x_n).$$

Now,

$$p_i(G) = p_i(\bigcup\{S(x,r): x \in G \text{ and } r \text{ (depending on } x) > 0\})$$
$$= \bigcup p_i(\{S(x,r): x \in G \text{ and } r \text{ (depending on } x) > 0\})$$
$$= \bigcup\{S_i(x_i,r): x_i \in X_i \text{ for which } x \in G \text{ and } r > 0\}$$

by Proposition 6.2.4. Since an arbitrary union of open sets is open (see Theorem 2.1.7(ii)), it follows that $p_i(G)$ is open. So p_i carries open subsets of X to open subsets of X_i and hence, is an open map.   □

Remark 6.2.9. The projection map is not a closed map. Consider the projection mapping $p_1: \mathbf{R}^2 \to \mathbf{R}$ of the plane into the x-axis, i.e., $p_1((x,y)) = x$. The map p_1 is not a closed map, for the set $F = \{(x,y): xy \geq 1, \ x > 0\}$ is a closed set, but its projection $p_1(F) = (0,\infty)$ is not closed, as illustrated in Figure 6.2.

Example 3.1.13(i) is a special case of the following result.

Proposition 6.2.10. A function $f: (Z, d_Z) \to (\prod_{j=1}^n X_j, d)$ from the metric space (Z, d_Z) into the product space is continuous if and only if for every projection $p_i: \prod_{j=1}^n X_j \to X_i$, the composition mapping

$$p_i \circ f : Z \to X_i$$

is continuous.

Proof. By Proposition 6.2.8, the projections p_i are continuous. So, if f is continuous, then $p_i \circ f$, being the composition of continuous maps, is continuous for $i = 1, 2, \ldots, n$ (see Theorem 3.1.11). On the other hand, let $p_i \circ f$ be continuous for $i = 1, 2, \ldots, n$ and let G_i be an open subset of X_i. Then

$$G_i = \bigcup\{S_i(x_i,r): x_i \in G_i, r \text{ (depending on } x_i) > 0\}.$$

(See Proposition 2.3.5). Since $p_i \circ f$ is continuous for $i = 1, 2, \ldots, n$, and

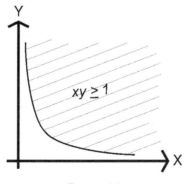

FIGURE 6.2

$$(p_i \circ f)^{-1}(G_i) = (p_i \circ f)^{-1}\left(\bigcup S_i(x_i, r(\text{depending on } x_i))\right)$$

$$= \bigcup f^{-1}(p_i^{-1}(S_i(x_i, r(\text{depending on } x_i))))$$

$$= \bigcup f^{-1}(S_i(x_i, r(\text{depending on } x_i)) \times \prod_{j \neq i} X_j)$$

$$= f^{-1}\left(\bigcup S_i(x_i, r(\text{depending on } x_i)) \times \prod_{j \neq i} X_j\right)$$

$$= f^{-1}(G_i \times \prod_{j \neq i} X_j),$$

it follows that $f^{-1}(G_i \times \prod_{j \neq i} X_j)$ is open in $\prod_{j=1}^{n} X_j$.

Let $G \subseteq \prod_{j=1}^{n} X_j$ be an open subset of the form $\prod_{j=1}^{n} G_j$, which is equal to $\bigcap < G_j >$ (see end of Section 6.1). Now,

$$f^{-1}(G) = f^{-1}\left(\bigcap < G_j >\right)$$

$$= \bigcap f^{-1}(< G_j >)$$

$$= \bigcap f^{-1}\left(G_j \times \prod_{i \neq j} X_i\right),$$

which is open, being a finite intersection of sets that are open in view of the preceding paragraph. The proof is now complete because such open sets constitute a base by Corollary 6.2.5.   □

Proposition 6.2.11. Let $F_i \subseteq X_i$, where (X_i, d_i), $i = 1, 2, \ldots, n$, are metric spaces. Let $X = \prod_{i=1}^{n} X_i$ be the product space with the metric d given by

$$d(x, y) = \max \{d_i(x_i, y_i) : 1 \leq i \leq n\}.$$

Then

$$\overline{F_1 \times F_2 \times \ldots \times F_n} = \bar{F}_1 \times \bar{F}_2 \times \ldots \times \bar{F}_n.$$

Proof. If $x = (x_1, x_2, \ldots, x_n) \in \prod_{i=1}^{n} \bar{F}_i$ for any $\varepsilon > 0$, there is, by assumption, $y_i \in F_i$, $i = 1, 2, \ldots, n$, such that $d_i(y_i, x_i) < \varepsilon$, $i = 1, 2, \ldots, n$. Hence, if $y = (y_1, y_2, \ldots, y_n)$, then $y \in \prod_{i=1}^{n} F_i$ and $d(y, x) < \varepsilon$. On the other hand, if $x = (x_1, x_2, \ldots, x_n) \notin \prod_{i=1}^{n} \bar{F}_i$ then there is a j such that $x_j \notin \bar{F}_j$. Observe that $\prod_{i \neq j} X_i \times (X_j - \bar{F}_j)$ is an open set in X containing x (by Proposition 6.2.6). Moreover,

$$(X_j - \bar{F}_j) \times \prod_{i \neq j} X_i \cap \prod_{i=1}^{n} F_i = \varnothing.$$

Hence,

$$x \notin \overline{\prod_{i=1}^{n} F_i}.$$ □

Proposition 6.2.12. The mapping $f: (Z, d_Z) \to (\prod_{j=1}^{n} X_j, d)$ from the metric space (Z, d_Z) into the product space is uniformly continuous if and only if for every projection $p_i: \prod_{j=1}^{n} X_j \to X_i$, the composition mapping

$$p_i \circ f: Z \to X_i$$

is uniformly continuous.

Proof. Observe that,

$$d_i(p_i(x), p_i(y)) \leq d(x, y)$$

where $x = (x_1, x_2, \ldots, x_n)$ and $y = (y_1, y_2, \ldots, y_n)$ are in X. On choosing $\varepsilon = \delta$, it follows that each p_i is uniformly continuous. Since the composition of uniformly continuous maps is uniformly continuous (see Theorem 3.4.6), it follows that so is $p_i \circ f$, $i = 1, 2, \ldots, n$.

On the other hand, assume that $p_i \circ f$ is uniformly continuous, $i = 1, 2, \ldots, n$, and let ε be any positive number. There exist δ_i, $i = 1, 2, \ldots, n$, such that

$$d_Z(z, \ w) < \delta_i \quad \text{implies } d_i(p_i \circ f(z), \ p_i \circ f(w)) < \varepsilon.$$

Let $\delta = \min\{\delta_1, \delta_1, \ldots, \delta_n\}$. Then

$$d_Z(z, \ w) < \delta \quad \text{implies } d_i(p_i \circ f(z), \ p_i \circ f(w)) < \varepsilon$$

for $i = 1, 2, \ldots, n$. Consequently,

$$d_Z(z, \ w) < \delta \qquad \text{implies } d(f(z), f(w)) < \varepsilon,$$

and so f is uniformly continuous.   □

6.3. Infinite Metric Products

Let (X_n, d_n), $n = 1, 2, \ldots$, be metric spaces with $d_n(X_n) \leq 1$ for each n. For $x, y \in \prod_{n=1}^{\infty} X_n$, define

$$d(x, \ y) = \sum_{n=1}^{\infty} 2^{-n} d_n(x_n, \ y_n), \tag{6.1}$$

where $x = \{x_n\}_{n \geq 1}$ and $y = \{y_n\}_{n \geq 1}$.

Observe that the series on the right in (6.1) converges. In fact,

$$2^{-n} d_n(x_n, y_n) \leq 2^{-n}$$

since $d_n(X_n) \leq 1$. The series $\sum_{n=1}^{\infty} 2^{-n}$ converges, and so, by the Weierstrass M-test (see Theorem 3.6.12), $\sum_{n=1}^{\infty} 2^{-n} d_n(x_n, y_n)$ converges.

Proposition 6.3.1. (X, d) is a metric space.

Proof. It is immediate that $d(x, y) \geq 0$ and that $d(x, y) = 0$ if and only if $x = y$. Also, $d(x, y) = d(y, x)$. For $x = \{x_n\}_{n \geq 1}$, $y = \{y_n\}_{n \geq 1}$ and $z = \{z_n\}_{n \geq 1}$,

$$d_n(x_n, z_n) \leq d_n(x_n, y_n) + d_n(y_n, z_n)$$

since d_n is a metric on X_n. Therefore, for every $k \geq 1$,

$$\sum_{n=1}^{k} 2^{-n} d_n(x_n, z_n) \leq \sum_{n=1}^{k} 2^{-n} d_n(x_n, y_n) + \sum_{n=1}^{k} 2^{-n} d_n(y_n, z_n)$$

$$\leq \sum_{n=1}^{\infty} 2^{-n} d_n(x_n, y_n) + \sum_{n=1}^{\infty} 2^{-n} d_n(y_n, z_n).$$

Since the left hand side of the above inequality is monotonically increasing and bounded above, we obtain on letting $k \to \infty$,

$$d(x, z) \leq d(x, y) + d(y, z).$$

Thus d satisfies all the requirements (MS1)–(MS4) of Definition 1.2.1 and so is a metric on X. This completes the proof. $\square$

The ball $S(x, r)$ of radius $r > 0$ centred at $x \in X = \prod_{n=1}^{\infty} X_n$ is the set

$$\{y = \{y_n\}_{n \geq 1} \in X : \sum_{n=1}^{\infty} 2^{-n} d_n(x_n, \ y_n) < r\}.$$

Then for $2r \leq 1$,

$$S(x, r) \subseteq \prod_{n=1}^{n_0} S_n(x_n, 2^n r) \times \prod_{n > n_0} X_n,$$

where n_0 is a suitably chosen positive integer. Let $y \in S(x, r)$. Then $d_n(x_n, y_n) < r2^n$ for $n = 1, 2, \ldots$. Let n_0 be the largest positive integer such that $2^{n_0} r \leq 1$. The integer n_0 exists because $2r \leq 1$. For $n > n_0$ and any $y_n \in X_n$, $2^{-n} d_n(x_n, y_n) < r$ since $d_n(X_n) \leq 1$, whereas for $n \leq n_0$, the admissible $y_n \in X_n$ are those which satisfy the inequality $2^{-n} d_n(x_n, y_n) < r$, that is, the y_n lie in $S_n(x_n, 2^n r)$. Thus, $y \in \prod_{n=1}^{n_0} S_n(x_n, 2^n r) \times \prod_{n > n_0} X_n$.

For $x = \{x_n\}_{n \geq 1} \in X = \prod_{n=1}^{\infty} X_n$, any integer $m \geq 1$ and any $r > 0$, let

$$S_m(x, r) = \{y = \{y_n\}_{n \geq 1} \in X : d_n(y_n, x_n) < r \text{ for } n \leq m\},$$

that is,

$$S_m(x, r) = \prod_{n=1}^{m} S_n(x_n, r) \times \prod_{n = m+1}^{\infty} X_n.$$

Let

$$\mathcal{B} = \{S_m(x, r) : m \text{ is any positive integer}, \ r > 0 \text{ and } x \in X\}.$$

Proposition 6.3.2. $\mathcal{B}$ is a base for the open subsets of the metric space X.

Proof. Clearly, the union of all members of $\mathcal{B}$ is X. Let $S_{m_1}(x^{(1)}, r_1)$ and $S_{m_2}(x^{(2)}, r_2)$ be any two members of $\mathcal{B}$ and let

$$x \in S_{m_1}(x^{(1)}, r_1) \cap S_{m_2}(x^{(2)}, r_2).$$

Then $S_m(x, r)$, where

$$m = \max(m_1, m_2) \text{ and } r = \min\left(r_1 - \max_{1 \leq n \leq m_1} d_n(x_n, x_n^{(1)}), r_2 - \max_{1 \leq n \leq m_2} d_n(x_n, x_n^{(2)})\right),$$

is a member of the collection $\mathcal{B}$ and is contained in the intersection $S_{m_1}(x^{(1)}, r_1) \cap S_{m_2}(x^{(2)}, r_2)$. In fact, if $y \in S_m(x, r)$, then $d_n(y_n, x_n) < r$ for $n \leq m$. Since

$$d_n(y_n, x_n^{(1)}) \leq d_n(y_n, x_n) + d_n(x_n, x_n^{(1)})$$
$$< r + d_n(x_n, x_n^{(1)}) < r_1$$

for $n \leq m_1$, we have $y \in S_{m_1}(x^{(1)}, r_1)$. Similarly, it can be shown that $y \in S_{m_2}(x^{(2)}, r_2)$. As $y \in S_m(x, r)$ is arbitrary, we have

$$S_m(x, r) \subseteq S_{m_1}(x^{(1)}, r_1) \cap S_{m_2}(x^{(2)}, r_2)$$

and hence $\mathcal{B}$ is a base for the open subsets of (X, d). $\square$

Remark 6.3.3. Let $x \in X$ be arbitrary. The class $\{S_m(x, r) : m \geq 1, \ r > 0\}$ is a local base at x.

Proposition 6.3.4. Let $\{x^{(k)}\}_{k \geq 1}$ be a sequence of points $x^{(k)} = \{x_n^{(k)}\}_{n \geq 1}$ of $X = \prod_{n=1}^{\infty} X_n$. Then $\{x^{(k)}\}_{k \geq 1}$ converges to a point $x \in X$ (respectively is Cauchy in X) if and only if for each n, the sequence $\{x_n^{(k)}\}_{k \geq 1}$ converges to x_n (respectively is a Cauchy sequence in X_n).

Proof. Suppose $\{x^{(k)}\}_{k \geq 1}$ is such that $x^{(k)} \to x$. Let p_n denote the projection of X onto X_n. It may be checked using the argument of Proposition 6.2.8 that p_n is continuous. Therefore, $p_n(x^{(k)}) \to p_n(x)$.

Conversely, suppose that $p_n(x^{(k)}) \to p_n(x)$ for every projection p_n. In order to prove that $x^{(k)} \to x$, it is sufficient to show that, if $x \in B$, where B is a member of the defining local base at $x \in X = \prod_{n=1}^{\infty} X_n$, then there exists $k_0 \in \mathbf{N}$ such that $k \geq k_0$ implies $x^{(k)} \in B$.

By definition of the defining local base at x,

$$B = S(x, \ r) = \prod_{n=1}^{j} S_n(x_n, r) \times \prod_{n=j+1}^{\infty} X_n$$
$$= \bigcap_{n=1}^{j} p_n^{-1}(S_n(x_n, r)).$$

By hypothesis, $p_n(x^{(k)}) \to p_n(x)$ and since $x \in B$, $p_n(x) \in S_n(x_n, r)$, $n \leq j$. There exists $k_n \in \mathbf{N}$ such that $k \geq k_n$ implies $p_n(x^{(k)}) \in S_n(x_n, r)$, i.e., $k \geq k_n$ implies $x^{(k)} \in p_n^{-1}(S_n(x_n, r))$. Let

$$k_0 = \max(k_1, k_2, \ldots, k_j).$$

For $k \geq k_0$, we have

$$x^{(k)} \in \bigcap_{n=1}^{j} p_n^{-1}(S_n(x_n, r)).$$

So, $x^{(k)} \to x$ in the product space.

The proof of the statement that $\{x^{(k)}\}_{k \geq 1}$ is Cauchy if and only if $p_n(x^{(k)})$ is Cauchy in X_n, $n = 1, 2, \ldots$, is left as Exercise 10. $\qquad\square$

Proposition 6.3.5. Let (X_n, d_n), $n = 1, 2, \ldots$, be metric spaces. Then $X = \prod_{n=1}^{\infty} X_n$ with the metric d defined by

$$d(x, y) = \sum_{n=1}^{\infty} 2^{-n} d_n(x_n, y_n),$$

where $x = \{x_n\}_{n \geq 1}$ and $y = \{y_n\}_{n \geq 1}$ are in X, is a complete metric space if and only if each (X_n, d_n), $n = 1, 2, \ldots$, is complete.

Proof. Let $\{x^{(k)}\}_{k \geq 1}$ be a Cauchy sequence of points $x^{(k)} = \{x_n^{(k)}\}_{n \geq 1}$ in X. Then $\{x_n^{(k)}\}_{k \geq 1}$ is a Cauchy sequence in X_n (see Proposition 6.3.4). Since X_n is complete, there exists $x_n \in X_n$ such that $\lim_k x_n^{(k)} = x_n$. Let $x = \{x_n\}_{n \geq 1}$. Then $x \in X$. It follows from Proposition 6.3.4 that $\lim_k x^{(k)} = x$.

On the other hand, assume that (X, d) is a complete metric space. First observe that if $a_n \in X_n$ then $< \{a_n\} >$ is a closed subset of the product space. In fact, $\{a_n\}$ is closed in X_n because a single point always forms a closed subset in a metric space, and therefore, the inverse image $< \{a_n\} >$ by the continuous map p_n must be closed. Hence, $\bigcap_{n \neq j} < \{a_n\} >$ is closed, being the intersection of closed subsets.

Consequently, $X_j \times \prod \{a_n : n \neq j\}$, being a closed subset of a complete metric space, is complete (see Proposition 2.2.6). The mapping $\varphi \colon X_j \to X_j \times \prod \{a_n : n \neq j\}$ defined by

$$\varphi(x_j) = (a_1, a_2, \ldots, a_{j-1}, x_j, a_{j+1}, a_{j+2}, \ldots)$$

is clearly one-to-one and onto. Moreover,

$$d(\varphi(x_j), \varphi(y_j)) = \sum_{n=1}^{\infty} 2^{-n} d_n(\varphi(x_j)(n), \varphi(y_j)(n)) = 2^{-j} d_j(x_j, y_j),$$

since $\varphi(x_j)(n) = \varphi(y_j)(n)$, $n \neq j$. Let $\{x_j^{(n)}\}_{n \geq 1}$ be a Cauchy sequence in X_j. Since

$$d\left(\varphi(x_j^{(n)}), \varphi(x_j^{(m)})\right) = 2^{-j} d_j\left(x_j^{(n)}, x_j^{(m)}\right),$$

it follows that $\{\varphi(x_j^{(n)})\}_{n \geq 1}$ is a Cauchy sequence in the complete space $X_j \times \prod\{a_n : n \neq j\}$ and hence, converges. As convergence in the product space is coordinatewise (see Proposition 6.3.4), it follows that $\{x_j^{(n)}\}_{n \geq 1}$ converges. Hence, X_j is a complete metric space. This completes the proof. $\qquad\square$

Proposition 6.3.6. Let (X_n, d_n), $n = 1, 2, \ldots$ be metric spaces and (X, d) be the product space, where $X = \prod_{n=1}^{\infty} X_n$ and $d(x, y) = \sum_{n=1}^{\infty} 2^{-n} d_n(x_n, y_n)$, whenever $x = \{x_n\}_{n \geq 1}$ and $y = \{y_n\}_{n \geq 1}$ are elements in X. The product space (X, d) is totally bounded if and only if each X_i is totally bounded.

Proof. Suppose (X, d) is totally bounded. Let $\varepsilon > 0$ be arbitrary. Fix $n \in \mathbf{N}$ and let ε_1 be such that $0 < 2^n \varepsilon_1 < \varepsilon$. Consider a finite ε_1-net $\{x^{(1)}, x^{(2)}, \ldots, x^{(k)}\}$ in X. We shall show that $\{x_n^{(1)}, x_n^{(2)}, \ldots, x_n^{(k)}\}$ is an ε-net in X_n. In fact, $d(y, x^{(j)}) < \varepsilon_1$, where $y \in X$ and $j \in \{1, 2, \ldots, k\}\}$, since $\{x^{(1)}, x^{(2)}, \ldots, x^{(k)}\}$ is an ε_1-net in X. It now follows from the definition of the metric on X that

$$2^{-n} d_n(y_n, x_n^{(j)}) \leq d(y, x^{(j)}) < \varepsilon_1,$$

that is, $d_n(y_n, x_n^{(j)}) < 2^n \varepsilon_1 < \varepsilon$. As y varies over X, y_n varies over X_n. So, $\{x_n^{(1)}, x_n^{(2)}, \ldots, x_n^{(k)}\}$ is an ε-net in X_n.

Conversely, assume that X_n is totally bounded for each n. Let φ be a sequence in X. We shall show that φ has a Cauchy subsequence. Now $p_1 \circ \varphi$ is a sequence in X_1, and X_1 is totally bounded. Therefore, we can extract subsequence φ_1 such that $p_1 \circ \varphi_1$ is Cauchy in X_1. Now consider $p_2 \circ \varphi_1$; for the same reason as before, we can extract a subsequence φ_2 of φ_1 such that $p_2 \circ \varphi_2$ is Cauchy in X_2. Proceeding by induction, we can obtain a sequence $\{\varphi_n\}_{n \geq 1}$ of subsequences of φ such that φ_{n+1} is a subsequence of φ_n for each $n \in \mathbf{N}$ and each sequence $p_n \circ \varphi_n$ is Cauchy in X_n. Now let $\hat{\varphi}$ be the subsequence $n \to \varphi_n(n)$ of φ; then for each fixed k, we have

$$\{\hat{\varphi}(m) : m \geq k\} \subseteq \{\varphi_k(m) : m \geq k\}.$$

Since $p_k \circ \varphi_k$ is a Cauchy sequence in X_k, it follows that $p_k \circ \hat{\varphi}$ is Cauchy in X_k. By Proposition 6.3.4, $\hat{\varphi}$ is Cauchy in X. $\qquad\square$

Theorem 6.3.7. (Tichonov) Let (X_n, d_n), $n = 1, 2, \ldots$, be metric spaces and (X, d) be the product space, where $X = \prod_{n=1}^{\infty} X_n$ and $d(x, y) = \sum_{n=1}^{\infty} 2^{-n} d_n(x_n, y_n)$, whenever $x = \{x_n\}_{n \geq 1}$ and $y = \{y_n\}_{n \geq 1}$ are elements in X. The product metric space (X, d) is compact if and only if each (X_n, d_n) is compact.

Proof. This is a consequence of Propositions 6.3.5, 6.3.6 and Theorem 5.1.16. $\quad\square$

6.4. Cantor Set

Recall that the Cantor set P is the part of the closed interval $[0, 1]$ that is left after the removal of a certain specified countable collection of open intervals described in

Example 2.1.40. If P_1 denotes the remainder of points in $[0,1]$ on deleting the open interval $(1/3, 2/3)$, then

$$P_1 = \left[0, \frac{1}{3}\right] \bigcup \left[\frac{2}{3}, 1\right].$$

If P_2 denotes the remainder of the points in P_1 on deleting the open intervals $(1/9, 2/9)$ and $(7/9, 8/9)$, then

$$P_2 = \left[0, \frac{1}{9}\right] \bigcup \left[\frac{2}{9}, \frac{1}{3}\right] \bigcup \left[\frac{2}{3}, \frac{7}{9}\right] \bigcup \left[\frac{8}{9}, 1\right].$$

Continuing in this manner, we obtain a descending sequence of sets

$$P_1 \supseteq P_2 \supseteq P_3 \supseteq \cdots$$

and P_n consists of the points in P_{n-1} excluding the "middle thirds". Observe that P_n consists of 2^n disjoint closed intervals. The Cantor set P is the intersection of all these sets, that is,

$$P = \bigcap_{n=1}^{\infty} P_n;$$

and hence, P is closed, being the intersection of closed sets P_n. Moreover, P is compact. In fact, P is a closed bounded subset of $\mathbf{R}$ with the usual metric.

Proposition 6.4.1. Let $X = \prod_{i=1}^{\infty} X_i$, where each $X_i = \{0, 2\}$ with the discrete metric. Then X is compact.

Proof. Observe that X_i is compact, being a finite discrete space (see Example 5.1.2(iii)). So, by the Tichonov Theorem 6.3.7, X is also compact. □

The set P_n consists of 2^n disjoint closed intervals and if we number them sequentially from left to right, we can speak of odd or even intervals in P_n.

We define a function f on the Cantor set P as follows:

$$f(x) = \{a_n\}_{n \geq 1},$$

where

$$a_n = \begin{cases} 0 & \text{if } x \text{ belongs to an odd interval of } P_n, \\ 2 & \text{if } x \text{ belongs to an even interval of } P_n. \end{cases}$$

The above sequence corresponds exactly to the decimal expansion of x to the base 3, that is, where

$$x = \sum_{n=1}^{\infty} \frac{a_n}{3^n}.$$

Proposition 6.4.2. Let $X = \prod_{i=1}^{\infty} X_i$, where each $X_i = \{0, 2\}$ with the discrete metric. The function

$$f: X \rightarrow P$$

defined by $f(\{a_n\}_{n \geq 1}) = \sum_{n=1}^{\infty} a_n/3^n$ is continuous. Moreover, f is a homeomorphism of X onto P.

Proof. Let $x = \{a_n\}_{n \geq 1}$ be in X and $\varepsilon > 0$. We need to show that there is an open subset U of X containing x such that

$$y \in U \text{ implies } |f(y) - f(x)| < \varepsilon.$$

Since the series $\sum_{n=1}^{\infty} (2/3)^n$ converges, there exists n_0 such that $n > n_0$ implies $\sum_{n=n_0+1}^{\infty} (2/3)^n < \varepsilon$. Consider the subset

$$U = \{a_1\} \times \{a_2\} \times \ldots \times \{a_{n_0}\} \times \prod_{n=n_0+1}^{\infty} X_n$$

of X. Observe that $x \in U$ and U is a member of the defining base for the open sets in X and is, therefore, open. Furthermore,

$$y = \{a_1, a_2, \ldots, a_{n_0}, b_{n_0+1}, b_{n_0+2}, \ldots\} \in U$$

implies

$$|f(y) - f(x)| = \left| \sum_{n=n_0+1}^{\infty} (b_n - a_n) \cdot \frac{1}{3^n} \right| \leq \sum_{n=n_0+1}^{\infty} (2/3)^n < \varepsilon.$$

Thus, f is continuous.

The function $f: X \rightarrow P$ is a one-to-one continuous function from the compact metric space X onto the space P. By Theorem 5.3.8, f is a homeomorphism. $\qquad \square$

Proposition 6.4.3. The Cantor set P has the cardinality of the continuum.

Proof. See Examples 2.3.14 (vi). $\qquad \square$

Definition 6.4.4. A subset A of a metric space (X, d) is said to be **perfect** if A is closed and every point of A is a limit point of A.

Proposition 6.4.5. The Cantor set P is perfect.

Proof. Hint. Let x_0 be a point of the Cantor set. In ternary representation,

$$x_0 = 0 \cdot a_1 a_2 a_3, \ldots, a_n \ldots,$$

where $a_n = 0$ or 2. Let $\{x_n\}_{n \geq 1}$ be a sequence of points where

$$x_n = 0 \cdot a_1 a_2 a_3 \ldots a_{n-1} a_n' a_{n+1} \ldots,$$

where $a_n' = 0$ if $a_n = 2$, and $a_n' = 2$ if $a_n = 0$, $n = 1, 2, \ldots$. The sequence $\{x_n\}_{n \geq 1}$ consists of distinct points all belonging to the Cantor set such that x_n differs from x_0

in the nth place in the ternary expansion. But $x_n \to x_0$ as $n \to \infty$, and so x_0 is a limit point of P. $\qquad\qquad\qquad\qquad\qquad\qquad\qquad\qquad\qquad\qquad\qquad\qquad\quad \square$

6.5. Exercises

1. Let X and Y be metric spaces and $A \subseteq X$ and $B \subseteq Y$. Then show that

   (a) $(A \times B)^\circ = A^\circ \times B^\circ$,
   (b) $(A \times B)' = (A' \times \overline{B}) \cup (\overline{A} \times B')$.

   Hint: (a) $S((x_1, x_2), r) \subseteq A \times B \Leftrightarrow S(x_1, r) \times S(x_2, r) \subseteq A \times B \Leftrightarrow S(x_1, r) \subseteq A$ and $S(x_2, r) \subseteq B$.

   $$(b)(S(x_1, r) \times S(x_2, r) \backslash \{x_1, x_2\}) \cap A \times B \neq \varnothing \Leftrightarrow [(S(x_1, r) \backslash \{x_1\}) \times S(x_2, r)] \cup [S(x_1, r)$$
   $$\times (S(x_2, r) \backslash \{x_2\})] \cap (A \times B)$$
   $$\neq \varnothing \Leftrightarrow [(S(x_1, r) \backslash \{x_1\} \cap A) \times (S(x_2, r) \cap B)]$$
   $$\cup [(S(x_1, r) \cap A) \times (S(x_2, r) \backslash \{x_2\}) \cap B] \neq \varnothing.$$

2. Let X be a compact metric space, Y be a metric space and $p: X \times Y \to Y$ be the projection. Show that p is a closed map.
   Hint: Suppose $F \subseteq X \times Y$ is closed. Let $y_0 \in Y \backslash p(F)$; then $(X \times \{y_0\}) \cap F = \varnothing$, so that each point (x, y_0) is contained in $S(x, r(x)) \times S(y_0, r(x))$, where $(S(x, r(x)) \times S(y_0, r(x))) \cap F = \varnothing$. From this open covering of $X \times \{y_0\}$, extract a finite subcovering $S(x_i, r(x_i)) \times S(y_0, r(x_i)), i = 1, 2, \ldots, n$. Then $\bigcap_{i=1}^{n} S(y_0, r(x_i))$ is an open ball with centre y_0 that does not intersect $p(F)$.

3. (a) Let f be a continuous mapping from a metric space X into a metric space Y. Show that $\{(x, f(x)): x \in X\}$ is a closed subset of $X \times Y$.
   (b) If $\{(x, f(x)): x \in X\}$ is a closed subset of $X \times Y$ and Y is compact, then show that f is continuous.
   Hint: (a) Let $(x, y) \in \overline{\{(x, f(x)): x \in X\}}$. Then there exists a sequence $\{(x_n, y_n)\}_{n \geq 1}$ in $\{(x, f(x)): x \in X\}$ such that $\lim_n (x_n, y_n) = (x, y)$. Now $y = \lim_n y_n = \lim_n f(x_n) = f(\lim_n x_n) = f(x)$, using continuity of f.
   (b) Let F be a closed subset of Y. Then $p_Y^{-1}(F) \cap \{(x, f(x)): x \in X\}$ is closed in $X \times Y$. Now $p_X: X \times Y \to X$ is a closed map (see Exercise 2) and

   $$p_X(p_Y^{-1}(F) \cap \{(x, f(x)): x \in X\}) = f^{-1}(F)$$

   is closed in X.

4. If d is a metric on X, then show that d is a continuous mapping of $X \times X$ into $\mathbf{R}$.
   Hint: $d: X \times X \to \mathbf{R}$ is defined by $(x, y) \to d(x, y)$. Now,

$$|d(x,y) - d(x_0, y_0)| = |d(x,y) - d(x, y_0) + d(x, y_0) - d(x_0, y_0)|$$
$$\leq |d(x,y) - d(x, y_0)| + |d(x, y_0) - d(x_0, y_0)|$$
$$\leq d(y, y_0) + d(x, x_0) \leq 2 d_{X \times X}((x, y), (x_0, y_0)).$$

5. Show that for metric spaces, the following properties are invariant under finite products: (a) boundedness; (b) total boundedness; (c) completeness.
Hint: (a) If $d_1, d_2, \ldots, d_n$ are metrics on $X_1, X_2, \ldots, X_n$, which are bounded, then

$$d((x_1, x_2, \ldots, x_n), (y_1, y_2, \ldots, y_n)) = \max \{d_i(x_i, y_i) : i = 1, 2, \ldots, n\}$$
$$\leq \max \{\operatorname{diam}(X_i) : i = 1, 2, \ldots, n\}.$$

(b) Let $\varepsilon > 0$ be given and let $\{x_i^{(1)}, x_i^{(2)}, \ldots, x_i^{(m_i)}\}$ be a finite ε-net in $X_i, i = 1, 2, \ldots, n$. Then

$$\{(x_1^{(i_1)}, x_2^{(i_2)}, \ldots, \ldots, x_n^{(i_n)}) : 1 \leq i_j \leq m_j, \ j = 1, 2, \ldots, n\}$$

is a finite ε-net in $X_1 \times X_2 \times \ldots \times X_n$.

(c) See Proposition 6.3.5.

6. Let X and Y be metric spaces. Show that $X \times Y$ is connected if and only if X and Y are connected.
Hint: Since the projection mappings are continuous and onto, if $X \times Y$ is connected, so are X and Y. Now suppose that X and Y are connected. Let (x, y) and (x^*, y^*) be any two points of $X \times Y$. Then $\{x\} \times Y$ and $X \times \{y^*\}$ are homeomorphic to Y and X, respectively, and hence, are connected. They intersect in (x, y^*) and so their union, which contains the two points (x, y) and (x^*, y^*), is connected. Thus, $X \times Y$ is connected.
Remark: If $(X_i, d_i), i = 1, 2, \ldots$, is a family of connected spaces, then their product $(\Pi X_i, d)$ is also connected, where $d(x, y) = \sum_{n=1}^{\infty} 2^{-n} d_n(x_n, y_n), x = (x_1, x_2, \ldots)$ and $y = (y_1, y_2, \ldots)$.

7. Prove: ΠA_i is dense in ΠX_i if and only if each $A_i \subseteq X_i$ is dense.

8. Let X and Y be metric spaces. $X \times Y$ is locally compact if and only if X and Y are locally compact.
Hint: Projections on coordinate spaces are continuous open maps. Let $(x, y) \in X \times Y$. There exist $S(x, r)$ and $S(y, r)$ such that $\overline{S(x, r)}$ (respectively, $\overline{S(y, r)}$) is compact in X (respectively, Y). Then

$$\overline{S((x, y), r)} = \overline{S(x, r)} \times \overline{S(y, r)}.$$

9. Let p_i be the projection of $\Pi_{i=1}^{\infty} X_i$ onto X_i. Prove that each p_i is uniformly continuous and open.

10. Show that $\{x^{(k)}\}_{k\geq 1}$ is a Cauchy sequence in the product space $\Pi_{i=1}^{\infty}X_i$ if and only if each $\{p_i(x^{(k)})\}_{k\geq 1}$ is Cauchy in X_i, where p_i is the projection of $\Pi_{i=1}^{\infty}X_i$ onto X_i.

References

1. Ahlfors, LV. *Complex Analysis* 2nd ed. McGraw-Hill, 1966.
2. Brown, AL, Page, A. *Elements of Functional Analysis*. Van Nostrand Reinhold Company, London, 1970.
3. Boas, RP. *A Primer of Real Functions*. Mathematical Association of America, 1960.
4. Bryant, V. *Metric Spaces, Iteration and Applications*. Cambridge University Press, Cambridge (UK), 1985.
5. Copson, ET. *Metric Spaces*. Cambridge University Press, Cambridge (UK), 1968.
6. Cohen, LW, Ehrlich, G. *The Structure of the Real Number System*. Van Nostrand Reinhold/ East West Press Pvt. Ltd., New Delhi, 1963.
7. Dieudonné, J. *Foundations of Modern Analysis*. Academic Press, 1960.
8. Dugundji, J. *Topology*. Allyn and Bacon, Inc., Boston, 1966.
9. Goffman, C, Pedrick, G. *A First Course in Functional Analysis*. Prentice-Hall, Inc., Englewood Cliffs, New Jersey, 1965.
10. Goldberg, R. *Methods of Real Analysis*. Blaisdell Publishing Company, Waltham, Massachusetts. Toronto, London, 1964.
11. Jain, PK, Ahmad, K. *Metric Spaces*. Corrected edition. Narosa Publishing House, New Delhi, Madras, Bombay, Calcutta, 1996.
12. Klambauer, G. *Mathematical Analysis*. Marcel-Dekker Inc., New York, 1975.
13. Kolmogorov, AN, Fomin, SV. *Elements of the Theory of Functions and Functional Analysis*. (Translated from the Russian) Graylock Press, Rochester, New York, 1957.
14. Limaye, BV. *Functional Analysis*, 2nd ed., New Age International Limited, New Delhi, 1996.
15. Liusternik, LA, Sobolev, VJ. *Elements of Functional Analysis*. Frederick Ungar Publishing Company, New York, 1961.
16. Pitts, CGC. *Introduction to Metric Spaces*. Oliver and Boyd, Edinburgh, 1972.
17. Pontriagin, LS. *Ordinary Differential Equations*. (Translated from the Russian) Addison-Wesley, Reading, Massachusetts, 1962.
18. Royden, HL. *Real Analysis*. The Macmillan Company, New York, 1963.
19. Rudin, W. *Principles of Mathematical Analysis*. McGraw-Hill, International Edition, 1976.
20. Sierpinski, W. *General Topology*. University of Toronto Press, Toronto, 1952.
21. Simmons, GF. *Introduction to Topology and Modern Analysis*. McGraw-Hill, 1963.
22. Sutherland, WA. *Introduction to Metric and Topological Spaces*. Clarendon Press, Oxford, 1975.

Index

AM-GM inequality 24
Arc 147, 165
Archimedean 20, 22, 97
Arcwise connected 156, 165–169
   subset 165
Arithmetic-Geometric Mean Inequality 24
Arzelà-Ascoli 189, 199

Baire 89, 152
   Category Theorem 88–90
Base 64, 83
   countable 82
   local 82
Boundary
   of a set 100
   point 150
Bounded 5
   above 5
   below 5
   function 31
   interval 1, 10
   pointwise 199
   subset of a metric space 76
      of $\mathbf{R}$ 77
      of $\mathbf{R}^2$ 77
      of $\mathbf{R}^n$ 170
   sequence 6, 10
   uniformly 189

Cartesian product of
   countable number of sets 201
   finite number of sets 201
   finite number of spaces 203
   infinite number of spaces 208

Cantor 78
   set 76, 101, 212
Category
   Baire's Theorem 88, 89
   set of first 88
   set of second 88
Cauchy Convergence Criterion (*or* Principle
      of Convergence) 7, 44, 127
Cauchy sequence 7, 45
Cauchy-Schwarz Inequality 25
Characterisation of
   arcwise connectedness 165
   closed subset 74
   compactness 178
   compact subset of $\mathbf{R}$ 10
   connected subset of $\mathbf{R}$ 158
   continuity 104, 105, 107
   limit point 71
   open set 66, 69
Closed
   ball 64
   disc 120
   map (*or* mapping) 146
   subset of a metric space 64, 71
      of $\mathbf{C}$ 71
      of $\mathbf{R}$ 9, 10, 71
      of $\mathbf{R}^n$ 170
Closure of a set 72
Collection of sets 2, 201
Compact sets in special metric spaces 188
Compact space 10, 171
   continuous functions on 182
   countably 180
   examples of 171

Compact space (*Continued*)
  locally 185
  other characterisations of 178
Complement 2, 68, 76
Complete metric space 22, 23, 47
Completion of a metric space 1, 13,
    55
  uniqueness 57
Component
  connected 161–162
  mapping 146
Composition of functions 4, 107, 117
Connected
  space 157
  arcwise (*or* pathwise) 165
  locally 163
  subset of 156–157
  subset of **R** 158
  subset of **C** 167
Continuous function (*or* mapping) 8, 103,
    104
  nowhere differentiable 93
  on a compact set 182
  on a connected set 159, 160, 166
  real and complex valued 112
  topological characterisation of 105, 106,
      107
  uniformly 114
Contraction mapping 132
  applications of 135–143
  Principle 133
  extension of 135
Convergent sequence 7
  in a product space 211
  in ℓ_p 41
  in **R**^{n} 39
  pointwise 123
  uniformly 125
Covering (*or* Cover)
  countable 84
  finite 10
  open 10, 84

Dense subset 84
Derived set 70
Diameter of a set 76
Distance
  between nonempty subsets 77
  of a point from a subset 76

Differential equation 132, 137
Dini's Theorem 194
Disconnected space 156
  totally 168
Disconnection 162
Discontinuity of a real-valued function
    152
Disjoint 4
  family 4

Empty set 2
Epsilon-net 173
Equicontinuous 188
Equivalence class 4, 14
  relation 4, 14, 37, 55
Equivalent metrics 120
Euclidean metric 29
Everywhere dense 84
Extended complex plane 33
Extension theorems 109, 118, 131

Finite
  intersection property 171
  metric product 202
  subcover (*or* subcovering) 10, 171
First countable space 83
Fixed point 133
Function (*or* mapping)
  bijective 3
  continuous *see* Continuous function
  contraction *see* Contraction mapping
  everywhere continuous, nowhere
      differentiable 93
  extension of 109
  identity 120
  injective 3
  integrable 11
  left (*or* right) hand derivative of 92
  limit of 104
  maximum or minimum of 10
  monotonic sequence of 194
  projection 201
  restriction of 109
  surjective 3
  uniformly continuous 114

Heine-Borel Theorem 10
Hölder's Inequality 25
Homeomorphism 119

Induced metric 28
Inseparable 86
Integral equation 132, 151
Interior of a set 69
    point 69
Intermediate Value Theorem 9, 160
Isometric imbedding 55
Isometric metric spaces 55
Isometry 55

Lebesgue number 181
Limit
    of a function 104–105
    of a sequence 7, 38
Limit point 70
Lindelöf 84
Linear equation 135
Locally compact 185
Locally connected 163

Mapping (or map) see function
Metric 28
    discrete 30
    equivalent 120
    Euclidean 29
    induced 28
    pseudometric 36
    standard 28
    supremum 32
    uniform 32
Metric space 27
    base for open sets of 83
    compact 171
    compact sets in special 188
    complete 47
    connected see Connected
    locally compact 185
    ℓ_p 31, 41, 50
    product of 203, 208
    separable 86
    sequence in 38
    totally bounded 174
Minkowski's Inequality
    for finite sums 26
    for infinite sums 27
M-test 129

Neighbourhood 8, 66
Nowhere dense 88
Nowhere differentiable 92

Open and closed 163
Open
    ball 64
    disc 147
    map (or mapping) 146
    subset of a metric space 66
    of $\mathbf{R}$ 10, 68
Oscillation over (or at) 95
Osgood's Theorem 90

Path 165
Pathwise connected 165
Picard's Theorem 137
Pointwise convergence 9, 123
Product space
    finite 203
    infinite 208
Projection 201

Relation 4
    equivalence 4
    reflexive 4
    symmetric 4
    transitive 4
Relatively
    closed 79
    open 79

Second countable space 83
Separable space 86
Sequence 6, 38
    Cauchy 7, 45
    convergent 7, 38
    diagonal 190
    equivalent 55
    subsequence 48
    uniformly convergent
        125
Series
    convergent 127
    uniformly convergent 128
    sum of 9

Set
  boundary of 100
  bounded 76
  closed 71
  equivalent characterisation of 72, 74
  closure of 72
  compact 171
  complement of 2
  connected 157
  countable 85
  dense 85
  diameter of 76
  disconnected 156
  everywhere dense 84
  exterior of 100
  interior of 69
  of first category 88
  of second category 88
  open 66
  separated 156
  totally bounded 174

Space
  ℓ_p 31, 41, 50
  of bounded functions 31
  of continuous functions 32
  of bounded sequences of numbers 30
Standard metric 28
Supremum metric 32

Tietze's Extension Theorem 131
Totally bounded
  space 174
  subset 174
Triangle inequality 28

Uniform metric 32
Uniformly continuous function 114
Urysohn's lemma 116

Weierstrass approximation theorem 12
Weierstrass M-test 129